Simplified Truss Design

Simplified Truss Design

Michele Melaragno
Professor of Architecture
University of North Carolina at Charlotte

VAN NOSTRAND REINHOLD COMPANY
NEW YORK   CINCINNATI   ATLANTA   DALLAS   SAN FRANCISCO
LONDON   TORONTO   MELBOURNE

Van Nostrand Reinhold Company Regional Offices:
New York   Cincinnati   Atlanta   Dallas   San Francisco

Van Nostrand Reinhold Company International Offices:
London   Toronto   Melbourne

Copyright © 1981 by Litton Educational Publishing, Inc.

Library of Congress Catalog Card Number: 78-16536
ISBN: 0-442-25129-7

All rights reserved. No part of this work covered by the copyright hereon may be reproduced or used in any form or by any means—graphic, electronic, or mechanical, including photocopying, recording, taping, or information storage and retrieval systems—without permission of the publisher.

Manufactured in the United States of America

Published by Van Nostrand Reinhold Company
135 West 50th Street, New York, N.Y. 10020

Published simultaneously in Canada by Van Nostrand Reinhold Ltd.

15 14 13 12 11 10 9 8 7 6 5 4 3 2 1

Library of Congress Cataloging in Publication Data

Melaragno, Michele G
  Simplified Truss Design

  Bibliography: p.
  Includes index.
  1. Trusses.   I. Title.
  TA660.T8M44   624'.1773   78-16536
  ISBN 0-442-25129-7

To My Mother and Father

Preface

Trusses are among structural systems with traditions reaching far back into many centuries of history. Yet they are as significant today as they have ever been, even though we now have a much larger spectrum of structural systems from which to choose. And if we project into the future of the truss, we can see that its potential has not yet been exhausted. As a matter of fact, exploration of the possible uses of reinforced concrete is a substantial area for future development. Also, speculation based on present research reveals that pneumatic structures may be a viable area of truss application.

Therefore, based on present truss status, I think it desirable for one involved in everyday design to have an understanding of trusses, as well as a quick method for selecting a truss suiting one's specific needs. Dealing with trusses usually requires the use of a structural engineer. Thus, in practical situations, a designer may have the tendency to bypass the use of trusses, either because of the inconvenience of consulting with an engineer at the beginning of a design idea, or because he unconsciously tends to eliminate something complicated or has a use which is unclear. With this book, I have attempted to solve these problems.

Of the five chapters, the first is concerned with the historical development of the truss, thereby contributing to the explanation of what a truss is. In the second chapter, I present those elements essential to the selection of a truss, including structural concepts, engineering data, and the description of the most common types of trusses and their common nomenclature. The third chapter contains a series of tables considering the most common types of trusses. Using these tables, the reader will be able to compute, almost immediately, the axial forces in the selected truss. By observing the magnitude of such forces, he can then change the configuration of the truss (by changing the pitch, the number of panels, or even the type) to satisfy his needs. Finally, the fourth chapter presents a step-by-step procedure on how to use the STRUDL II computer program for the solution of truss problems. The reader can apply this procedure to his own practical problem, and the procedure can be used to resolve statically determinate problems as well as most kinds of indeterminate problems, whether caused by internal or external conditions. The reader can also determine elastic distortions of the members (elongation or contraction) and the deflection of joints. Indeed, once he practices with STRUDL II on trusses, he can then work without difficulty on any other type of structure in a plane or in space. On the basis of what has been presented in the previous chapters, the reader is now ready for the final design of a truss. Chapter 5, therefore, furnishes basic design procedures and tables for the design of tension and compression members of wood, steel, or reinforced concrete. Also included are common types of connections for members built of the above materials.

MICHELE MELARAGNO
Charlotte, North Carolina

Contents

Preface	vii
1. THE HISTORICAL DEVELOPMENT OF THE TRUSS	1
The Greek Truss	1
Roman Trusses	7
Medieval Trusses	9
Truss Development in the Orient	18
Trusses During the Renaissance	20
Long-Span Timber Trusses	32
American Truss Development	33
European Truss Development through the Nineteenth Century	43
Truss Development in the Twentieth Century	52
2. A SELECTIVE EVALUATION OF TRUSSES	81
How Do We Select a Truss?	81
Truss Forms in Nature	81
Definitions	84
Interpretation	86
Concept of Economy	86
Structural Comparison	88
Statical Determinacy	88
Geometrical Instability	97
Statically Indeterminate Trusses	100
Teaching Aids	101
Cremona's Diagram	102
Weight of Trusses	104
Prestressed Steel Trusses	106
Concrete Trusses	109
Types of Trusses	115
Simple Triangle	115
King Post	116

Reinforced Triangles	118
Scissors of German Truss	120
Palladian	120
Hammer-Beam	122
Polonceau of Fink	123
Fan	124
Howe	125
Pratt	126
Belgian	127
Warren	128
Tridimensional Trussing	128
Structures of Expediency	136

3. COEFFICIENTS FOR SIMPLIFIED TRUSS ANALYSIS — 139

Use of Tables	139
Determination of Axial Forces	140
Determination of Length	140
Tables for Determining Axial Force Coefficients, Length Coefficients, and Angles	142

4. ANALYSIS BY COMPUTER — 243

Problem 1	244
Problem 2	261
Comparison of Methods	275
Problem 3	276
Problem 4	290
Prbolem 5	398
Problem 6	310
Problem 7	312
Problem 8	319

5. MEMBER AND CONNECTION DESIGN 327

Wood Compression Members 327
Steel Compression Members 337
Steel Tension Members 348
Latticework or Lacing 349
Reinforced Concrete Compression Members 351
Connections 355

APPENDIX A 383
APPENDIX B 386
APPENDIX C 392
APPENDIX D 397

BIBLIOGRAPHY 397

INDEX 401

1
The Historical Development of the Truss

How has the truss evolved throughout history? If we consider the concept of triangulation as the basic principle of trussing, we can trace the beginning of trussing from the first primitive shelters built by prehistoric man. Guided only by his intuition, ancient man adopted the inherent stability of the triangle in his buildings to solve the problem of connecting wood members. Continuity and fixed ends were difficult to achieve in working with wood components alone; thus, the geometric stability of triangulation was a necessary ingredient in primitive man's shelters (see Figures 1-1 and 1-2).

THE GREEK TRUSS

The historic evolution from the concept of triangular bracing to that of trussing is still, for the architectural historian, a matter of speculation, since wooden roof structures of ancient buildings have not survived. However, the large, roof-covered spans existing in these buildings suggest that some sort of trussing system must have been used. Examples such as the temples of Concord in Agrigentum, Sicily, and Theseion in Athens seem to verify this hypothesis. Was the truss then known to Greek civilization? There is evidence that the Greek truss has existed since the Mycenaean period although the design criteria of such structure was quite different from that of the modern truss. It is more accurate, in fact, to refer to the ancient Greek truss as a reinforced beam consisting of two diagonal rafters bearing at midspan over a main beam, which carries the load as a flexural member. This concept is symbolically illustrated by the Lions' Door in Mycenae, where the bodies of the two lions represent the diagonal rafters, and the lions' front legs, bearing on a block at midspan over the architrave below, complete the scheme (see Figures 1-3a and 1-3b). Similar details for the construction of the Greek

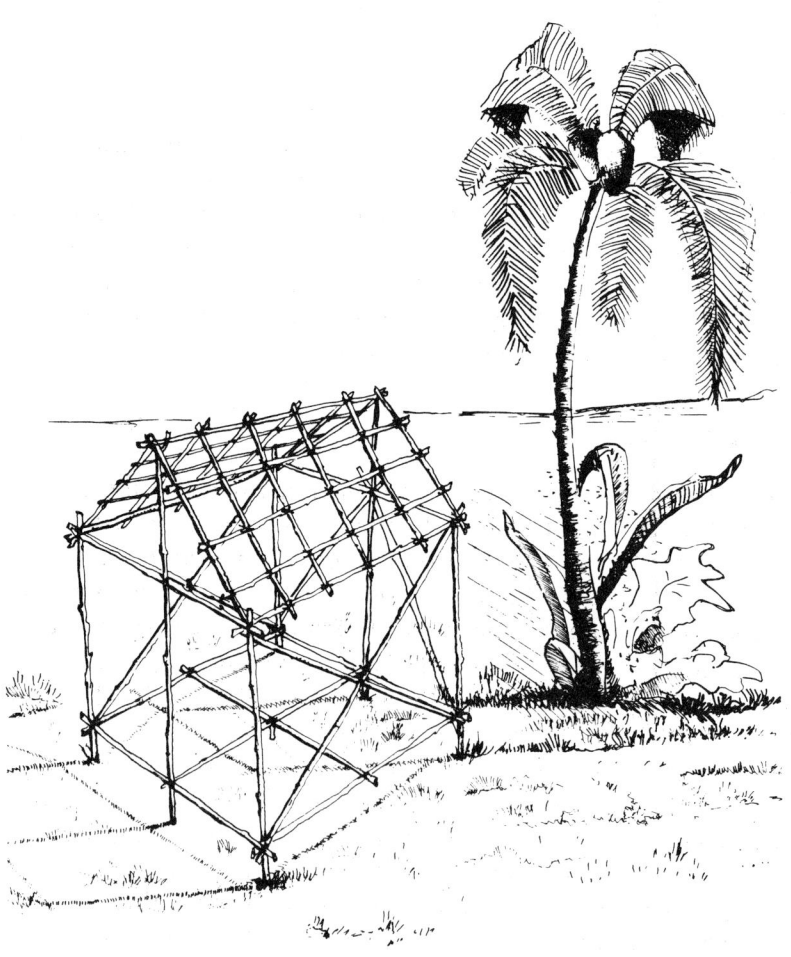

Figure 1-1. A typical primitive hut where triangulation of the wooden members provides the structural integrity.

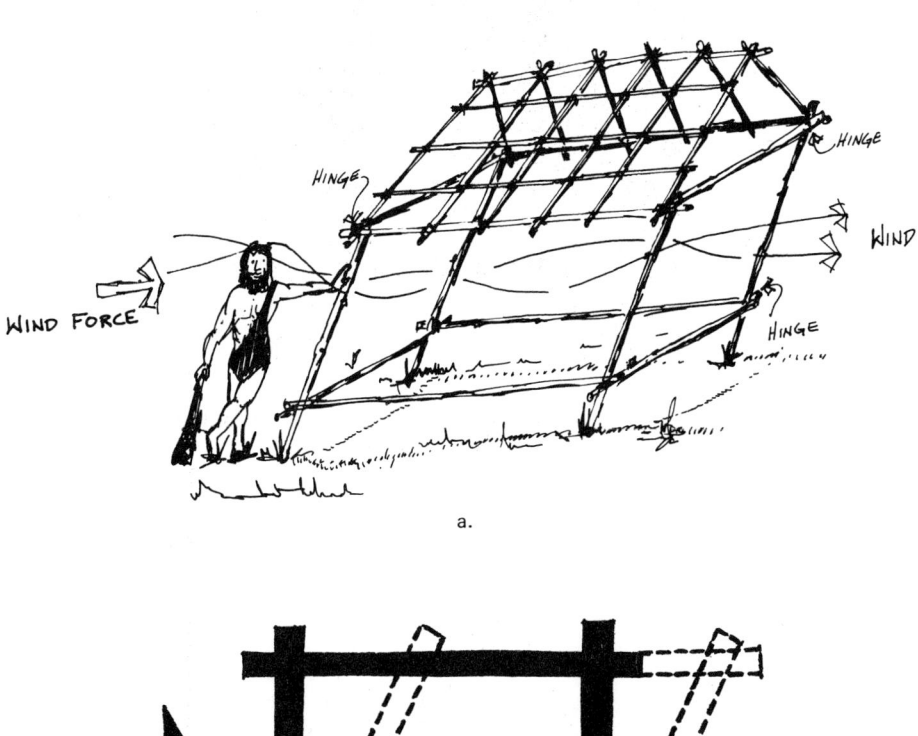

Figure 1-2a, b. Even a small lateral force such as could be produced by relatively low speed wind can make such structures collapse.

4 SIMPLIFIED TRUSS DESIGN

Figure 1-3a. The Lions' Door in Mycenae has been considered by some to symbolize the concept of the truss as understood by early Greek civilization.

Figure 1-3b. The inclined bodies of the lions represent the loads directed toward the supports, and the lions' vertical front legs represent that portion of the load at midspan on the architrave.

THE HISTORICAL DEVELOPMENT OF THE TRUSS 5

truss appear in inscriptions found on a marble slab. These inscriptions, illustrating the specifications for the construction of the arsenal at Piraeus, indicate that in structures built between 340 and 330 B.C., the Greek truss was still quite different from our modern interpretation of trusses. The inscription describes a truss 22 ft in length, consisting of a main beam 32 ft wide and 29 in. high. This beam carries a wood block at midspan, on top of which a ridge beam ($22\frac{1}{2}$ in. × $17\frac{3}{4}$ in.), perpendicular to the truss, carries the ends of two rafters (12 in. wide and 8 in. high) (see Figure 1-4).

Figure 1-4. The construction details of the arsenal of Pireaus (fourth century B.C.), according to the inscriptions found on a marble slab (drawing of restoration by Dörpfeld).

6 SIMPLIFIED TRUSS DESIGN

Thus, the Greek truss prior to the Hellenistic period could hardly be considered a real truss in the modern sense. After this period, however, the truss in its modern version appeared in Greece, as we learn from the tratise' *De Architectura*, supposedly written by Vitruvius in the first century A.D. (see Figure 1-5). Whether this new type of truss was imported from Rome or Egypt is not really clear.

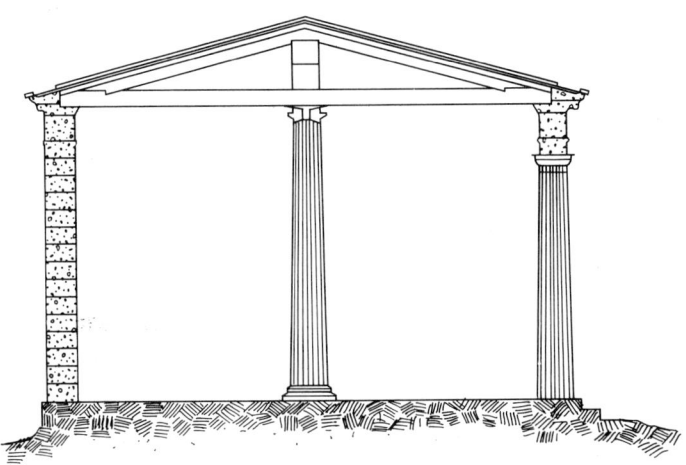

Figure 1-5. Reconstruction of the Greek truss, according to the description by Vitruvius.

ROMAN TRUSSES

The existence of modern trusses in Roman building technology is accepted. Only efficient, real trusses could have spanned over the large spaces of Roman basilicas. Even more significant proof that modern trusses were used in Roman times is provided, once again, by Vitruvius. There is no doubt that Roman builders knew the distribution of the forces carried by triangular trusses—compressive forces in the sloping members of the top chord and tensile forces in the bottom chord (see Figure 1-6). The geometrical configurations and the construction details of the Roman trusses reached the Renaissance, as demonstrated by Palladio.

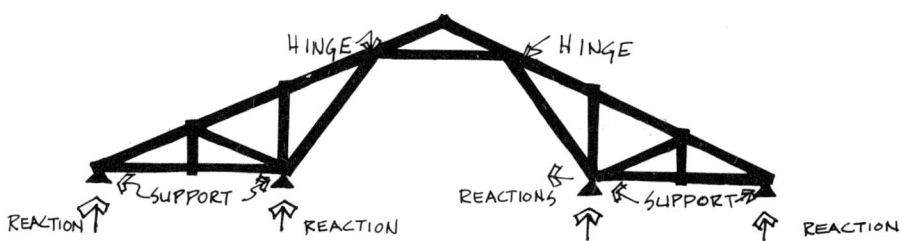

Figure 1-6. Schematic interpretation of the bronze truss shows that the structure is a "Gerber" and therefore statically determinate: basically, two trusses simply supported and having a cantilevered support for the third truss. Can this be just intuition or another example that the statics of trusses was a Roman achievement?

8 SIMPLIFIED TRUSS DESIGN

A significant example of a Roman truss which employed exactly the same design and construction criteria of present technology was the bronze truss supporting the portico in front of the Pantheon in Rome. Built about 130 A.D., with riveted connections, the truss lasted until 1625, when Pope Urban VIII removed it and melted it to reuse the bronze. Fortunately, the drawings by Palladio, made before the destruction of the truss, preserve its memory (see Plate 1-1). Still another example of wood trusses used in Roman

Plate 1-1. The bronze truss supporting the roof in the portico in front of the Pantheon. The drawing, made by Palladio, preceded the structure's removal in 1625 by Pope Urban VIII. This drawing is tangible proof of the degree of truss development in Roman times.

THE HISTORICAL DEVELOPMENT OF THE TRUSS 9

carpentry is in the bridge over the Danube, built by Apollodorus in 104 A.D., and illustrated on the Trajan column in Rome. The absence of scale in the high-relief does not allow for a presentation of construction details of the bridge in a realistic sense, but it is evident that the structure included some type of trusses. The 120 ft. spans from pier to pier are sufficient evidence that a wood structure of such length must have employed a structural system based on trusses (see Plate 1-2 and Figures 1-7a and b).

MEDIEVAL TRUSSES

With the decline of the Roman Empire, Europe entered the Dark Ages. What, then, happened in the development of trusses? Medieval wood carpentry flourished considerably in Northern Europe, as indicated by typical medieval house structures, which consisted of half-timber and plaster, with

Plate 1-2. A standing document on Roman trusses. Apollodorus' bridge over the Danube (104 A.D.) sculptured on a Trajan column. That a trussing system was used in the bridge is evident through this relief.

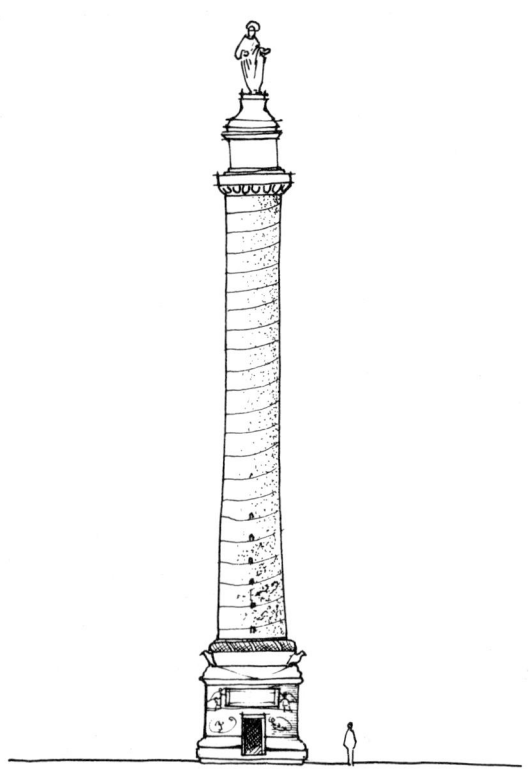

Figure 1-7a. The Trajan column still standing in Rome: history written on stone.

roofs carried by trusses (see Figures 1-8 through 1-12). At a higher level of architectural values, even the ecclesiastical buildings that culminated in the cathedrals of the Gothic period employed trusses for roof supports (see Figure 1-13). These trusses must have been derived from the original Roman prototypes that covered many Roman basilicas and other buildings (see

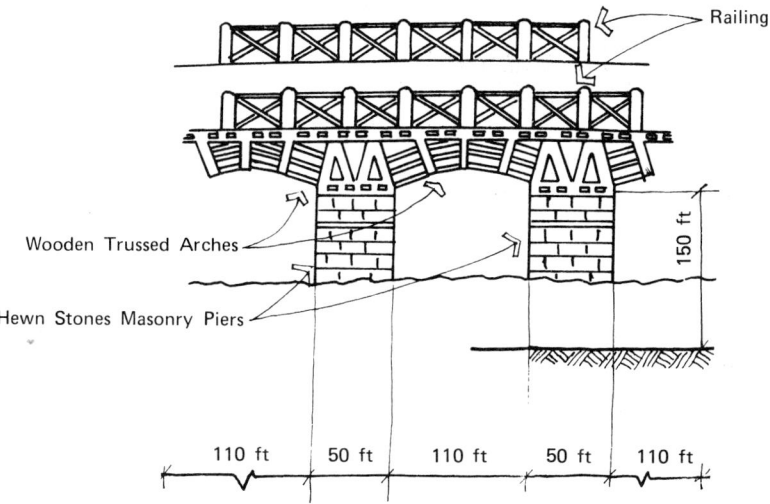

Figure 1-7b. Details of the bridge are obviously out of scale. Dimensions indicated on the drawing are taken from historical accounts written by Dion Cassius a century after the construction.

Plate 1-3). Preservation of the classical culture by the Church and by word of mouth through generations of craftsmen must have saved the tradition of truss construction for centuries, although the real concept on which trussing was based was partially lost, as medieval structures prove. The craftsmen of the Middle Ages had unclear ideas of the mechanics of real trusses. Many members were not really necessary, but had been added by erroneous assumptions. Yet even if some of these redundant members did not relieve others of their axial force, they at least provided some lateral support for those members that were structurally active. Especially for long compression members, these redundant additions strengthened their load-carrying ability against buckling. Static calculations at that time were not possible, as the knowledge needed to carry out such calculations was lacking, and the evaluation of static stability was also

Figure 1-8. Typical Northern European half-timber framing systems of the Middle Ages, showing trussing usage.

Figure 1-9. Example of mature stage in the development of trussing in the wood framing of Northern European building technology. Dwelling structure in Schweigern.

THE HISTORICAL DEVELOPMENT OF THE TRUSS   13

Figure 1-10. The Old Hall at Moreton in Cheshire (1550–1559). The stiffening diagonals are fully exposed on the facade, as well as with the vertical and horizontal members. The braced structure incorporating a series of patterns becomes the dominant design feature of this building type.

Figure 1-11. The Adam House in Angiers, France, shows the dominance of the diagonal members in the exposed wood structure. Their number and density are such that they give the observer the effect of latticework.

Figure 1-12. Example of delicately carved wood trusses in the framing of an eighteenth century farm building in Zernez, Switzerland.

14 SIMPLIFIED TRUSS DESIGN

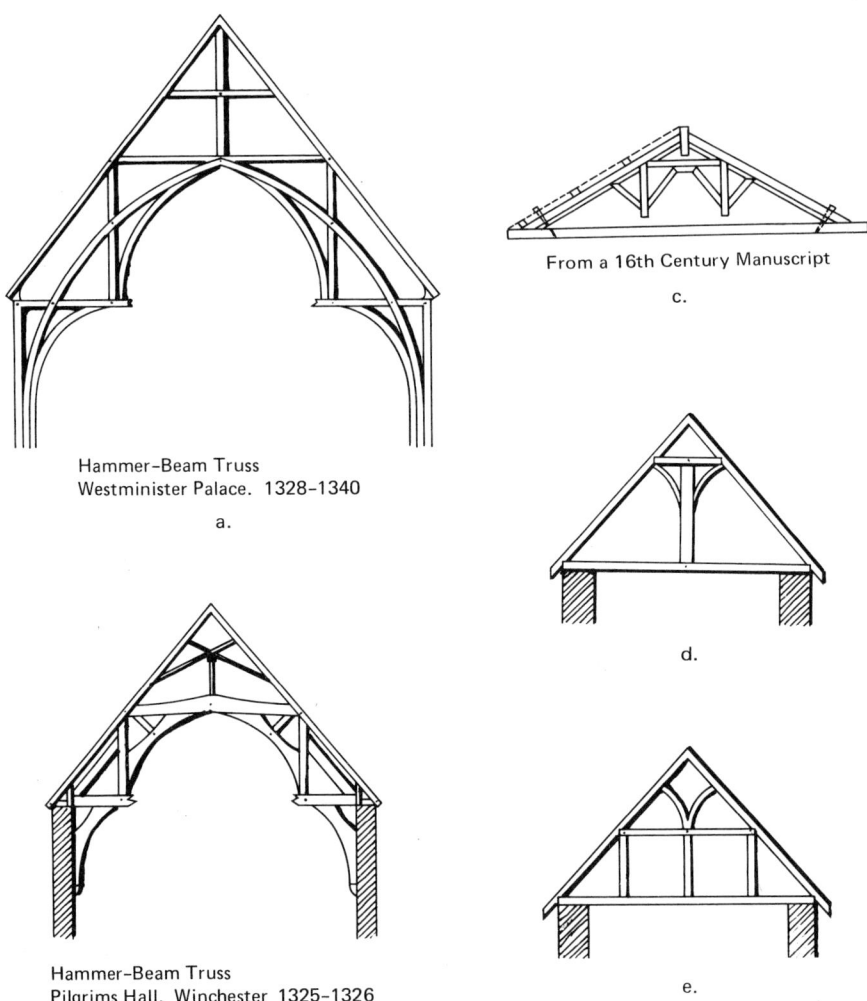

Hammer-Beam Truss
Westminister Palace. 1328-1340
a.

Hammer-Beam Truss
Pilgrims Hall. Winchester 1325-1326
b.

From a 16th Century Manuscript
c.

d.

e.

Figure 1-13. Typical examples of medieval trusses.
    a. Deep roof trusses over Gothic cathedrals accentuate the verticality of the buildings, typical of Northern Europe. The steep slope of the top members reduces snow loads, although wind loads increase. The high slope, however, is also a means to reduce the horizontal truss so that the bottom chord extending horizontally from one support to the other could be eliminated.
    b. The hammer-beam, like the scissor truss, brings to mind the behavior of a two-hinge arch with a deep section at the crown (where moments are maximum) and thin section at the supports (where moments are zero).
    c. From a sixteenth century manuscript.
    d. King post.
    e. More elaborate than queen post, the intra-structure subdivides the chords in shorter spans.

THE HISTORICAL DEVELOPMENT OF THE TRUSS 15

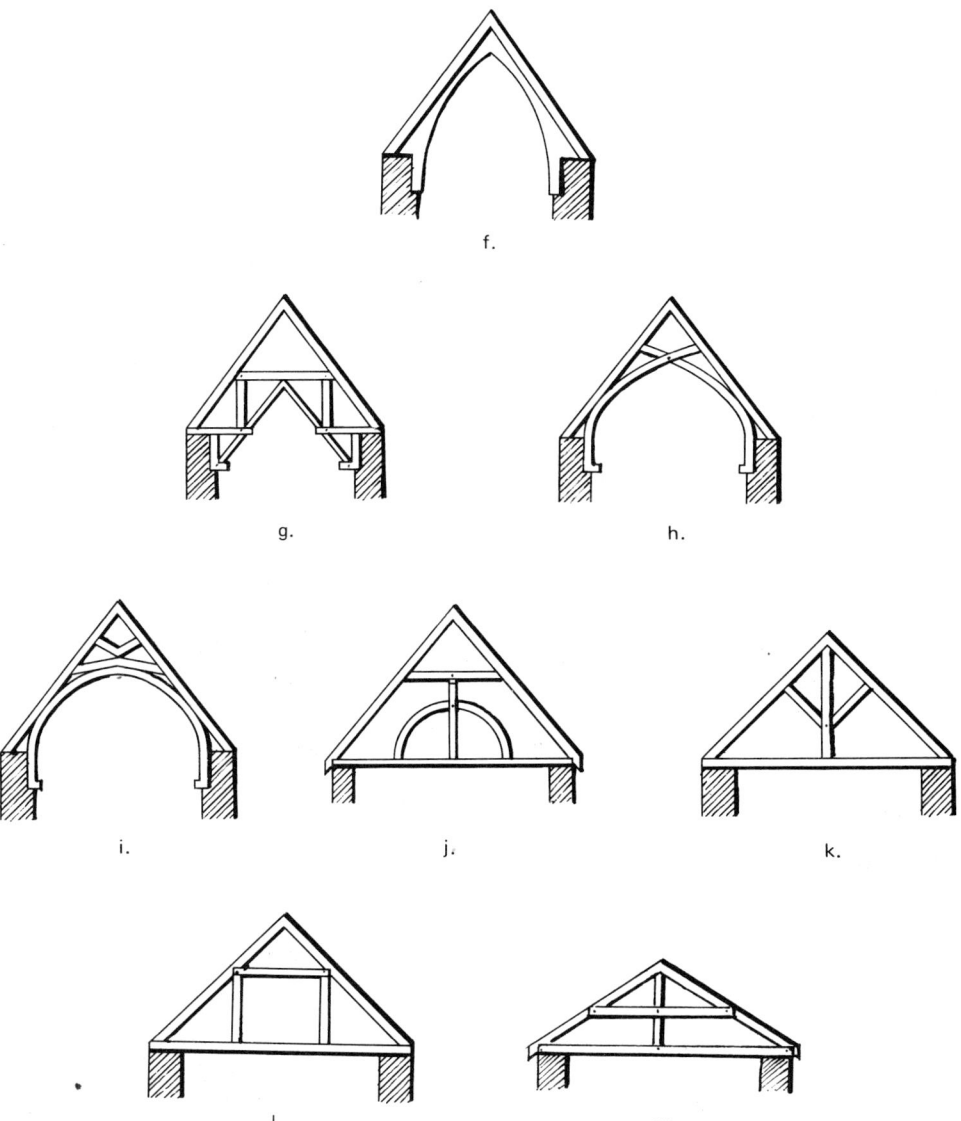

f. A structure that is hard to include among trusses. Horizontal reactions at supports and long unsupported members negate basic truss concepts.
g. Hammer-beam with all rectilinear members.
h. Scissor truss with curvilinear member reflecting the Gothic arch forms.
i. A structural shape performing as a whole like a scissor truss or a two-hinged arch, deeper at the crown, where moments are larger.
j. The elaborate geometrical pattern of intra-structure capitalizes on straight lines and arched members.
k. King post (almost a Palladian type).
l. Queen post.
m. Popular scheme found in many early churches (see Plate 1-4).

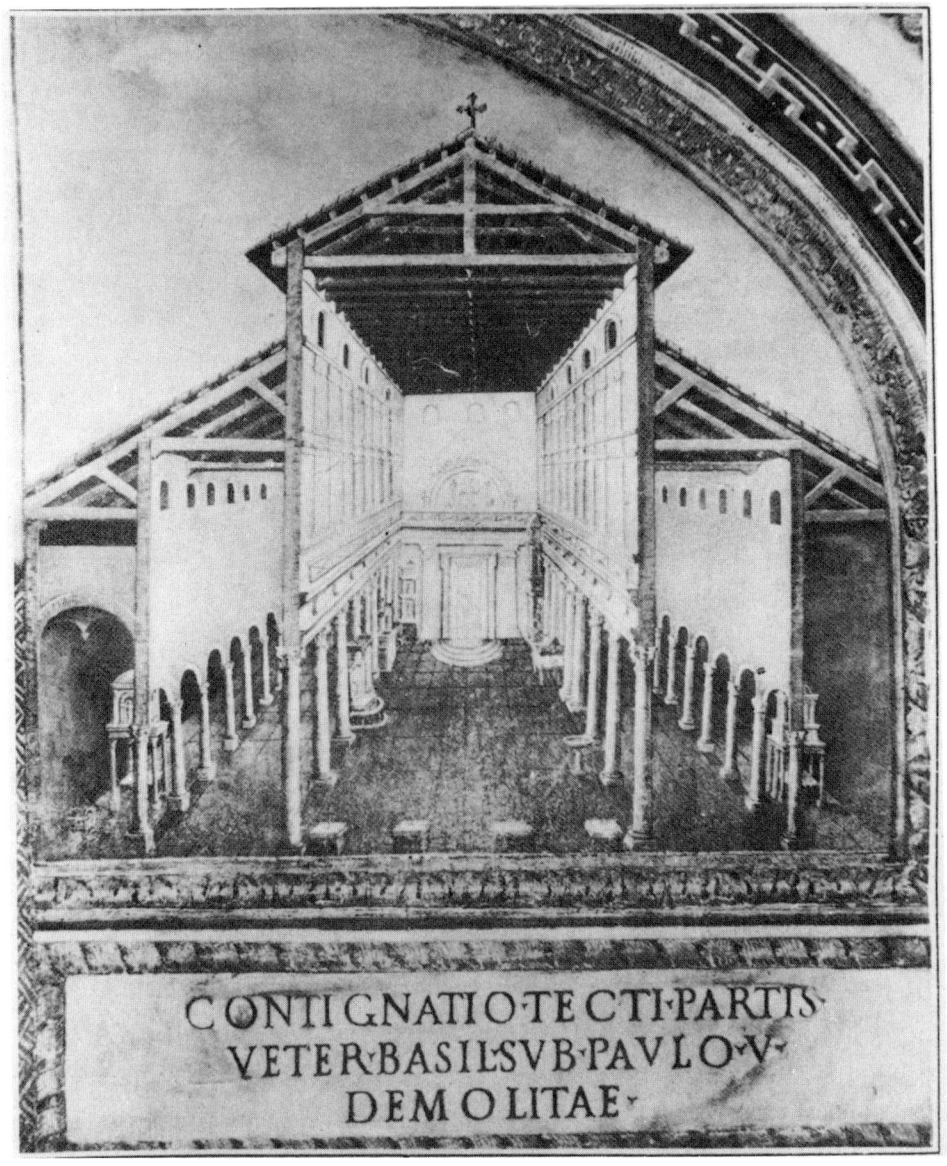

Plate 1-3. Example of old truss systems used in early Christian basilicas. Fresco in the Vatican grottos illustrating the partial reconstruction of the roof of the old Vatican basilica under the papacy of Paul V (1552–1621).

THE HISTORICAL DEVELOPMENT OF THE TRUSS 17

difficult, especially because some of the connections were fixed (rather than hinged, as joints in modern trusses are assumed to be). In fact, for practical construction reasons, members at a given joint were continuous with some of the members and hinged to others.

Apart from structural efficiency, medieval trusses had a certain inherent beauty in the care of the details and in the overall shape. Such features show the human expression of the craftsman (see Plate 1-4) as well as his dedication to the artifact. The mystical feelings of the time seem to be reflected even in the vernacular work of medieval carpenters. Curved members were often incorporated into the truss. They, of course, were in contradiction with the axial nature of the tension and the compression forces they carried from one joint to another. The connections are also particularly intriguing in their elaborate configurations and in the use of metal fasteners of various types.

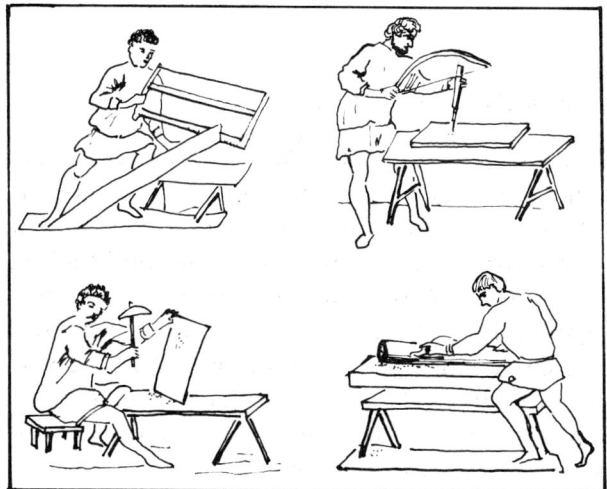

Plate 1-4a.   Carpenters at work (from a stained glass in the Vatican).

18 SIMPLIFIED TRUSS DESIGN

TRUSS DEVELOPMENT IN THE ORIENT

In the Orient, contrary to all expectations, the truss did not develop as it did in the West. In the cultures of China and Japan, the dominance of wood as a building material, especially in dwelling structures, was so widely spread that it made wood carpentry a basic component of the building technology in these countries. And since wood construction seems to induce spontaneously the use of triangulation for bracings, one would expect to find trussing an integral part of Chinese and Japanese building systems. Surprisingly enough, this was not so (see Figure 1-14). L. Sprague de Camp confirms this fact in *The Ancient Engineers,* where he writes, "Despite their skillful use of wood, [the Chinese] never discovered the truss."*

---

*L. Sprague de Camp, *The Ancient Engineers* (New York, Doubleday and Company, 1963) p. 290.

THE HISTORICAL DEVELOPMENT OF THE TRUSS   19

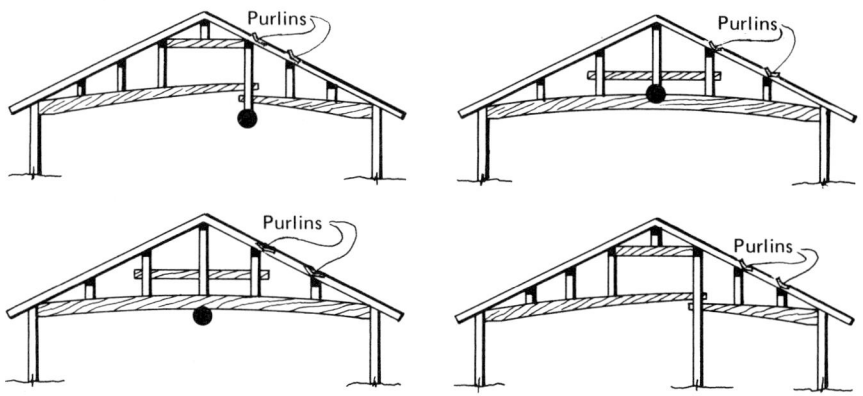

Figure 1-14. Typical roof framing systems in Japan. Note the absence of trussing that seems purposely avoided in situations where it would have been a logical solution.

In the Western tradition, the development of the morphology of the truss derived from a roof system consisting of a pair of inclined straight rafters that were the main members. This system, in essence, tries to counteract the rafters' horizontal thrust, whereas the Eastern approach to roof structures assumed the horizontal purlins were the main structural elements. The purlins are supported with a system of queen posts and tie-beams over columns surmounted by corbel brackets. (See Figure A-4 in Appendix A). This system lacks in lateral stability that is eventually provided by other nonstructural building components. Why the concept of trussing was not used, despite Japan's highly organized building technology, which codified the most elaborate construction details for wood structures, is difficult to understand. Either the culture never discovered the concept or purposely decided to ignore it. In the latter case the question still remains: Why? Triangular framing components were clearly used in the overhanging parts of Oriental roofs, but there was no further development of triangular forms into real trusses.

TRUSSES DURING THE RENAISSANCE

With the rediscovery of classical culture during the Renaissance, even the truss reappeared in the building art as it probably existed at the end of the Roman Empire. Bypassing the experiences of the Middle Ages, the new truss was now rediscovered and scientifically understood. Moreover, its evolution to present day forms can be traced directly from the Renaissance interpretations. Responsible for the new interpretation and its modern forms was Andrea Palladio (1518–1580) from Vicenza. His trusses, used for roof structures and for bridges, were statically stable in accordance with present engineering criteria. The geometry of the Palladian truss still in use today signalled the beginning of the trussing system on a scientific basis. Whether Palladio rediscovered this principle during his period of research on classical architecture, or whether he invented trussing criteria *ex novo*, he nonetheless remains the inventor of the modern truss, even if he modestly denied such a claim in his writings. Some of his basic truss types are included here (see Plates 1-5 and 1-6, and 1-11 through 1-18).

THE HISTORICAL DEVELOPMENT OF THE TRUSS 21

Plate 1-5. The Palladian truss must have been a conventional structural type during the Italian Renaissance, as demonstrated by its representation in the details of Sandro Botticelli's paintings. "The Adoration of the Magi" (1482), in fact, predates Palladio (1518–1580).

Plate 1-6. Detail from "The Adoration of the Kings," London, National Gallery, also by Sandro Botticelli (1444?–1510). As does the previous painting, it shows the Palladian truss as a conventional structure even before Palladio's birth (1518).

THE HISTORICAL DEVELOPMENT OF THE TRUSS 23

Plate 1-7. Leonardo da Vinci (1452–1519); self-portrait, red chalk (about 1512–1515).

During the Renaissance, Leonardo da Vinci, too (see Plate 1-7), was interested in trusses, as indicated by his sketches illustrating the proposal for a wood truss bridge (see Plate 1-8; also see Plates 1-9a and b). As the truss was skillfully used by the master builder in creating architectural forms of daring virtuosity for the Gothic cathedrals, the Renaissance architect also used the truss as part of the space composition, rather than as a structural means alone. The truss in S. Miniato in Florence is an example of this, but even more convincing is the representation of the truss as a space composition in Botticelli's 1482 painting, "The Adoration of the Magi." After the new trussing concept had been established by Palladio's* treatise on architecture and through the prototypes he built, the new technique spread throughout Europe, then expanded to the New World. (See Plates 1-10 through 1-18.)

24 SIMPLIFIED TRUSS DESIGN

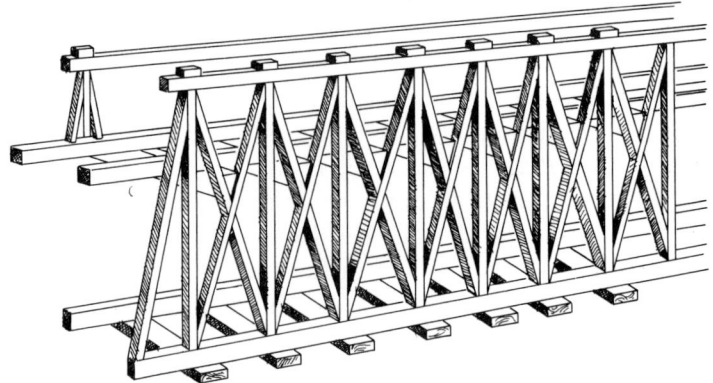

Plate 1-8. Model of a bridge truss by Leonardo da Vinci. This is another example of the mature stage of truss development reached during the Renaissance.

THE HISTORICAL DEVELOPMENT OF THE TRUSS 25

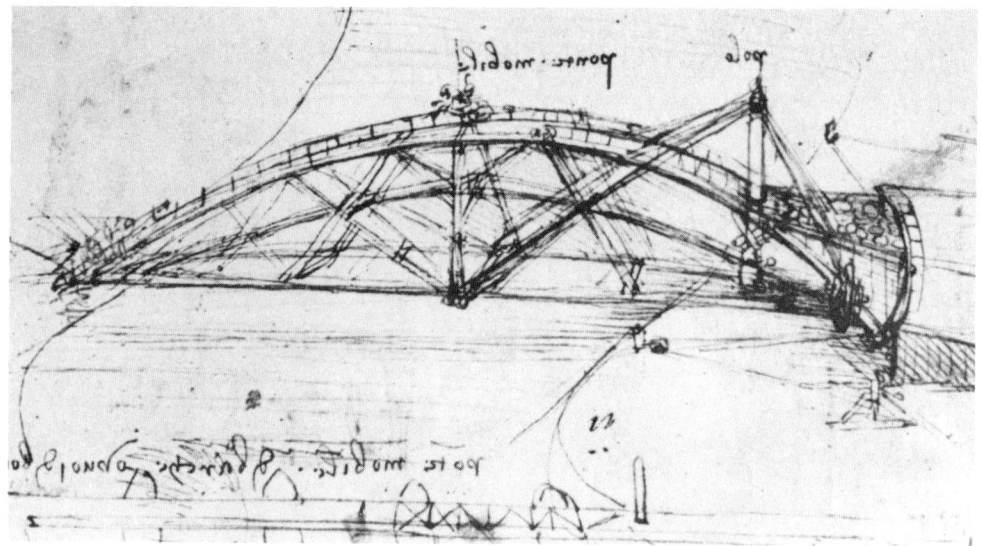

Plate 1-9a. Drawing by Leonardo da Vinci (about 1480–1485) of a temporary rotating bridge, consisting of an arched cantilevered truss. Agile, light and avant-garde design at the time.

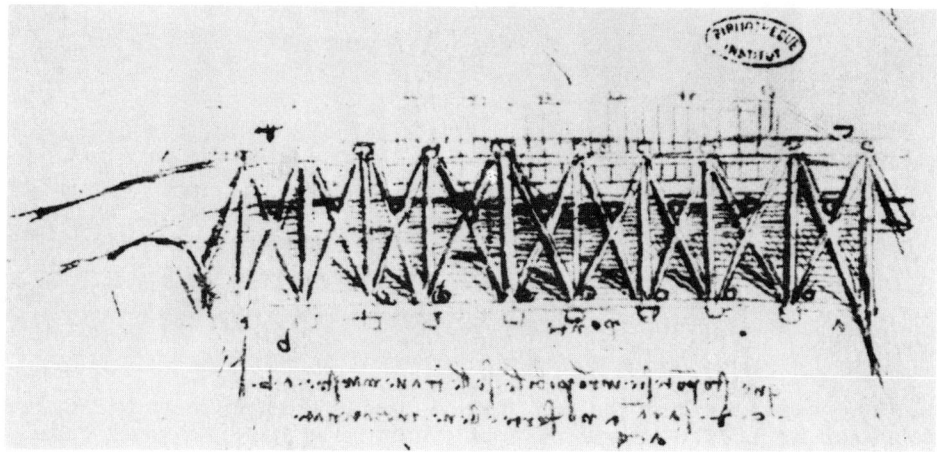

Plate 1-9b. Schematic sketch by Leonardo da Vinci, illustrating a light, slender truss for a mobile bridge.

Plate 1-10. Andrea di Pietro, known as Palladio, born in Vicenza in August 1580. Portrait engraved by B. Picart and derived from a painting by Paolo Veronese. Palladio, through his studies in classical architecture and intense research on Roman monuments, is the rediscoverer of the truss and interpreter of its structural significance in terms of present concepts of statics.

Plate 1-11. Palladio's drawing of the Egyptian Halls, illustrating the typical Palladian truss (Plate XXVIII, Book II, *The Four Books of Architecture* by Andrea Palladio, 1570).

28 SIMPLIFIED TRUSS DESIGN

Plate 1-12. A truss type (a reinforced triangle) as illustrated by Palladio in the drawings of the house of Count Iseppo de Porti (Plate IV, Book II, *The Four Books of Architecture*).

Plate 1-13. A typical truss type, indicated by Palladio in his drawing describing the Tuscan Atrio (Plate XVII, Book II, *The Four Books of Architecture*).

Plate 1-14. Truss type illustrated by Palladio in his drawing reconstructing the temple of Jupiter in Rome (Plate XXVII, Book IV, *The Four Books of Architecture*).

30 SIMPLIFIED TRUSS DESIGN

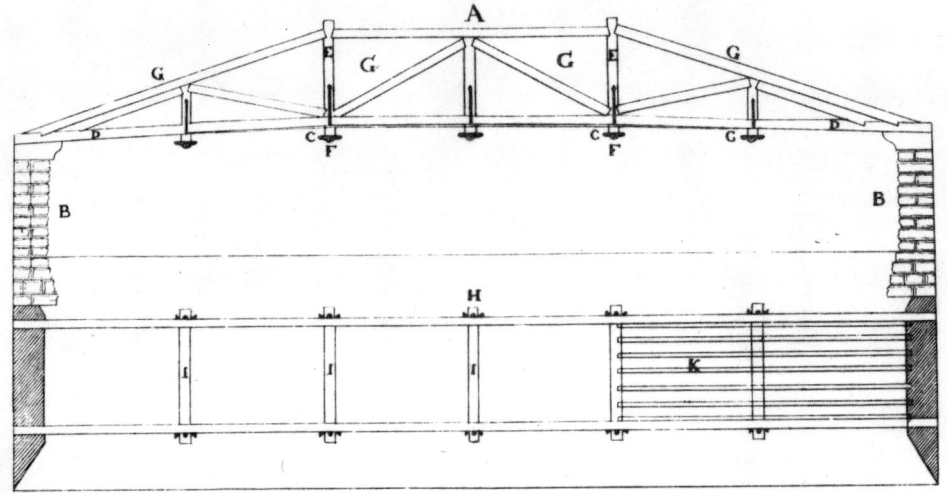

Plate 1-15. Illustration by Palladio of a truss for the bridge built over the Cismone River near Bassano (Plate III, Book III, *The Four Books of Architecture*).

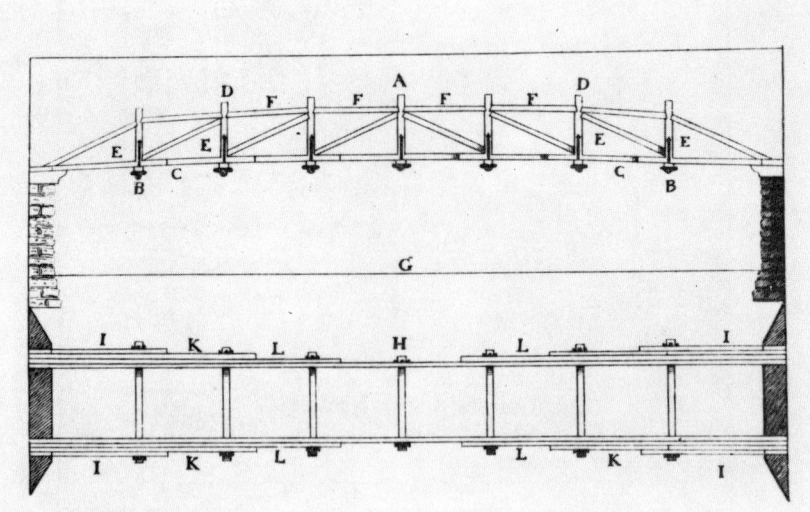

Plate 1-16. Illustration of a truss by Palladio, representing a bridge type he claimed was popular in Germany, according to the description by Alessandro Picheroni (top of Plate IV, Book III, *The Four Books of Architecture*).

THE HISTORICAL DEVELOPMENT OF THE TRUSS 31

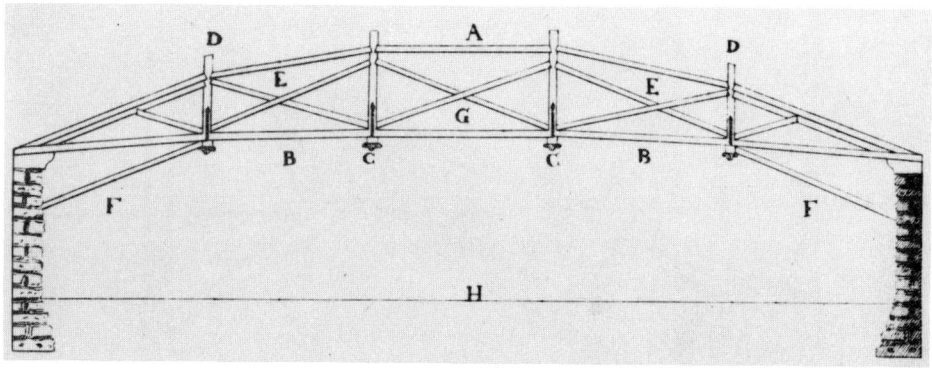

Plate 1-17. Illustration of a truss bridge by Palladio (bottom of Plate IV, Book III, *The Four Books of Architecture*).

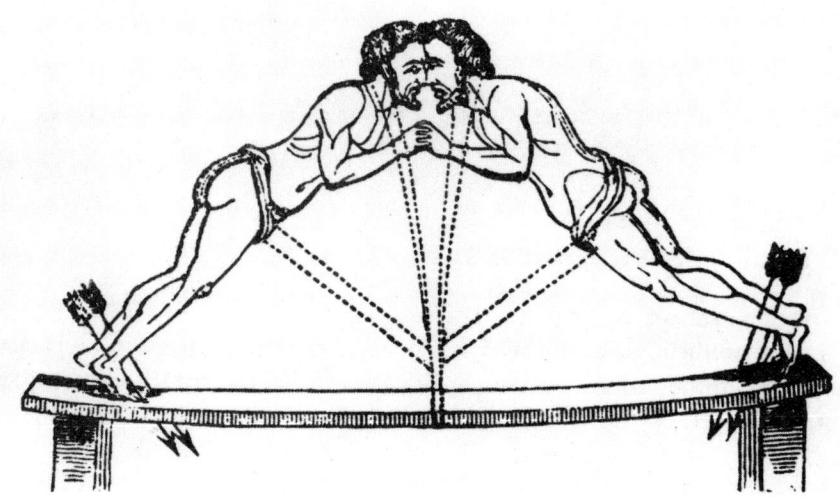

Plate 1-18. Graphical explanation of the structural behavior of the Palladian truss.

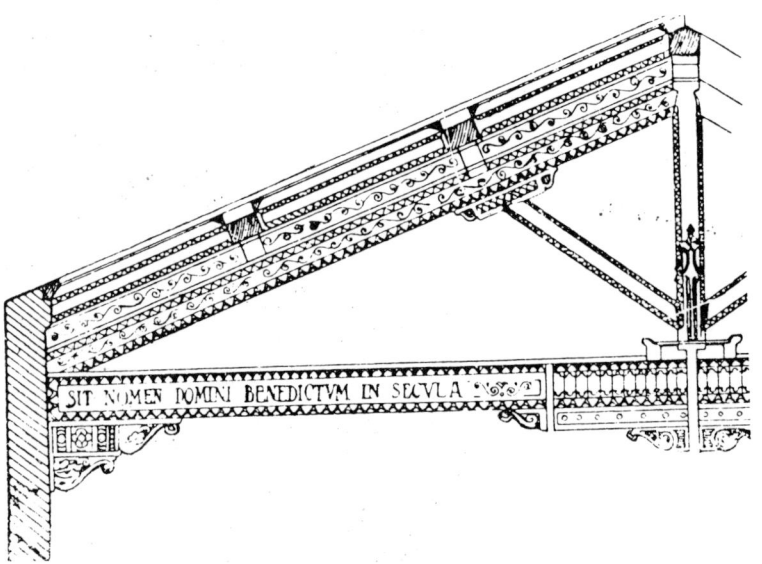

Plate 1-19. A roof truss in Saint Miniato in Florence, showing the elaborate decoration added to trusses during the Renaissance.

LONG-SPAN TIMBER TRUSSES

The first generation of the new trusses was, of course, of timber construction. The Kpellbrücke and the Spreuerbrücke in Lucerne, Switzerland, are two early examples of wood trusses applied to bridges. Other examples are the bridges built by the Swiss brothers, Johann and Hans Goubermann, in the eighteenth century. One such bridge was built in 1758, over the Limmat at Baden. It displayed a clear span of 390 ft.

It is important to notice that the development of the truss primarily depended on the necessity of building longer and longer bridges, rather than supporting roof structures in buildings, although the roof truss takes advantage of the bridge truss's development. Of course, roof structures differ from trusses used in bridges in many substantial ways. Loading conditions, for instance, which are basically static in nature for roof trusses, include, for bridges, dynamic loads, which are more complex to analyze. Moving loads and vibrations from vehicular loads and certain wind actions are conditions involving bridge trusses that do not apply to trusses in buildings.

AMERICAN TRUSS DEVELOPMENT

Trusses in the New World evolved in a different manner than European truss development. The expansion of the New World, which required emphasis on the practical aspects of technology, allowed pioneering work without the restriction of the European conservative traditions. The urgent need for bridges over wide rivers demanded the construction of long-span structures which at that time could be realized only with wood trusses. The men involved usually had little formal training in engineering, thus their methods were still empirical and based on the intuitive approach. This development was contrary to that of the European system, which relied heavily on scientific principles. It is amazing to see the extraordinary accomplishments achieved in the first period of timber truss construction, even with such great limitations. One of the first names we should remember is that of Timothy Palmer of Newburyport, Massachusetts, a self-taught man who started his career of bridge building in the 1780's, with a background in wood carpentry. His work is characterized by the covered timber truss, which protected its structural members from weather deterioration. Examples of his work are the Essex-Merrimac Bridge over the homonymous river (1792), and the Pisca Tagua Bridge at Portsmouth, New Hampshire (1794), in which he used a Palladian arched truss for the main span.

Following in the vernacular architecture of timber bridge building was Theodore Burr of Harrisburg, Pennsylvania. A self-taught engineer, he patented a form of truss (named after himself) in 1817, which consisted of a Palladian truss reinforced with an arched wood element, forming a structural unity.

Louis Wernwag, a German-born carpenter and immigrant of Philadelphia, continued the American tradition of empirical engineering for timber truss bridge building. Among the several bridges he built, the Colossus over the Schuylkill River at Fairmont Park, Philadelphia (1812), was the most famous, as it was the longest-span wood bridge in the United States at that time. A significant contribution which he introduced to timber truss development was wrought iron diagonal tension members.

Through formal training, the American bridge designers' knowledge of scientific principles gradually increased; yet these men still lacked the professional training of their European counterparts. Ithiel Town was one of the first formally trained American engineers to continue timber truss bridge design. His truss was patented in 1820. A basic innovation of the design was the absence of the arch form and the adoption of lattice work.

In 1830, Stephen L. Long patented a timber truss design referred to as the "assisted truss," a rectilinear Palladian truss reinforced with additional members at midspan and at both supports (see Figure 1-15). Two other patents in 1836 and 1839 followed the first one and modified it. Contrary to the American empirical designers of the earlier period, it is evident that Long could precisely calculate the stress in individual members. Long also patented, in 1858, a modification of the Burr arch truss in iron.

William Howe (1803–1852) born in Spenser, Massachusetts, one of the most significant contributors to the evolution of trusses, built his first bridge structure for the Baltimore and Albany Railroad at Warren, Massachusetts, in 1838, and patented this structure in 1840. Its major characteristics consisted of a timber truss with vertical wrought iron tension members and diagonal timber members extending across two consecutive panels (see Figure 1-16). An application of such a patent can be found in the Western Railroad of Massachusetts Bridge across the Connecticut River at Springfield (1841).

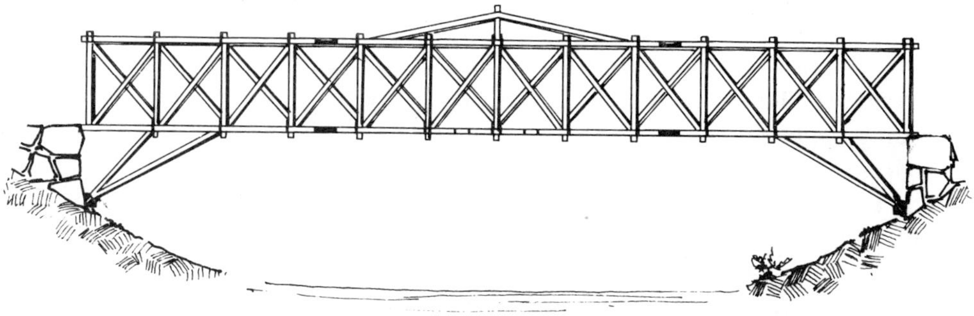

THE HISTORICAL DEVELOPMENT OF THE TRUSS 35

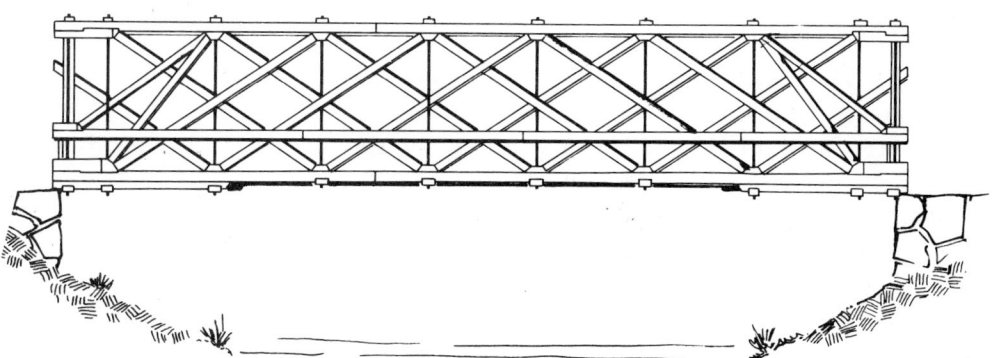

Figure 1-15. The timber truss patented in the United States in 1830 by Stephen H. Long. It was named the "assisted truss."

Figure 1-16. The bridge truss patented by William Howe in the United States in 1840. Basically, it is a timber truss with wrought iron vertical tension members.

The famous Howe truss that was adopted for bridge construction throughout the nineteenth century was, in reality, a modification of Howe's patent made by Harris and Stone, who applied Howe's design to the numerous bridges they built. The basic configuration of this Howe truss consisted of two diagonals crossing each panel. The vertical members between panels were usually wrought iron, while the top and bottom chords and the vertical end posts were heavy timber. A third patent of less impact, obtained by Howe in 1846, had a timber arch on the side.

As America began to enter the age of iron, and as iron began to be industrially produced, its application to trusses was natural. Although its first construction was in combination with wood, very quickly the entire truss was built in metal. Due to the different characteristics of cast iron and wrought iron, the first was usually used for compression members; the second, for tension members. As the new material could afford more ambitious projects, expanding the limits of the wood truss even further, and since truss design was still dictated by empiricism, the danger of structural collapses of trusses became a reality. In the United States, the first patent for iron trusses applied to bridges was by August Canfield of Patterson, New Jersey, in 1831. In 1841, Earl Trombull patented another iron truss, which modified Canfield's.

Thomas Pratt (1812-1875), a graduate of the Rensselaer Polytechnical Institute in New York and a railroad engineer, designed several truss prototypes that gained long lasting popularity. His first contribution to innovative truss design was a modification of the Howe truss, which Pratt patented in 1842. This truss consisted of parallel chords and vertical struts in wood, and X wrought iron diagonals in tension (see Figure 1-17). Later modifications, which removed one of the diagonals in each panel, except in the middle, were widely accepted. In 1844, Thomas and his father, Caleb Pratt, patented two other trusses (see Figure 1-18), but Pratt's first patent remained one of the most famous types adopted in the nineteenth century for railroad bridges.

THE HISTORICAL DEVELOPMENT OF THE TRUSS   37

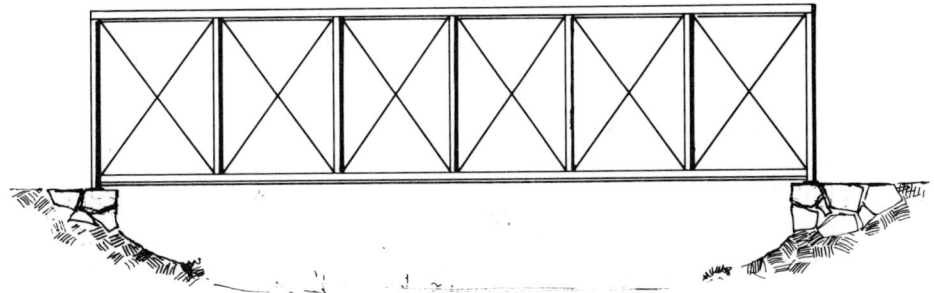

Figure 1-17. Thomas Pratt's bridge truss, patented in the United States in 1842, consisted of parallel chords and vertical struts in wood, and wrought iron diagonals in tension. This was one of the most popular trusses for railroad bridges in nineteenth century America.

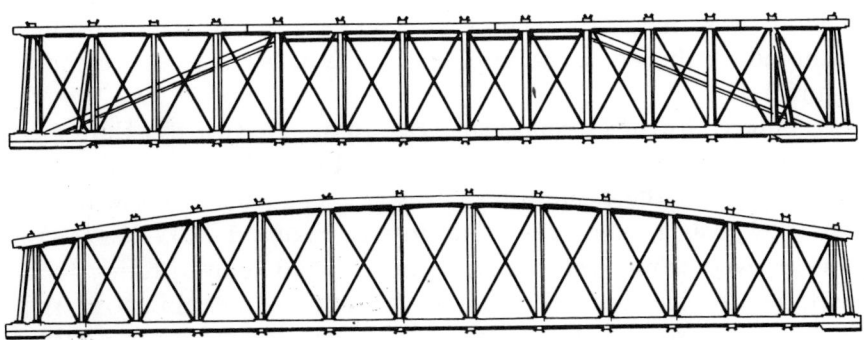

Figure 1-18. In 1844, Thomas Pratt and his father patented these two bridge trusses in the United States.

Squire Whipple, a railroad engineer born in Hardwich, Massachusetts in 1804, patented his first truss, named the "bowstring," in 1841. Using wrought iron for the compression members, the truss was characterized by a polygonal top chord acting as an arch in compression. But Whipple's most important contribution was the patent of 1847, for the truss named after him. Its original version consisted of a horizontal top chord and diagonal tension members in wrought iron. Modified in 1863 by John W. Murphy, this truss became one of the most popular types used for the long-span railroad bridges at that time. A further contribution by Whipple to the progress of trusses was a rudimental treatise on bridge design, published in 1851.

Wendell Bollman, a German-born engineer and bridge designer for American railroads, contributed to the development of the truss with a patent in 1850 for a new type, which was an original improvement of the Pratt. Its main feature was the addition of a number of wrought iron tension members anchored at the ends and supporting the truss as a suspension bridge (See Figure 1-19).

Albert Fink, born in Germany in 1827, emigrated to the United States, where he worked as a railroad engineer and bridge designer. He became a prominent figure in the evolution of trusses. The Fink truss, which he patented in 1851, was a major contribution (see Figure 1-20). Using concepts which he derived from Pratt's and Whipple's, but reinterpreting them in a new light, he produced something new. Without the bottom chord, the truss consisted of the horizontal top chord and vertical posts in timber, and wrought iron diagonals in tension. Later modification of the truss included the addition of a bottom chord. Notice that some confusion arises by attributing to Fink a truss type developed by Polonceau.

THE HISTORICAL DEVELOPMENT OF THE TRUSS 39

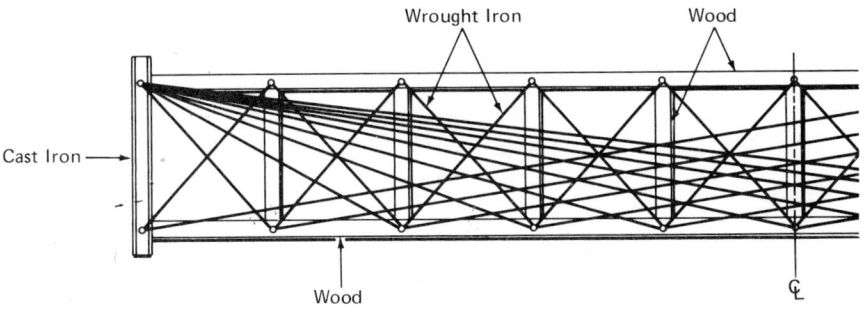

Figure 1-19. A bridge truss patented by Wendell Bollman in the United States in 1850.

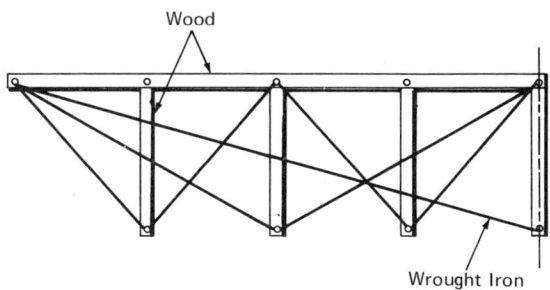

Figure 1-20. The original Fink bridge truss, patented by Albert Fink in the United States in 1851. Note the absence of the bottom chord, an original characteristic of this design.

In the second half of the nineteenth century, in the United States, many patents of minor importance added significant refinements to the development of the truss, which had already reached a mature stage. Among the many contributors, some of those worth remembering are presented in the list below.

Edwin Stanley (lenticular truss), 1851

J. B. Gridley, 1852

George W. Thayer, 1845, 1854

Samuel and Thomas Champion, 1854

Abraham S. Swartz (modification of Whipple's truss), 1857

George S. Avery, 1857

Lewis Eikenberry (compensating truss), 1859

George H. Pegram, 1887

W. E. Stearns (modified Fink), 1892

THE HISTORICAL DEVELOPMENT OF THE TRUSS 41

Plate 1-20. Published in the late 1860's, this illustration is one of the earliest examples of integration of vertical trussing with the facade of buildings. Proposal of Eugene Emmanuel Viollet-le-Duc (1814–1879) for an apartment building.

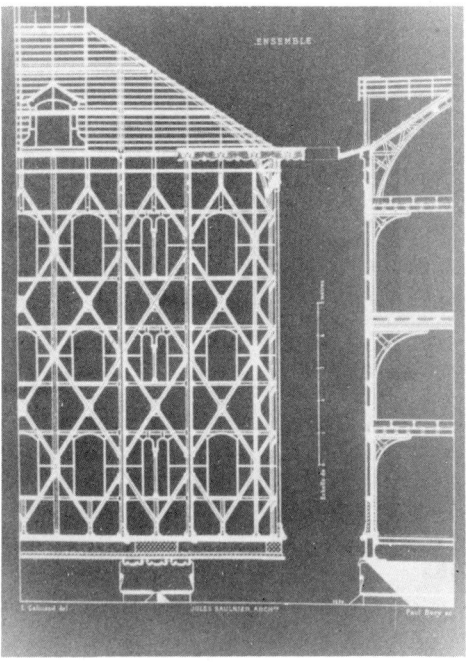 

Plate 1-21. Early example of vertical truss system exposed on the facade: the Chocolate Factory (1871–1872) at Noisiel-sur-marne near Paris, by Jules Saulnier. Infilled with wood tiles, the iron framework is part of the architectural composition of the facade. a. The structure itself. b. The facade with the integrated iron framework.

EUROPEAN TRUSS DEVELOPMENT THROUGH THE NINETEENTH CENTURY

With the progress of wood trusses through the evolution of bridge construction, the application of long-span wood trusses to buildings was a logical consequence. The best examples of this application are found in theaters that required large, unobstructed spaces. Examples of this include:

La Scala in Milano, Italy (1778), the world's most sophisticated theater structure of the time, with a span of 27 m or 89 ft.

The Farnese Theater in Parma, Italy (1618), designed by Gianbattista Aleotti, the prototype of the modern Western playhouse, with a seating capacity of 3,500 and a span of 32 m or 105 ft.

The theater in Darmstadt, Germany, built in the fifteenth century, with a span of 40 m or 131 ft.

The theater in Moscow, built toward the end of the nineteenth century, with a span of 46 m or 150 ft.

A new revitalization in the application of trusses occurred in the middle of the nineteenth century, when the development of graphic statics made calculations much easier than the algebraic methods previously used. By 1870, statical analysis of determinate trusses, either by analytical or graphical methods, was fully developed in Europe. By 1880, the analysis of statically indeterminate trusses was also fully developed. The heroic, pioneering efforts in America that had expanded the development of the truss to such a great extent were coming to an end, in this new era of modern science.

44 SIMPLIFIED TRUSS DESIGN

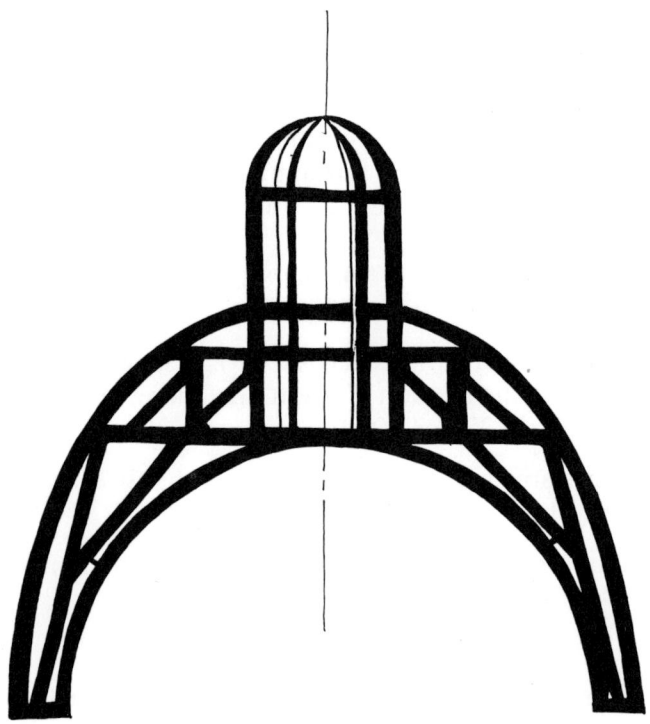

Figure 1-21. Trussed domes. When wood or metal, rather than masonry, are used for the construction of domical shapes, the structure does not work as a dome but in reality becomes a circular trussed frame supporting a non-structural skin cover. The economy in labor and material cost of wood or steel versus masonry in certain countries, along with the attachment to traditional architectural forms, produced such a hybrid type of truss application.

The most significant contributions in the field of graphic statics include the work of K. Culmann (1821–1881) and W. Ritter (1847–1906) in Zurich, L. Cremona (1830–1903) in Milano, and O. C. Mohr (1835–1918) in Dresden. Culmann, succeeded by his pupil, Ritter, is considered the father of graphic statics and its practical applications to truss analysis. Ritter, however, continued Culmann's work and carried it further with his own writings. Cremona's major work, in the analysis of trusses which became very popular throughout Europe just after its publication, was the formulation of a simplified graphical method for determining the stress diagram.

Figure 1-22. Luigi Cremona (1830–1903), engineer, architect, distinguished mathematician, professor at the University of Balogna and the University of Rome, originator of the famous methods for graphical determination of the axial forces in the members of statically determinate (isostatic) trusses. Although credited by some sources to Clerk Maxwell, the Cremonian diagrams, named for him, spread rapidly throughout Europe for their simplicity and practicality.

Just as iron gained prominence in American truss design, it captured European designs as well, particularly in the area of bridge building. In 1845, Neville, born in Belgium, built an iron bridge using a truss configuration then known as the Warren truss, which was made out of consecutive equilateral triangles. In 1846, James Warren and Willoughby Monzani built a truss which they patented in England in 1848. This truss was similar to Neville's of 1845, and had no vertical members, just diagonals of equal lengths sloping alternately in opposite directions. These diagonals formed equilateral triangles with parallel top and bottom chords. A major objective of this design was economy of construction, all members being of the same length (see Figure 1-23).

F. A. Pauli (1802–1883), born in Worms, Germany, designed a lens-shaped truss called the Pauli girder. Its particular characteristic was that its geometry tended to keep the axial forces constant throughout the length of the top chord (see Figure 1-24). Many bridges over the Rhine and the Danube were built using this design.

Jean Barthélemy Polonceau (1813–1859), a railroad engineer born in Chambery, France, who graduated from L'Ecole Centrale, had an important

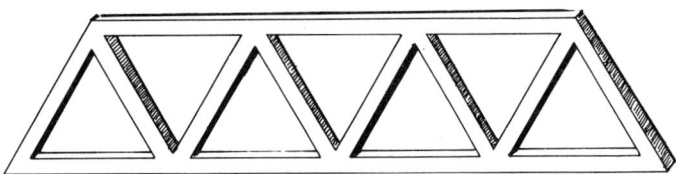

Figure 1-23. Original Warren truss as patented in 1848 by James Warren and Willoughby Manzani in England. Note that the original patent was based on a pattern of equilateral triangles. Later applications often used isosceles triangles.

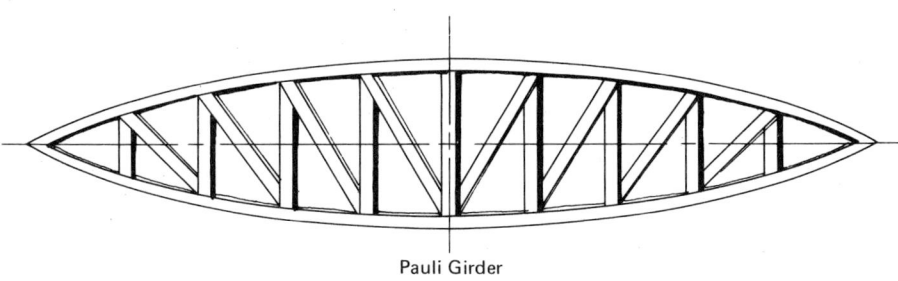

Pauli Girder

Figure 1-24. A lens-shaped truss, called the Pauli girder, patented in Germany by F.A. Pauli (1802–1883). Its particular characteristic is that its geometry tends to keep the axial forces constant throughout the length of the top chord.

role in the evolution of truss morphology in Europe. In truss design, his major contribution was the truss named after him, consisting of pitched top chords in wood, short diagonal struts in cast iron, and wrought iron tension members (see Figure 1-25). His truss, which was very popular as well as being a major achievement in truss design, is sometimes confused with the Fink truss.

The Mogine truss was named after its inventor, a Belgian engineer who patented it in Germany in 1858. It consisted of a Warren truss in which additional diagonal members in the central panel were added.

As iron structural elements were introduced into buildings, making large spans possible, the iron truss was one of the most popular applications because of its light weight. Long-span iron trusses appeared as early as the 1850's, as in the case of the Crystal Palace in London. The largest span used over the main nave reached 72 ft (see Figure 1-26). But the best known glorification of the iron trussing system of this period is the tower that Gustav Alexandre Eiffel designed and built in 1889 for the World Expo in Paris. It is 300 m tall (over 900 ft) and weighs over 8,000 lb (see Figure 1-30).

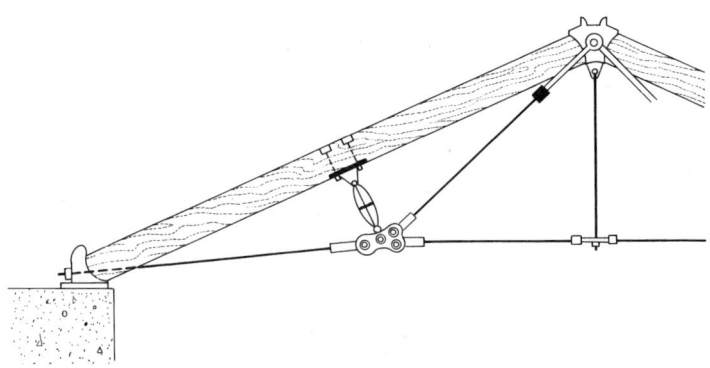

Figure 1-25. The truss designed by Jean Barthelemy Polonceau (1813–1859), with the top chord in timber, short compression struts in cast iron, and wrought iron tension members.

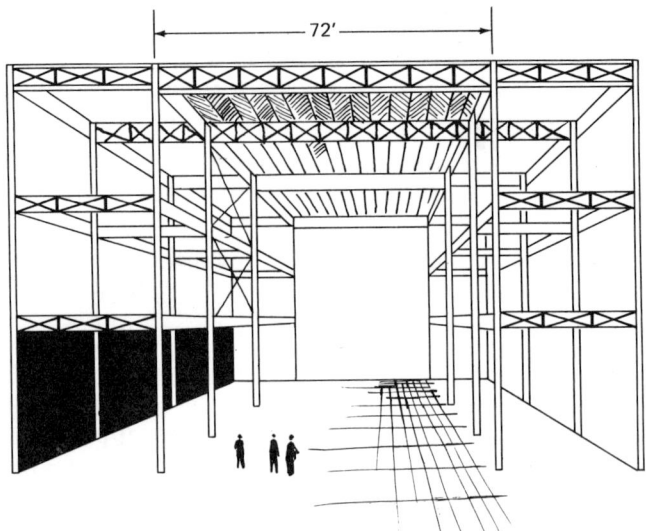

Figure 1-26. Lightweight iron trusses allowed long-span floors and roofs in buildings. Unobstructed large spaces could be achieved, as in the case of the Crystal Palace in London (1850).

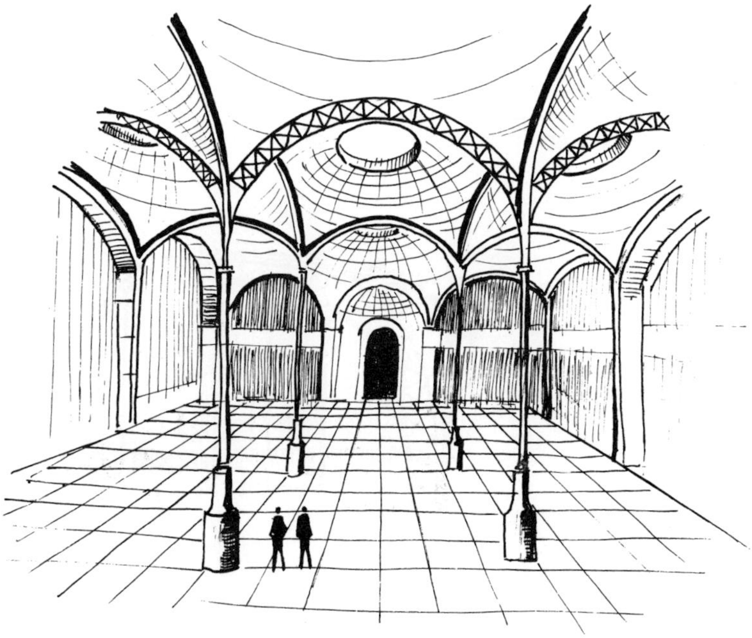

Figure 1-27. Trussed iron arches supporting the terra-cotta domes over the reading room of the National Library in Paris (1867-1868), by Henri Labrouste, architect.

50 SIMPLIFIED TRUSS DESIGN

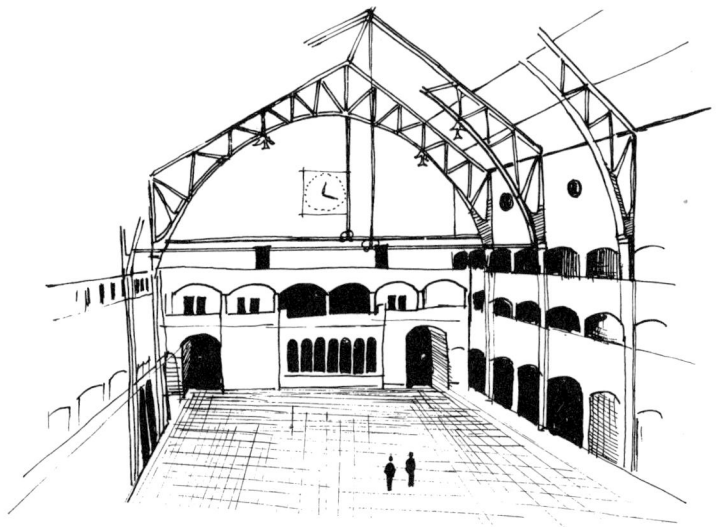

Figure 1-28. Arched iron trusses supporting the glass and metal roof of the Stock Exchange in Amsterdam (1898–1903), designed by H. P. Berlage. The trusses fully exposed in the hall are an integral part of the design of interior space.

Figure 1-29. Trussed iron arches, 375 ft span, supporting the roof of the Halle des Machines, built for the 1889 International Exhibition in Paris. Designed by F. Dutert, architect, and Contamin, engineer, the building capitalizes on the iron trussed structures just as did the Eiffel Tower.

Figure 1-30. A nineteenth century symbol that capitalized on the inherent structural aesthetics of the truss.

TRUSS DEVELOPMENT IN THE TWENTIETH CENTURY

As the industrial age spread more and more technology throughout the world, another material besides iron and wood began to be applied to long-span trusses. This was steel, industrially produced after the development of the Bessemer process in 1856. Steel trusses of great length, employed in building structures, during the 1930's reached clear spans up to 300 ft, as in the Martin Airplane Assembly Building in Baltimore (1938), designed by Albert Kahn. The trusses in this building were 30 ft deep and were spaced 50 ft on center. The span-depth ratio of 10, used in this case, is still approximately the most efficient, although for that span, other structural systems could probably be competitive nowadays.

With the evolution of the steel frame in skyscrapers, vertical truss systems have been used to stiffen the frame against the effects of lateral forces such as wind and earthquakes (see Figure 1-32). To increase the efficiency of slender steel frames, trusses have also been employed on top of the building structure to carry the floors below, which are suspended from them (see Figure 1-33). A further development of the previous method has introduced additional horizontal trusses at intermediate heights in order to distribute the floor loads over more than one level (see Figure 1-34). Another application of vertical trusses is the X bracing system used on the external frame of the John Hancock Building in Chicago (1968), designed by SOM (see Figure 1-39 and Plate 1-22). The structural efficiency of the vertical truss in reducing lateral sways is proven by the reduction of the ratio: (structure weight/unit floor area) of this 100-story building (29.5 psf, or 155 kg/m^2) compared with that of the 102-story-high Empire State Building (42.2 psf, or 98 kg/m^2). (See Figures 1-40 through 1-44.)

THE HISTORICAL DEVELOPMENT OF THE TRUSS 53

Figure 1-31. A twentieth century working structure that capitalizes on a truss system for its structural efficiency. Light, economical, easy to be transported, assembled and maintained, with minimal surface exposed to the wind, the trussed tower has a sculptural charm to many people.

54 SIMPLIFIED TRUSS DESIGN

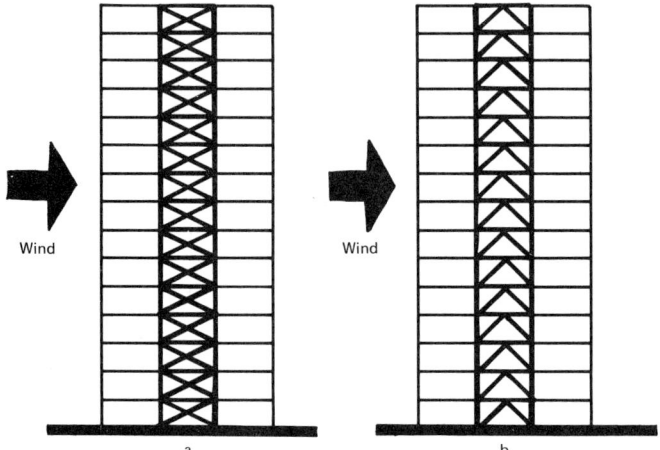

Figure 1-32. Steel frames with simple connections in high-rise buildings require vertical truss systems for stiffening to resist lateral forces. Typical geometrical patterns are the X bracing (a) and the K bracing (b).

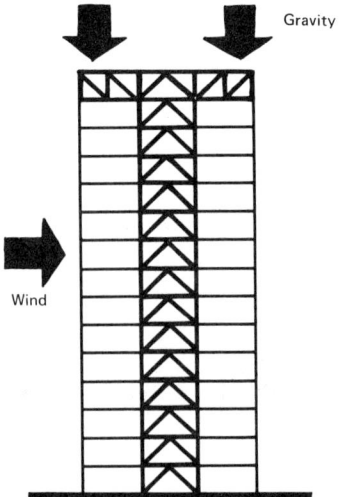

Figure 1-33. Topmost horizontal trusses, carrying the floors below in suspension, improve the lateral stiffness of tall steel frames.

THE HISTORICAL DEVELOPMENT OF THE TRUSS 55

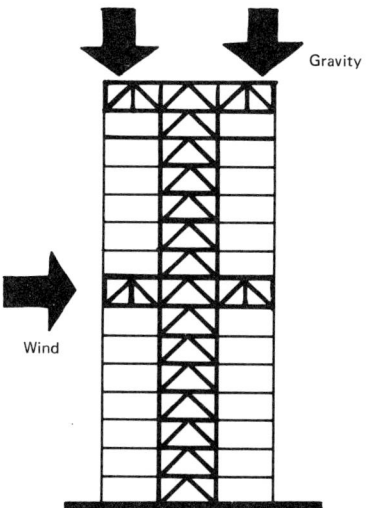

Figure 1-34. To improve the efficiency of the system in Figure 1-33, the addition of intermediate trusses carrying the floors below at various levels proves effective.

Figure 1-35. First Wisconsin Bank Building in Milwaukee, designed with belt trusses which stiffen the structure against side sways. Belt trusses at the top and at mid-height connect the exterior columns to the vertical interior trusses. The belt trusses are clearly expressed on the facade and made part of the architectural composition of the building.

56 SIMPLIFIED TRUSS DESIGN

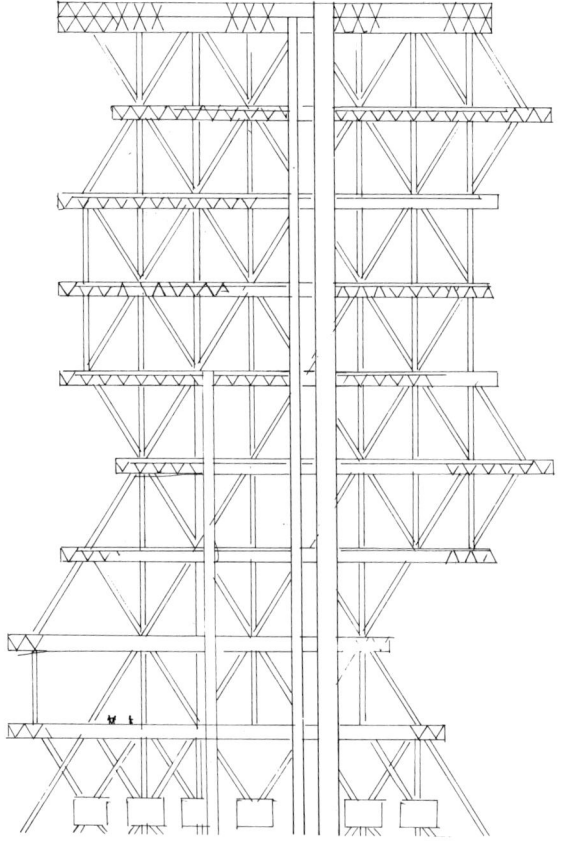

Figure 1-36. Louis I. Kahn's schematic models of two proposed tower buildings for the city of Philadelphia are spatial compositions of trusses on various scales. The largest trusses, in fact, include several stories within the depth of their triangular panels; others, at a low scale, instead support the floors. The theme developed by Kahn in this proposal was certainly a bold new statement in the use of truss systems in building design back in the 1950's. a. Precast concrete multi-story truss system that provides the primary frame for the whole structure (1952-1953). b. Concrete truss system for multi-story structure that resists wind forces, acting as a vertical truss cantilevering from the ground (1957).

THE HISTORICAL DEVELOPMENT OF THE TRUSS   57

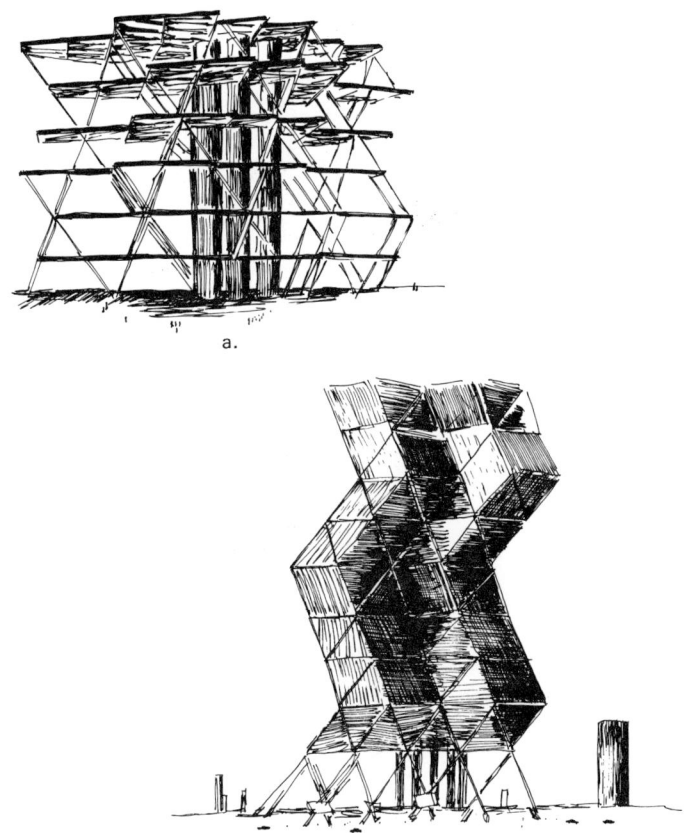

Figure 1-37. Section through the 616 ft high City Tower Municipal Building showing more details of the triangulated structure. Notice the vertical distance between nodal points of the primary frame equal to 66 ft.

58 SIMPLIFIED TRUSS DESIGN

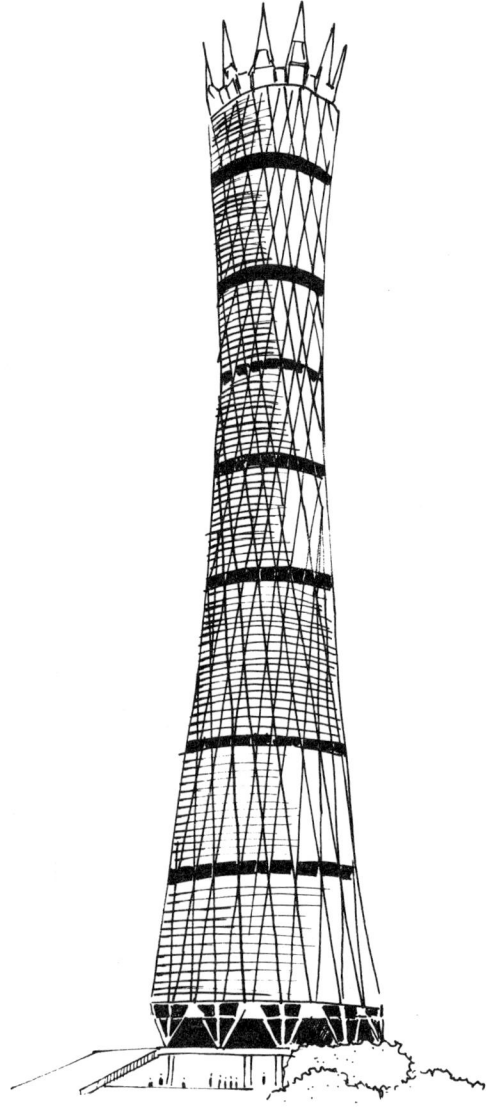

Figure 1-38. Intersecting diagonal members along the generators of the hyperboloid form the exposed structure of this proposed office building. Architects: I. M. Pei & Associates; Engineers: Roberts & Schaefer Co. (1956–1957). Such exterior lattice of tension and compression members is connected every 13 stories to the structural interior core, considerably increasing the stiffness.

THE HISTORICAL DEVELOPMENT OF THE TRUSS 59

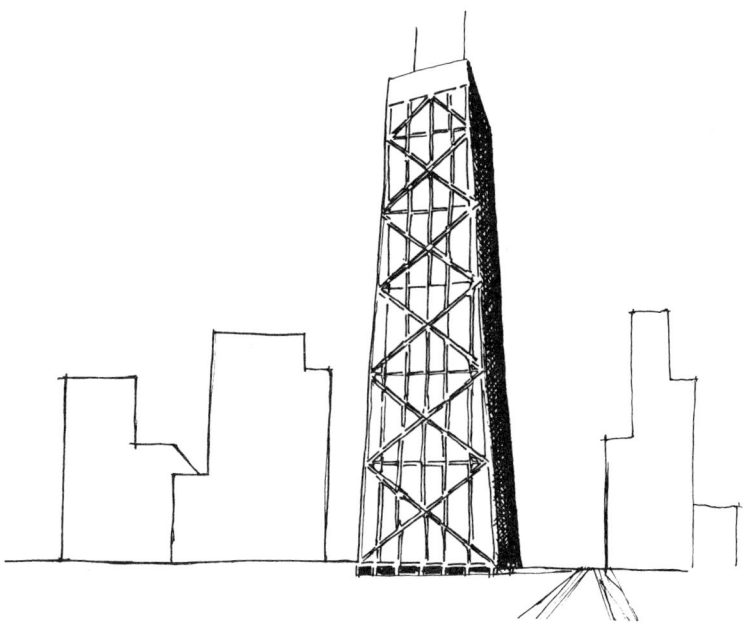

Figure 1-39. Perimetral bracing (vertical trusses) of large steel structures proves very effective for particularly tall steel structures, as illustrated by the John Hancock Building in Chicago (1968), designed by SOM.

Figure 1-40. The Prudential Town Center in Southfield, Michigan, by 3-D International: another example of an exposed trussing system that, like the SOM's Alcoa Building in San Francisco, is integrated with the rest of the structure and dominates the composition of the facade.

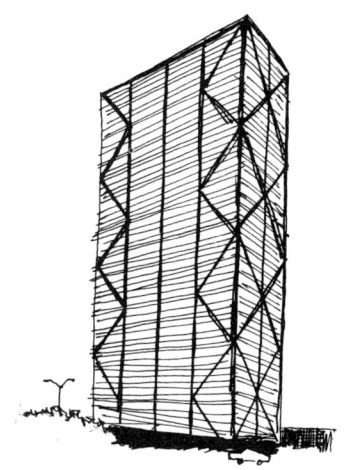

60 SIMPLIFIED TRUSS DESIGN

Plate 1-22. Although previously applied in other examples, the integration of steel braces in the architectural composition of the facade became a new trend when, in 1967, SOM applied this concept to the Alcoa Building at the Golden Gateway Center in San Francisco. Notice that the truss action of the braces absorbs 75% of the lateral load (wind and earthquake forces), leaving the remaining 25% to the 24-story frame.

THE HISTORICAL DEVELOPMENT OF THE TRUSS 61

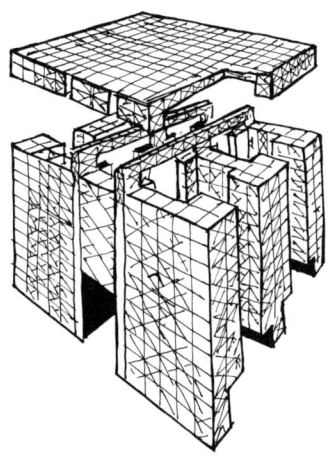

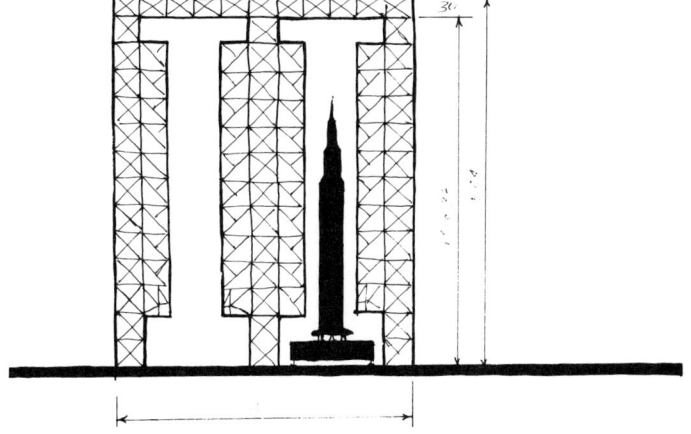

Figure 1-41a, b. The Vehicle Assembly Building at the NASA John F. Kennedy Space Center in Florida (1967). Architect: Max Urban; Structural Engineers: Roberts & Schaefer Co. It includes the largest volume ever recorded: 129 million ft^3 that can be expanded to 178 million ft^3 with future extension. At 550 ft high, with four doors 456 ft high and capable of admitting a 45-story skyscraper inside, the building consists of a giant space-truss system in steel. Designed and built in record time on the basis of 2,700 drawings and 20,000 shop drawings, this structure is an impressive example of the potential truss systems offer.

62 SIMPLIFIED TRUSS DESIGN

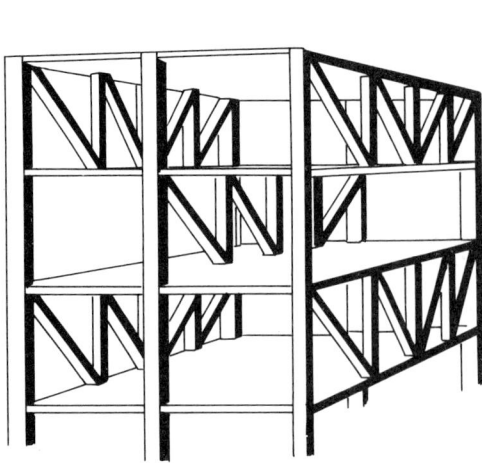 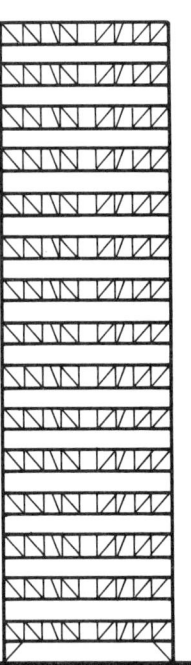

Figure 1-42. A conceptually new structural system for high-rise buildings, designed at M.I.T. in 1967. One-story high steel trusses, spanning the width of the building, carry the floor loads bearing on the top chord or suspended from the top chord.

THE HISTORICAL DEVELOPMENT OF THE TRUSS 63

Figure 1-43. Steel trusses are still used efficiently in the floor system of the World Trade Center in New York City (1972).

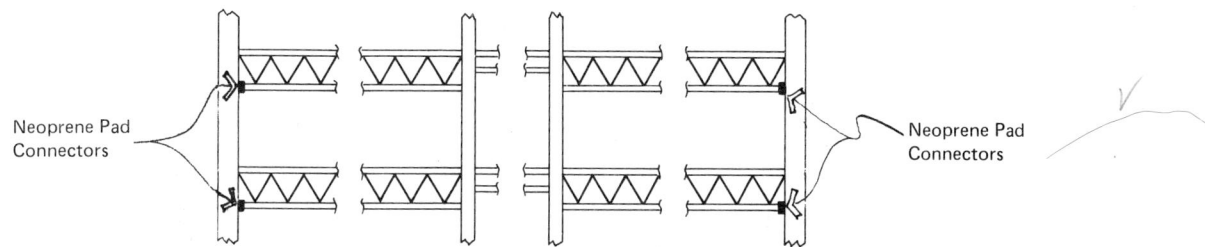

Figure 1-44. A peculiarity of the connections between the trusses and the columns in the World Trade Center is the interposition of neoprene pads that dissipate wind-induced vibration.

64 SIMPLIFIED TRUSS DESIGN

Exposed vertical trussed system on the facade of the Mercantile Center in St. Louis, Mo. (1975): dominant theme in the architecture of the tower. (Courtesy of: Sverdrup & Parcel and Associates, Inc., architect.)

THE HISTORICAL DEVELOPMENT OF THE TRUSS 65

An expression of the Russian "avant-garde" constructirism, the design for the monument to the Third International (1920) by V.E. Tatlin capitalizes on the new aesthetic values of trussed structures.

66 SIMPLIFIED TRUSS DESIGN

Another material applied to truss systems today is reinforced concrete. In fact, with the advent of reinforced concrete at the start of this century, trusses were also built with this new plastic material that could have expanded the configuration of the truss to new limits. However, the most common applications in the beginning were confined, for the most part, to industrial buildings (see Plates 1-23 and 1-24), and the concrete truss only occasionally emerged in works of architectural importance. Among those who have used concrete trusses in innovative and creative forms is Pier Luigi Nervi (1891-1979), born in Sondrio, Italy. Some of the most significant examples of his work with trusses include his airplane hangers at Orbetello, Orvieto, and Torre del Lago (1939-1941). In these buildings, Nervi used the trussed arched systems of reinforced concrete as the main components of the structure (see Plate 1-25). Another significant use of reinforced trusses in Nervi's work is in the bus station at the George Washington Bridge in New York, built in 1961-1962 (see Plate 1-26).

Plate 1-23. A typical example of reinforced concrete trusses in industrial applications. Voxon Plant (Rome, 1966). Architect: Del Bufalo. Engineer: Aldo Ghira.

THE HISTORICAL DEVELOPMENT OF THE TRUSS 67

Plate 1-24. A typical reinforced concrete truss in industrial applications. Soc. ICAR (Commissionaria Fiat, Latina, Italy). Engineer: Aldo Ghira. Span: 20 m or 66 ft ±; Height: 2 m or 6.6 ft ±; Spacing: 750 m or 25 ft ±.

Plate 1-25. Extremely elegant solutions in reinforced concrete, arched truss systems are the structures designed by Pier Luigi Nervi in his military hangars built in Orbetello, Orvieto and Torre del Lago (1939–1941).

68 SIMPLIFIED TRUSS DESIGN

a.

b.

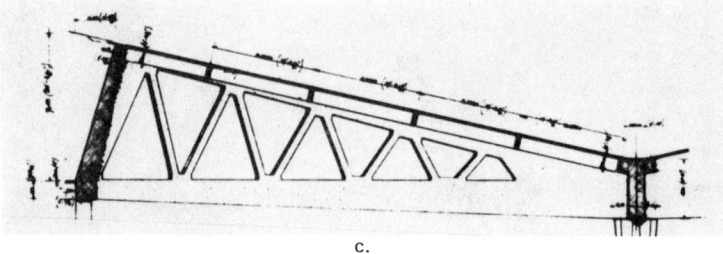

c.

Plate 1-26. A significant example of architecture and engineering in a reinforced concrete truss is the bus station at the George Washington Bridge in New York (1961–1962), designed by Pier Luigi Nervi: (a) top view; (b) side view; (c) truss detail.

THE HISTORICAL DEVELOPMENT OF THE TRUSS 69

Plate 1-27. Examples of reinforced concrete trusses are also found in the works of Eduardo Torroja. Concrete truss at Avila, Spain, during erection.

In countries rich in mineral resources, such as the United States, steel trusses are still among the most economical building systems used in conventional construction (see Figures 1-45 through 1-51). In fact, open web steel joists, usually of the Warren type, built presently with A36 low carbon steel or with high-strength steel (Fy = 50.000 psi), can span up to 128 ft, with a depth of 72 in. and a load-carrying capacity of approximately 500 lb/linear ft, and be very low in cost. For the sake of objectivity, however, it is fair to say that the present steel truss, although very competitive in various applications, including the steel joist, has recently been challenged by the new built-up steel girders. These girders have deep webs made of thin plates that recent AISC specifications allow to be built.

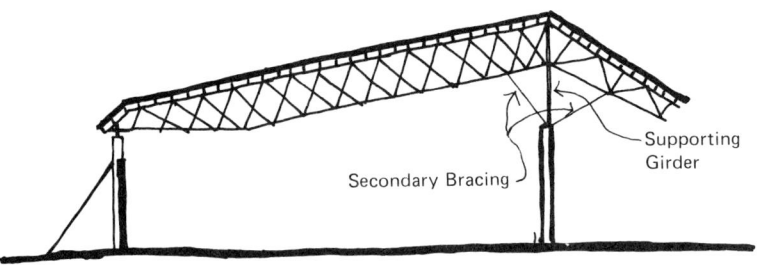

Figure 1-45. The dramatic aspect of cantilevering has a fascinating effect on which trusses can capitalize. Their relatively low weight allows longer span than those that could be reached by solid girders. The variety of conditions for using cantilevers is endless. However, an example of steel trusses simply supported with one cantilevered end span is the structure for the hangar at Torrejon, Madrid, by Eduardo Torroja.

Figure 1-46. A building that is all trusses. Roof, floor, walls, and columns are made of welded steel trusses clearly exposed through glass walls. Lightweight and efficient, the trusses allow long cantilevered spans and large column-free spaces. Phillis Wheatley Elementary School in New Orleans, Louisiana.

THE HISTORICAL DEVELOPMENT OF THE TRUSS   71

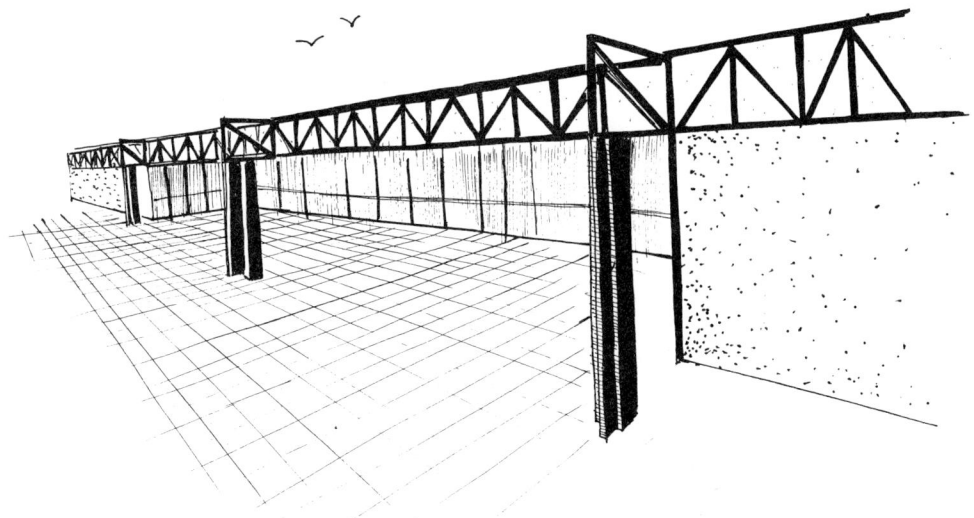

Figure 1-47. Example of steel trusses successfully incorporated in the architectural composition of a building. Painted charcoal brown, the trusses interact visually with the precast concrete panels with exposed aggregates and with the bronze-tinted glass. Factory and Administrative Building of Scientific Data Systems, El Segundo, near Los Angeles, California. Architect: Craig Ellwood Associates. Capitalizing on the aesthetic values of structural efficiency, trusses have come back recently with a new vocabulary. Completely different from the old image, the new generation of trusses encompasses tridimensional compositions flowing in space.

72 SIMPLIFIED TRUSS DESIGN

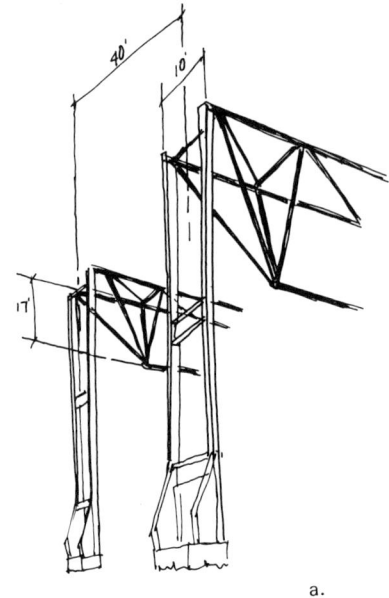

Figure 1-48a. Trussed portals 40 ft on center and spanning 135 ft. The horizontal truss, made of high-strength tubular steel, has a triangular cross section that includes two top chords 10 ft apart in compression and one bottom chord 17 ft below in tension. The two vertical columns, made of structural steel shapes 10 ft apart, are fixed at the base and hinged at the intersection with the truss. **b.** Exposed truss system that supports the roof structure of the Athletic Facility Building for Phillips Exeter, Exeter, New Hampshire, by Kallmann & McKinnell, architects, and Le Messurier Associates, structural engineers.

THE HISTORICAL DEVELOPMENT OF THE TRUSS 73

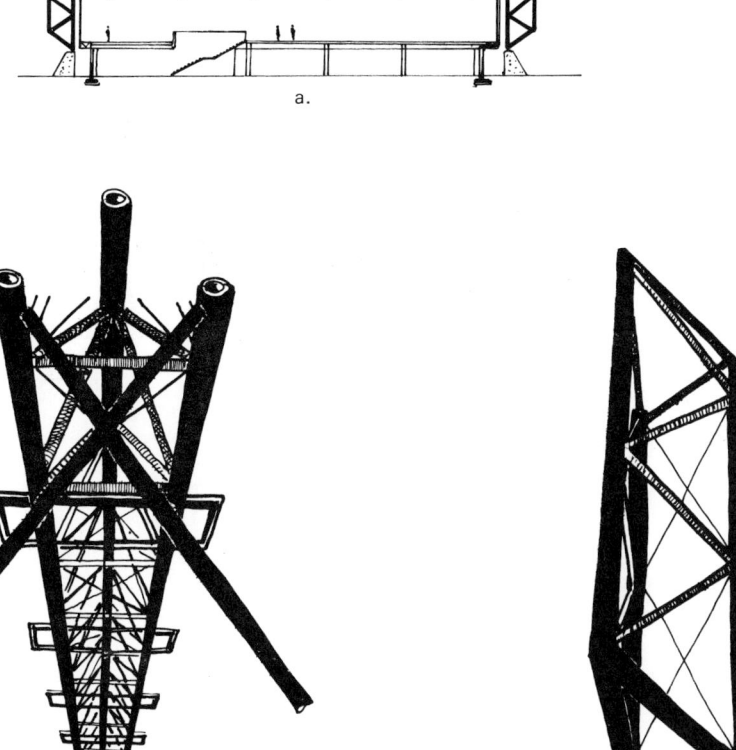

Figure 1-49a, b, c. New spatial forms and the variety of member size create sculptural compositions in the exposed trussed structure of this building. Designed by Angelo Mangiarotti, architect, and B.C.V. Studio, structural engineers, the industrial complex (1978) at Majano del Friuli, Italy presents an exciting new expression of trussing that expands and enriches the vocabulary of this structural typology.

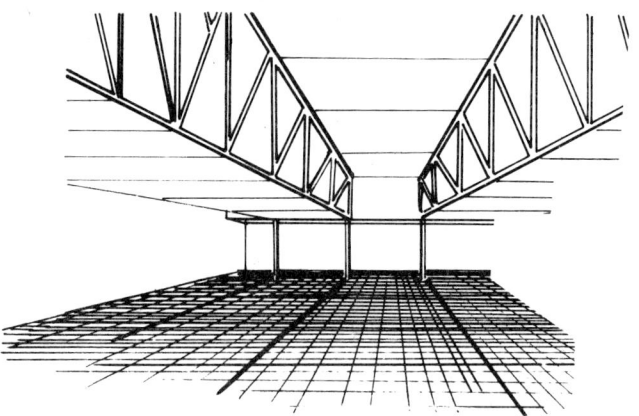

Figure 1-50. Open web steel joists.

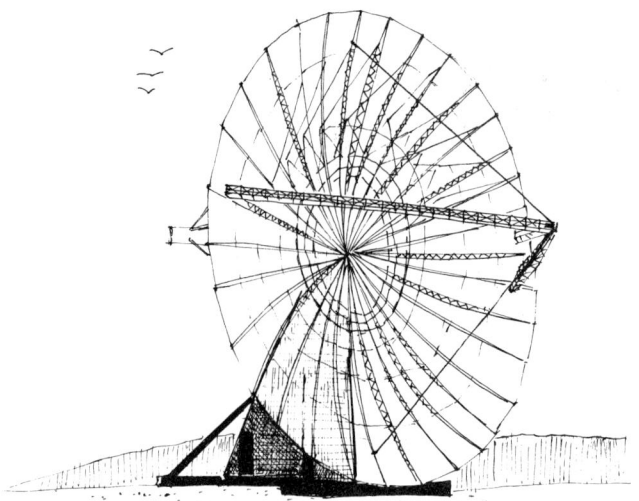

Figure 1-51. Tubular welded trusses of high-strength steel form the primary system for the parabolic antenna disk shown in this figure. Light, but exceptionally rigid, the structure can satisfy the specified minimum deflections under high wind loads because of the efficiency of the truss system.

A unique application of horizontal trusses in buildings is the "staggered trusses system" designed at M.I.T. through a research grant from USSC in 1967,* for high-rise apartment buildings up to 60 stories. The trusses, spanning across the width of the structure, are one story high, and combined with the external columns to form a multi-story frame, increase the stiffness against lateral forces.

The truss girders that are part of the main frame of the World Trade Center in New York City (1972) are a significant example of the fact that the steel truss still has an important role in present building technology. The ratio: (structure weight/unit floor area) of the 110-story-high steel structure, the second highest building in the world, is 37 psf, or 181 kg/m^2, as a result of the light-weight floor trusses. An innovative method adopted in this building to connect the trusses to the columns employs neoprene pads (10,000 for the whole building) that absorb and dissipate energy as the wind makes the building vibrate. Indeed, the application of trusses to high-rise buildings is commonplace. In fact, the world's tallest building, the Sears Tower in Chicago, uses trusses at the sixty-sixth and ninetieth floors, where the framing tubes drop off, and between the twenty-ninth and thirty-first floors, where trusses are two stores high. These trusses' presence, moreover, assists the building in behaving as a structural unity vertically cantilevered from the ground.

Where will future truss development lead? Capitalizing on the character of trusses and projecting into the near future, we think the application of pneumatic structures to trusses is a feasible and practical idea, particularly for buildings in places remote from civilization, or eventually, for structures to be erected on other planets. The concept is based on the adoption of tubular compression members filled with pressurized air or other gases, and tensile members made of light-weight fibers (see Figure 1-52).

76 SIMPLIFIED TRUSS DESIGN

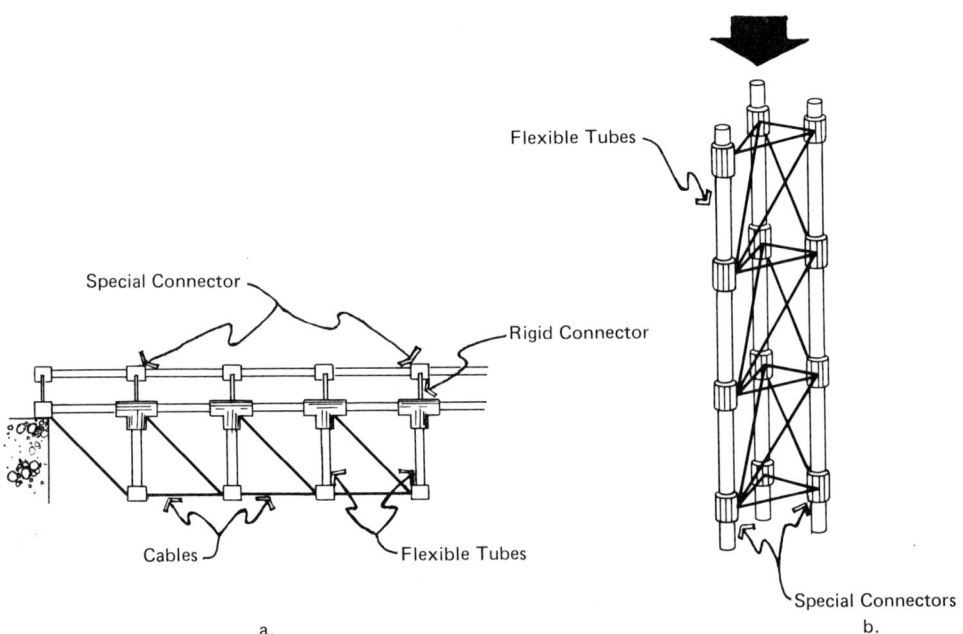

Figure 1-52. The pneumatic truss system, employing pressurized tubular compression members, is hypothetically feasible in the near future. Because of its exceptionally light weight, this system can be used where a conventional system cannot: (a) horizontal truss; (b) vertical trussed column.

Recent speculation on the origin of the truss proposes the theory that the truss was known in Persia at least from the third millennium B.C., on the basis of archeological findings.[1]

[1] Farshad, M., of Pahlavi University and D. Işfahanian of the University of Isfahan, "Iranian Plateau—The Homeland of Original Truss Structures," *Journal of the American Oriental Society* 98, No. 3: 248-250, July-September 1978.

THE HISTORICAL DEVELOPMENT OF THE TRUSS 77

From the excavations at Susa on the Persian plateau (early third millennium B.C.), one of the carved seals of the time reveals records of what appears to be a truss. Such a truss, of the same configuration as the modern type referred to as Warren, is part of a scene represented on the seal,[2] which includes a cylindrical building sitting on such a truss. The interpretation given is that the building is a grain storage raised above the ground in order to prevent contact with the moisture on the soil. Although the geometry of the truss is very clear, the fact that it was used as a foundation system leaves some doubts as to whether the system could be considered a real truss. Can we conclude just from this representation that the truss is a Persian invention 5,000 years old? A major question to such a hypothesis is why other cultures, such as the Egyptian and the Grecian, as well as those of the Far East, which interacted with the Persians, did not acknowledge this structural topology or, by the same token, why no other evidence of this truss has been found in Persia? (See Figure 1-53).

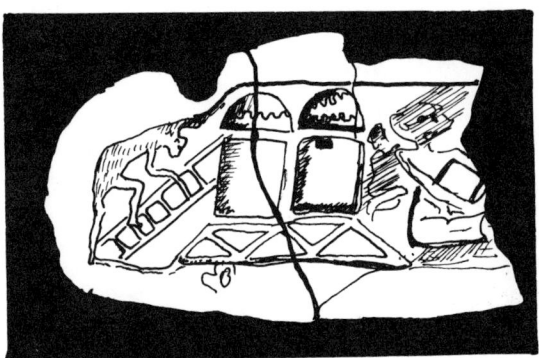

Figure 1-53. Carved representation on a seal of the third millennium B.C. found at Susa, Persia, showing what appears to be a truss supporting a cylindrical granary.

---

[2]Pope, Arthur U. and Phyllis Ackerman, *A Survey of Persian Art* I, Oxford University Press, London, p. 291, 1965-1965.

78 SIMPLIFIED TRUSS DESIGN

a.

SPACE ASSEMBLY OPERATIONS

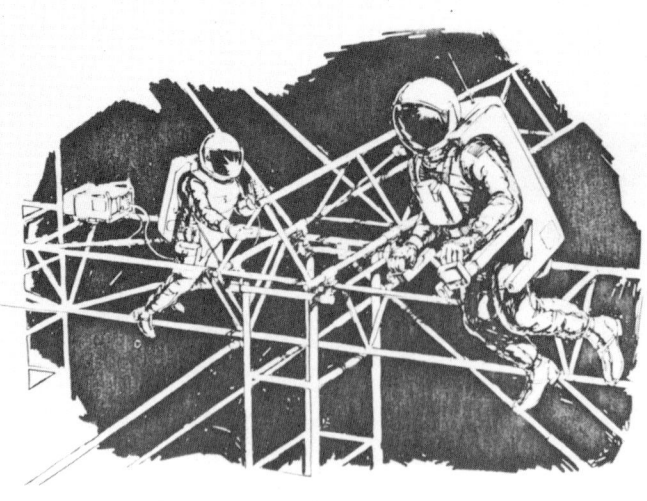

b.

Plate 1-28. Based on its high structural efficiency in terms of strength and low weight and for ease in transporting and assembling, the truss system is applicable to space structures as indicated by some projects that are under study at NASA.
    a. Large structure erection in space, using automatic fabrication module ("Beam Builder") from shuttle cargo bay.
    b. Space assembly operation of truss structures.

THE HISTORICAL DEVELOPMENT OF THE TRUSS 79

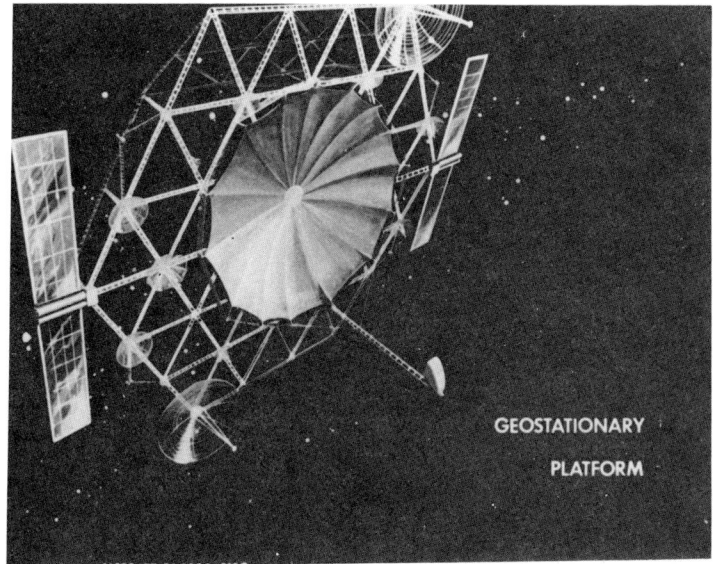

c.

d.

c. Geostationary platform for communications services.
d. Orbiting radiometer with 400 ft antenna (for weather, climate, and oceanography observations).

If the truss is conceptually an engineering achievement based on intuition of structural behavior and on the process of optimization of flexural members, its visual expression and its participation in the architectural unity of the building involve also the physical realization of the artifact. Humble, but vital and productive must have been the role of the craftsmen who shaped with their hands the final product. Especially during the Middle Ages and later in the Renaissance the work of these men inspired a mystic feeling first and a proud sense of human accomplishment in the latter period.

Moreover, if the truss is indeed a Roman achievement, as all evidences seem to prove — e.g. the bronze truss in the Pantheon, the bridge over the Danube River represented on the Trojan column, the large span roofs on the basilicas, etc.—the Roman carpenters may, most probably, be the inventors of the truss. It is not surprising, in facts, that this could be true if we consider how sophisticated was the skill of the Roman carpenters who built the wood falsework for arches, vaults and domes of Roman megastructures.

2
A Selective Evaluation of Trusses

HOW DO WE SELECT A TRUSS?

As we attack the problem of designing a truss, we must first gather some data on span, pitch, spacing, materials, etc. Then we must select some feasible geometrical configuration, either from among those already existing in pertinent literature, or from modifications we make as we see fit. We must then make the statical analysis for computing the axial forces in all members of the truss. Although we can use any of the many methods available, we suggest either the use of tables in Chapter 3 for one of the configurations therein, or the use of the computer program illustrated in Chapter 4. In the latter, there are no restrictions in geometry, degree of statical indeterminacy, or loading conditions. The classical graphical method by L. Cremona is also included.

We suggest that the reader follow the steps below.

1. Understand what the truss really is, through definitions and further interpretations of the concept.
2. Have an overall view of the many types of trusses already described in related literature.

Truss Forms in Nature

As we have seen up to this point, the fundamental concept that generated the truss is in essence a process that optimizes the efficiency of flexural members from the structural as well as construction standpoint. If we put aside the triangulation of bracing systems as truss generating factor (not very probable in fact), the process of evolution from the solid beam to the mature truss is a slow and laborious product of man's ingenuity and intuition without the benefit of inspiring examples from the natural world.

82 SIMPLIFIED TRUSS DESIGN

What is the actual counterpart in nature, we ask. How did nature refine and optimize flexural members in living organisms? Surprisingly enough, examples of trusses in nature are hard to find. Flexural members may have variable cross sections that follow the shape of the bending moment diagram (example: the tapering cross-section of tree branches which cantilever from the trunk), but almost nothing comes to mind that could be called a truss. The bone structure in wings of vultures is probably a unique example of beam refinement. It resembles the Warren truss. (See Figure 2-1).

Stretching our imagination a little we can also see a truss in the configuration of the bone in the pelvis of certain birds (see Figure 2-2). The king post is the truss shape that fits the image.

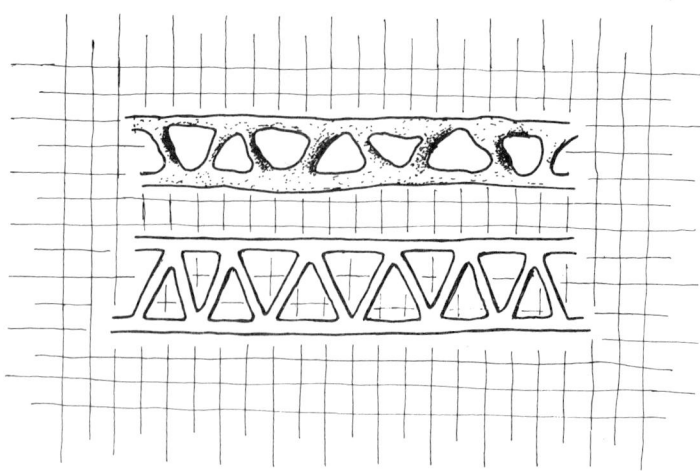

Figure 2-1. In spite of its high structural efficiency, trussing is not usually found in nature. The bone structure in the wings of vultures is one of the few examples of pseudo-truss (fixed joints) that resembles the shape of the Warren.

A SELECTIVE EVALUATION OF TRUSSES 83

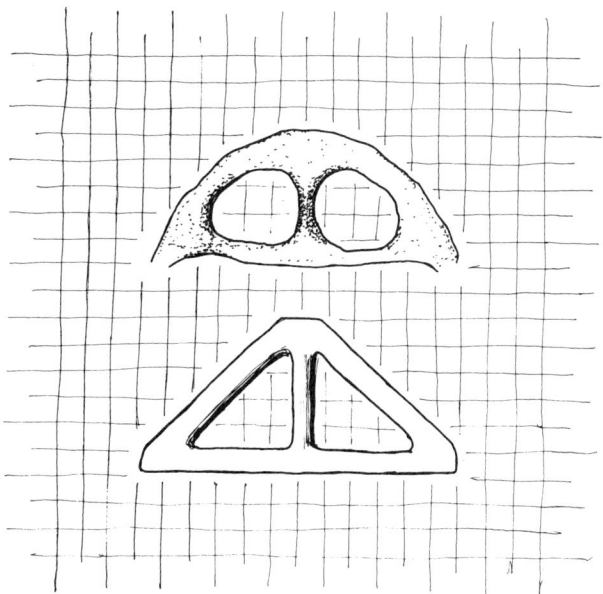

Figure 2-2. Another rare example of pseudo-trusses in nature: the pelvic structure of some birds that resembles to a certain extent the king post.

Definitions

A truss is a longitudinal structure which carries superimposed loads to the supports, just as a beam does. It is more efficient and lighter than an equivalent beam carrying the same load over an equal span. The structure consists of several members joined together. Their length, size, and material can vary from one truss to another. The loads travel through the members, stressing them with axial forces only—tension (+) or compression (−). All members of the truss lie in one plane, usually vertical, and all members are joined at their ends with other members, forming a series of consecutive triangles (triangles only). Joints, or points of connection, are assumed to be frictionless hinges, even if this is not true.

"Trussing systems" or "trussed structures" are terms with a larger scope than that of "trusses." They include all structures consisting of pin-connected members, axially stressed only in tension or compression, whether lying in the same plane or differently arranged in space, with or without regular geometrical order. These structures can have any primary function, such as beams, arches, slabs, columns, or walls. In fact, trussing systems include conventional planar trusses as well as tridimensional trusses, space frames, framed domes (geodesic domes), and structures of any configuration, such as the complex bridge structure in Figure 2-3, which contains a multitude of irregularly arranged struts and ties.

A SELECTIVE EVALUATION OF TRUSSES 85

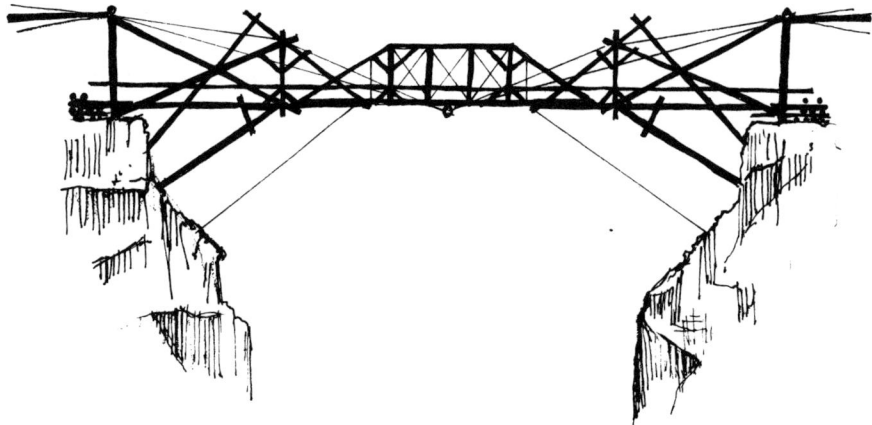

Figure 2-3. This structure, consisting of a multitude of members, either in tension or compression, and each performing a structural task, is a real truss. Its complicated and chaotic appearance at first glance can give to some observers a sense of confusion and distrust, but at the same time it can give to others a feeling of spontaneous artistic creation. Like modern "constructivism," the structure seems to take form gradually, moved by an intuitive sense of self-improving as symptoms of weakness arise. This light, primitive, and intriguing structure is the Ahwillgate Bridge in India.

Interpretation

A truss can be interpreted as the modification of a beam (see Figure 2-4a) such that its structural efficiency is improved. Thus, let us consider the two steps below, to illustrate the rationale for the generation of trusses:

1. Removing some of the material along the neutral axis (see Figure 2-4b) where the stresses are small, but leaving enough material to resist shear (horizontal and vertical).

2. Moving the remaining material farther away from the neutral axis to increase its flexural resistance (see Figure 2-4c).

A trussing system is an assembly of parts so connected and arranged that inefficient stresses of bending, torsion, and shear are eliminated and replaced with tension and compression. The resulting geometrical configuration of the whole embodies a process of optimization where mass is minimized by the rational use of form. Aesthetic values, as the product of rationality and creative design, are potentially associable to such forms.

Concept of Economy

If we want to improve the economical efficiency of a truss, we should follow the steps below.

1. Minimize the length of compression members.

2. Minimize the number of compression members, even if the number of tension members must be increased.

3. Increase the depth of the truss as much as is practical; this will reduce the axial forces.

4. Explore the possibility of using more than one material in the truss, one for compression and another for tension members.

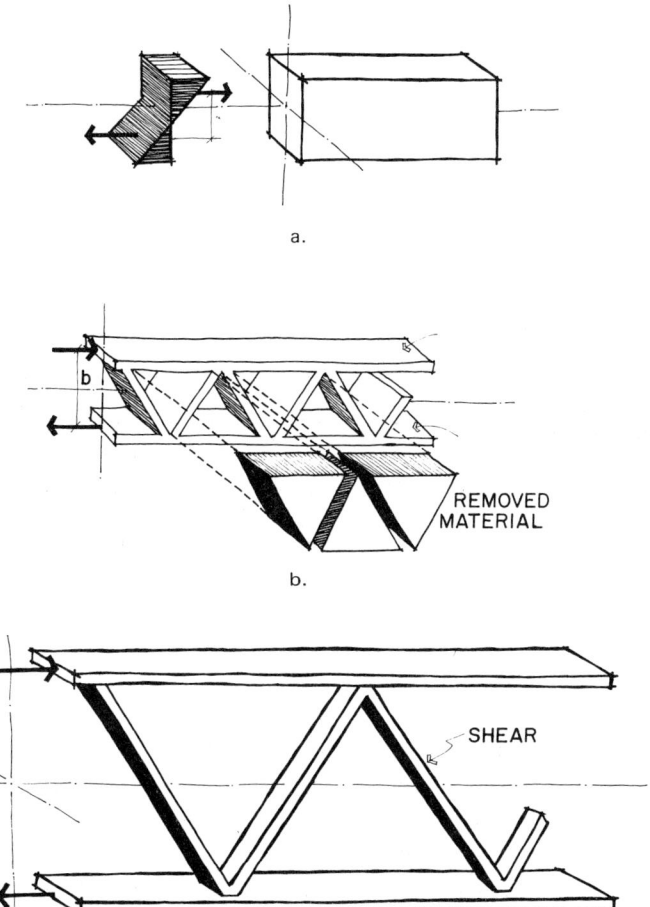

Figure 2-4. The genesis of the truss from the beam as a process of structural optimization.
  a. The beam action in bending.
  b. Removal of inefficient material in the zone near the neutral axis.
  c. Expanding the remaining material away from the neutral axis, thus increasing the resisting moment and generation of the truss.

Structural Comparison

A basic characteristic of trusses is their ability to carry loads over long spans. But other structural forms, such as plate girders, arches, cables, etc., share such an ability, although with different cost efficiency and span limitations.

A comparative cost analysis summarized in Figure 2-5 indicates that in the 100 ft span the truss is second in efficiency following the arch, and in the 200 ft span the truss is in third place after the arch and the cable.

Statical Determinacy

As we observe a given truss, we must determine whether it is unstable, statically determinate, or statically indeterminate. As we know, an unstable structure is not in equilibrium and therefore is unacceptable. A statically *determinate* structure is one that can be analyzed by means of statics alone, and it has the minimum number of members and supports required for its equilibrium. A statically *indeterminate* structure cannot be analyzed by means of statics, but only by means of the theory of elasticity. These structures, of course, have more members and/or supports than necessary.

A truss can be statically determinate or indeterminate for internal conditions (number of members) or for its external supports. There is no reason to discuss the latter, since a truss as a whole functions as a solid beam in terms of its supports, and it is assumed known.

A SELECTIVE EVALUATION OF TRUSSES

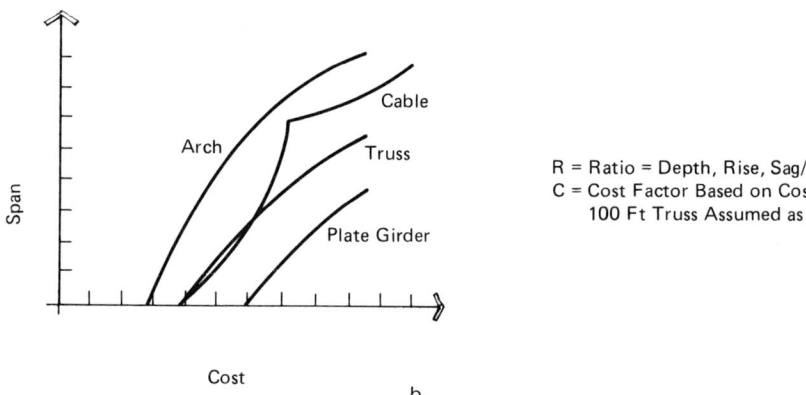

Structural System	Ratio R	Span (Ft)	Cost Factor C	Ratio R	Span (Ft)	Cost Factor C
Plate Girder	1/20	100	1.38		200	
Truss	1/10	100	1.00	1/10	200	1.35
Arch	1/5	100	0.69	1/5	200	0.88
Cable	1/10	100	1.04	1/10	200	1.23

a.

b.

R = Ratio = Depth, Rise, Sag/Span
C = Cost Factor Based on Cost of 100 Ft Truss Assumed as Unit

Figure 2-5. Comparison of the efficiency of trusses with that of other structural systems.

A quick method for assessing the internal stability of a truss, introduced by August Mobius of Germany, is to compute k as indicated below:

$$k = 2j - r$$

where:

k = minimum number of necessary members in the truss,
j = number of joints in the truss, and
r = number of reaction components at the exterior supports necessary for the stability of the truss (for example, $r = 3$ for trusses of one simple span, $r = 4$ for three-hinged arches, etc.).

Then, we must compute m, the actual number of members in the truss, and compare it with k. If $m < k$, the truss is unstable, since some members are missing. If $m = k$, the truss is statically determinate. In other words, the number of members is equal to the minimum necessary. If $m > k$, the truss is statically indeterminate, since the number of members is more than the minimum required. The number of members in excess of k, $(m-k)$ indicates the degree of indeterminacy.

A SELECTIVE EVALUATION OF TRUSSES 91

Examples. Determine whether the trusses in the accompanying illustrations are unstable, statically determinate, or statically indeterminate, and if statically indeterminate, to what degree.

Example 2-1

Solution:

$r = 3$
$j = 5$
$k = 2j - r = (2)(5) - (3) = 7$
$m = 6$
$m < k$.

Therefore, the truss is unstable.

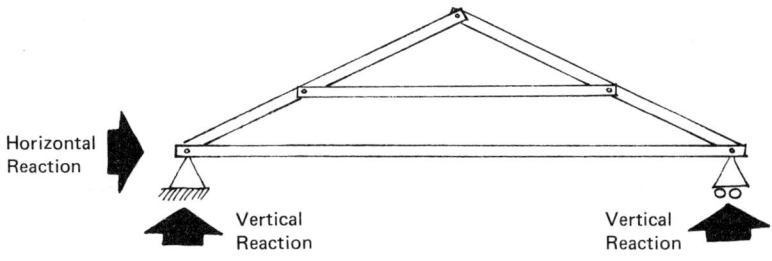

Figure 2-6. A truss that is assumed to be statically determined (isostatic) because the top chords can be considered continuous and the collar beam can be ignored in the statical equilibrium. Thus: $j = 3; m = 3; r = 3; k = 2j - r = (2)(3) - (3) = 3$. Therefore, $m = k$.

Example 2-2
Solution:

$r = 3$
$j = 14$
$k = 2j - r = (2)(14) - (3) = 25$
$m = 25$
$m = k.$

Therefore, the truss is statically determinate.

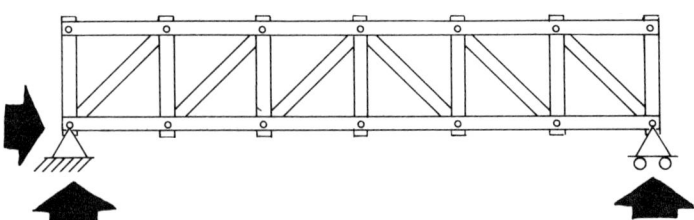

Figure 2-7. A statically determinate (isostatic) truss. $j = 14$; $m = 25$; $r = 3$; $k = 2j - r = (2)(14) - (3) = 25$. Therefore, $m = k$.

Example 2-3

Solution:

$r = 3$
$j = 14$
$k = 2j - r = (2)(14) - (3) = 25$
$m = 27$
$m > k.$

Therefore, the truss is statically indeterminate to the second degree.

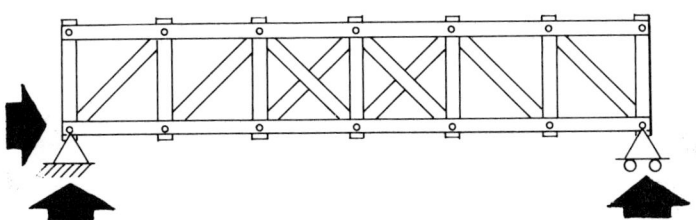

Figure 2-8. A statically indeterminate (hyperstatic) truss to the second degree, due to internal conditions. It is easy to spot the two redundant members. Consider: $j = 14$; $m = 27$; $r = 3$; $k = 2j - r = (2)(14) - (3) = 25$. Therefore, $m > k$.

Example 2-4

Solution:

   $r = 4$
   $j = 7$
   $k = 2j - r = (2)(7) - (4) = 10$
   $m = 10$
   $m = k.$

Therefore, the truss is statically determinate.

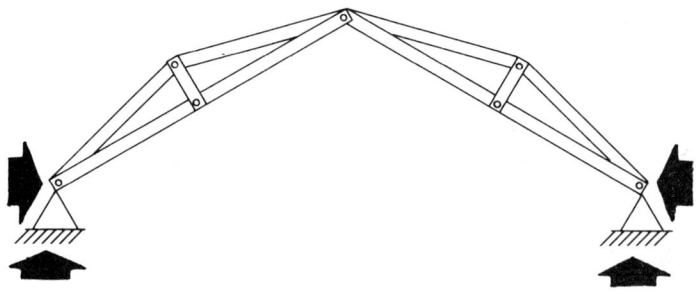

Figure 2-9. A statically determinate truss. $j = 7$; $m = 10$; $r = 4$; $k = 2j - r = (2)(7) - (4) = 10$. Therefore, $m = k$.

Example 2-5

Solution:

$r = 3$
$j = 4$
$k = 2j - r = (2)(4) - (3) = 5$
$m = 6$
$m > k.$

Therefore, the truss is statically indeterminate to the first degree.

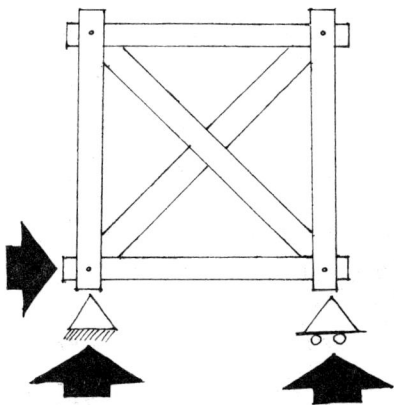

Figure 2-10. A statically indeterminate (hyperstatic) truss to the first degree that includes one redundant member. $j = 4$; $m = 6$; $r = 3$; $k = 2j - r = (2)(4) - (3) = 5$. Therefore, $m > k$.

Example 2-6

Solution:

   $r = 3$
   $j = 5$
   $k = 2j - r = (2)(5) - (3) = 7$
   $m = 8$
   $m > k.$

Therefore, the truss is statically indeterminate to the first degree.

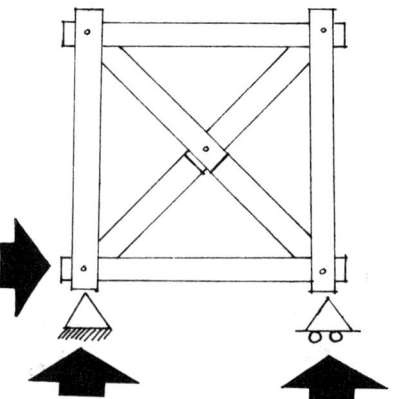

Figure 2-11. A truss that is statically indeterminate (hyperstatic) to the first degree because of one redundant member. $j = 5; m = 8; r = 3; k = 2j - r = (2)(5) - (3) = 7$. Therefore, $m>k$.

GEOMETRICAL INSTABILITY

The statical evaluation of trusses by means of the equation $k = 2j - r$ is not always possible. There are cases where the truss is unstable even though the equation's solution indicates otherwise. This phenomenon of so-called "geometrical instability" is hard to identify by mere visual inspection. For instance, the trusses in the following examples appear to be statically determinate according to the previous equation; however, the computer analysis in Chapter 4 proves their instability.

Examples. Determine the statical stability of the trusses shown in the accompanying illustrations.

Example 2-7

Solution:

$r = 3$
$j = 6$
$k = 2j - r = (2)(6) - (3) = 9$
$m = 9$
$m = k$.

Therefore, the truss should be statically determinate according to the above computations. Its geometrical instability, however, will be proven by computer analysis in Chapter 4.

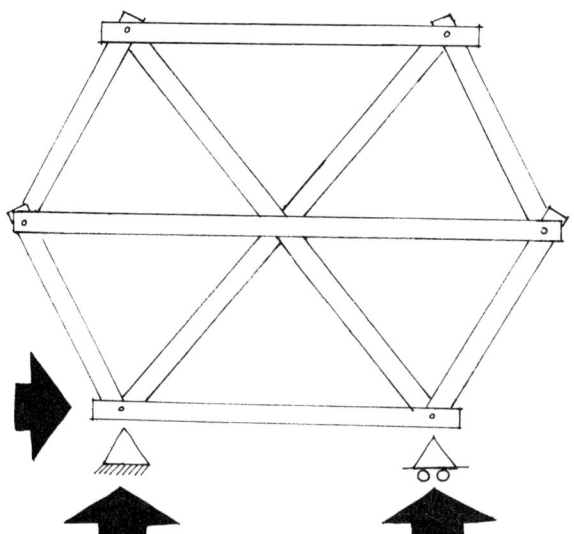

Figure 2-12. A case of geometrical instability. According to the condition of stability, the truss seems isostatic: $j = 6$; $m = 9$; $r = 3$; $k = 2j - r = (2)(6) - (3) = 9$; $m = k$. However, a computer analysis (see Chapter 4, Problem 6) shows the instability of the truss. Nevertheless, if the geometrical symmetry is removed (see Chapter 4, Problem 7), the truss becomes isostatic.

A SELECTIVE EVALUATION OF TRUSSES 99

Example 2-8

Solution:

$r = 3$
$j = 40$
$k = 2j - r = (2)(40) - (3) = 77$
$m = 77$
$m = k.$

Therefore, the truss should be statically determinate according to the above computations. Its geometrical instability, however, will be proven by computer analysis in Chapter 4.

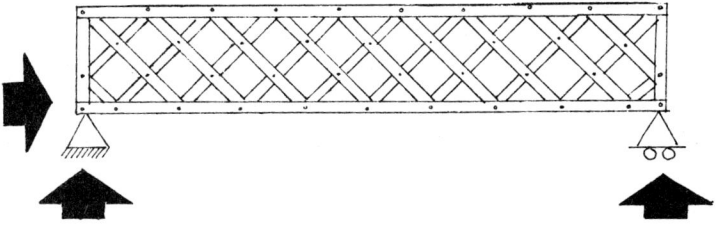

STATICALLY INDETERMINATE TRUSSES

In the definition of statically indeterminate (or hyperstatic) trusses, we say they are characterized by the inclusion of more members or supports than strictly required for statical equilibrium. The redundant members or supports seem to be unnecessary intruders which either do not work as do the others or, if they do, remove the clarity and purity from the structure. This is not necessarily because the redundancy is not intended as a factor of inefficiency but merely as a theoretical condition of equilibrium. A full discussion on the merits and drawbacks of static versus hyperstatic structures goes beyond the scope of the present work, but the reader should remember that hyperstatic trusses have among major advantages the following: increased rigidity, the redistribution of axial stresses in case of local failure in a member, and the possibility of more creative geometrical forms. Furthermore, the major drawback of long calculations has been removed by the accessibility of computer analysis (see Chapter 4).

TEACHING AIDS

Differentiating between tension and compression in the various members of a loaded truss is, in most cases, not a simple task without the benefit of a statical analysis. Especially for complex types, and particularly for statically indeterminate trusses with many redundant members, it would be convenient to establish immediately the nature of the stresses in each member. A visual aid particularly helpful in classroom teaching is a simple apparatus which, based on models with deformable members which emphasize the magnitude of elongations and contractions under load, differentiates therefore between members in tension and those in compression. To achieve this, the members of the models include, at midspan, a deformable plastic ring which, when not stressed, is a perfect circle. Under tension or compression, the ring elongates or contracts, assuming the shape of an ellipse and therefore giving a visual description of the stress. With sufficient members of various lengths, models of different geometrical configurations can be assembled. A simple pegboard is used to mount the model and to suspend it at the appropriate supporting points. Figure 2-14 shows a sample of the aid, particularly simple and inexpensive to build.

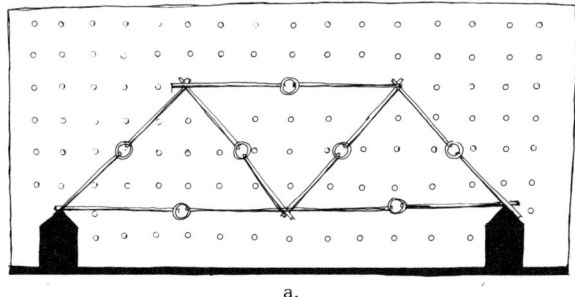

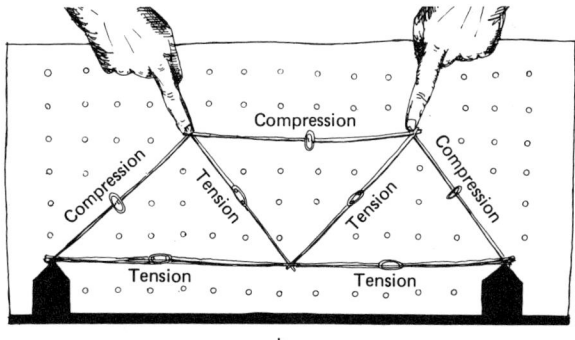

Figure 2-14. Differentiating between tension and compression in the members of trusses is not a simple task in many cases. A visual aid particularly useful in the classroom is the simple apparatus shown above. Each individual member includes at midspan a deformable ring that is circular when the member is not stressed. Under stress, the ring expands or contracts longitudinally to an elliptical shape, indicating tension and compression, respectively.
    a. Deformable model of a truss mounted on a vertical pegboard. Without any load, the ring in each member is circular.
    b. Model being loaded visually by manual action shows the nature of the axial forces in each member by the shape of the deformed rings.

CREMONA'S DIAGRAM

The following step-by-step instructions should help the reader in the construction of the Cremona's diagram for the determination of the axial forces in any given statically determinate truss (see Figure 2-15).

1. Draw accurately the outline of the truss in any convenient scale.

2. Identify each triangle in the truss and each space between external forces by placing arbitrary letters in such spaces. The reader will find that, after completing the lettering, any member of the truss or any external force can be identified by two letters, since each member or force should have a letter on each side. For instance, the left reaction can be identified by *AB*.

3. Although not essential to the construction of the diagram, it is simpler to number the joints in the order in which they must be examined. Notice that only joints with no more than two unknown member forces can be analyzed. In the example in Figure 2-15 the only joint from which we can begin is joint #1, where the only unknowns are the forces in members *MA* and *CM*. Note also that after solving such two-member forces, we can proceed to the analysis of joint #2, which originally had three unknowns and now has only two (member forces *DL* and *LM*), since the *MC* member force has been previously found.

4. Decide on the order in which the forces at a joint will be considered. For instance, choose a clockwise direction. Starting at joint #1, draw in any convenient scale the known forces at that joint, *AB* and *BC*, and complete the polygon by finding *CM* and *MA* in the order described.

5. The previous step is repeated at joint #2, where the unknown forces are now only two—*DL* and *LM*—because *MC* has previously been calculated in Step 4.

6. The procedure continues until the end. Note that due to the symmetry of the structure and its loads, the procedure could have been stopped at joint #4.

A SELECTIVE EVALUATION OF TRUSSES

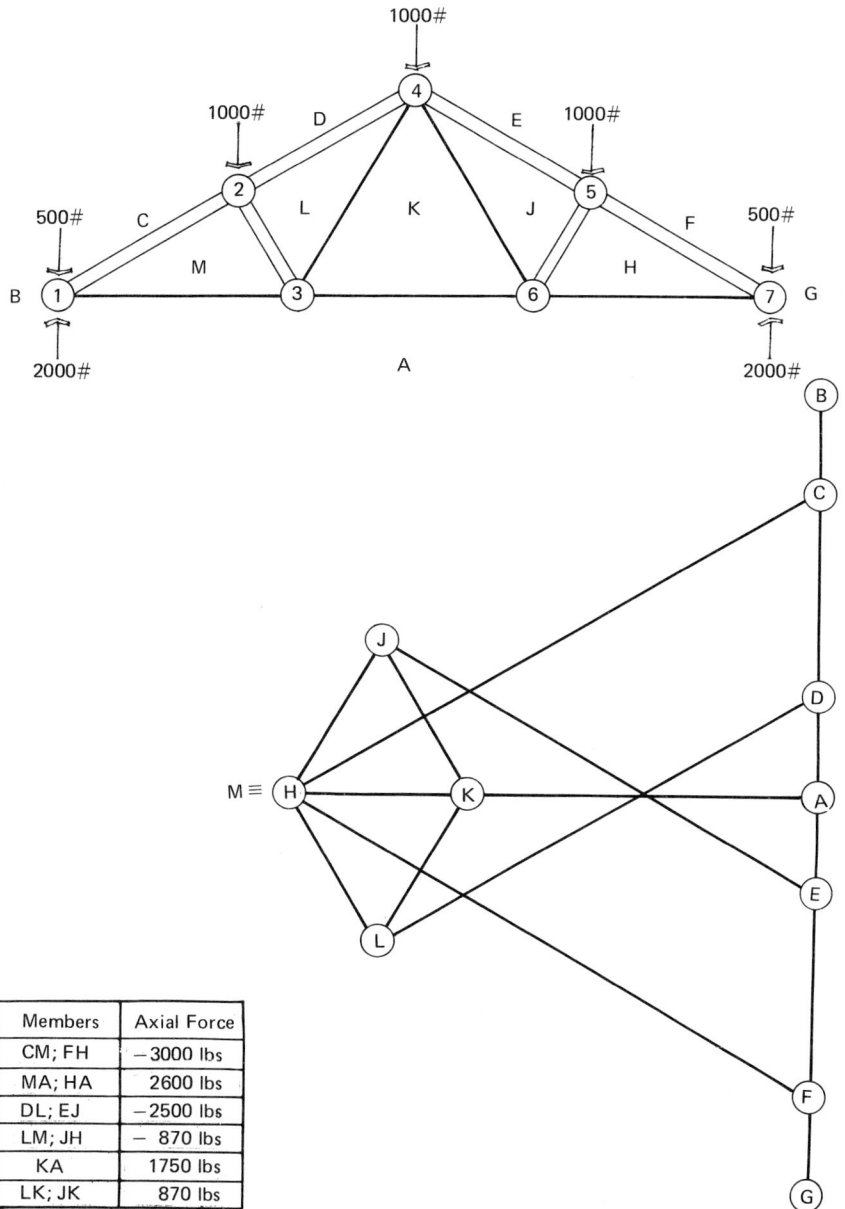

Members	Axial Force
CM; FH	−3000 lbs
MA; HA	2600 lbs
DL; EJ	−2500 lbs
LM; JH	− 870 lbs
KA	1750 lbs
LK; JK	870 lbs

Figure 2-15. Cremona's diagram.

WEIGHT OF TRUSSES

To establish the loads to be applied on a truss, the designer will find it necessary to make a preliminary estimation of the weight of the truss. It is useful, therefore, to have some rules of thumb or empirical formulas for this.

Steel. To obtain the total weight of steel trusses, use the following formula.*

$$W = cA \qquad (2.1)$$

where:

W = total weight of the steel truss in lb,
c = coefficient from Tables 2-1 or 2-2 for steel, and
A = total area in sq. ft of the roof carried by the truss.

Another formula for calculating the total weight of steel trusses is given in 2.2.

$$W = \sqrt{\frac{wa}{s}(4L^2 + 60L)} \qquad (2.2)$$

where

W = total weight of truss, in lb,
w = roof load in psf of horizontal projection of roof,
s = average allowable stress in psi used in design,
a = center to center spacing of trusses in ft, and
L = truss span in ft.

Wood. By following the same procedure as that for steel trusses, the weight of timber trusses can also be estimated, using the coefficients from Table 2-3 and the same formula 2.1 previously seen.

Table 2-1. c = Weights of Steel Trusses in lb/ft^2 of Roof Surface*

Span (in ft)	$\frac{1}{2}$ Pitch $\alpha = 27° \pm$	$\frac{1}{3}$ Pitch $\alpha = 18° \pm$	$\frac{1}{4}$ Pitch $\alpha = 14° \pm$	Flat $\alpha = 0°$
Up to 40	5.25	6.3	6.8	7.6
40–50	5.75	6.6	7.2	8.0
50–60	6.75	8.0	8.6	9.6
60–70	7.25	8.5	9.2	10.2
70–80	7.75	9.0	9.7	10.8
80–100	8.50	10.0	10.8	12.0
100–120	9.50	11.0	12.0	13.2

*Harry Parker, *Simplified Designs of Roof Trusses for Architects and Builders* (New York, John Wiley and Sons, Inc., 1953) p. 50.

Table 2-2. c = Weights of Steel Trusses in lb/ft^2 of Roof Surface*

Span (in ft)	Pitch from $\frac{1}{3}$ to $\frac{1}{4}$ $\alpha = 18° \pm \alpha = 14° \pm$		Pitch over $\frac{1}{3}$ over $18° \pm$		Pitch flat	
	min	max	min	max	min	max
40	2	$3\frac{1}{2}$	1	3	$2\frac{1}{2}$	$4\frac{1}{2}$
50	3	$4\frac{1}{2}$	2	4	$3\frac{1}{2}$	$5\frac{1}{2}$
60	4	$5\frac{1}{2}$	3	5	$4\frac{1}{2}$	$6\frac{1}{2}$
70	5	$6\frac{1}{2}$	4	6	$5\frac{1}{2}$	$7\frac{1}{2}$
80	6	$7\frac{1}{2}$	5	7	$6\frac{1}{2}$	$8\frac{1}{2}$

*Lenton E. Grinter, *Design of Modern Steel Structures* (New York, The MacMillan Company, 1964) p. 284.

Table 2-3. c = Weights of Timber Trusses in lb/ft² of Roof Surface*

Span (in ft)	½ Pitch $\alpha = 27° \pm$	⅓ Pitch $\alpha = 18° \pm$	¼ Pitch $\alpha = 14° \pm$	Flat $\alpha = 0°$
Up to 36	3	3½	3¾	4
36–50	3¼	3¾	4	4½
50–60	3½	4	4½	4¾
60–70	3¾	4½	4¾	5¼
70–80	4¼	5	5½	6

For cambered trusses, increase 30%.

*Harry Parker, *Simplified Designs of Roof Trusses for Architects and Builders* (New York, John Wiley and Sons, Inc., 1953) p. 49.

Weights of bracings. An estimated weight of the bracing required for steel trusses is proposed to be one pound per square foot of roof surface for short trusses, and two pounds for long trusses.

PRESTRESSED STEEL TRUSSES

For years, prestressing has proven very efficient in flexural and axially loaded concrete structures, including trusses. Tests on prestressing of steel trusses have also been made and its efficiency proven.

Prestressing is accomplished by applying a compressive force on the chord which, under normal loading, would be in tension. In other words, the truss is prestressed in such a manner that the induced axial forces are the reverse of those produced by normal loads. A simple example is illustrated in Figure 2-16.

*Robins Fleming, "Weight of Roof Trusses by Empiric Formulas," *Engineering New Record* (1919).

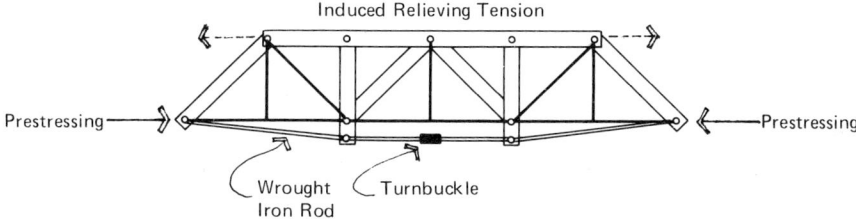

Figure 2-16. Early example of prestressing. With just a wrought iron rod and a turnbuckle, a compressive force can easily be applied on the bottom chord, creating stresses that will reduce those the loads generate. For instance, the top chord that the loads will stress in compression receives a relieving tensile force from the prestressing.

This concept, of course, is not new, but large, significant applications are relatively recent. In the early 1950's, for instance, 60 ft steel Warren trusses supporting the roof of an industrial building in Harlow, England, were prestressed with turnbuckles (Figure 2-17). The induced stresses were opposite those generated by dead and live loads, with consequent stress relief.

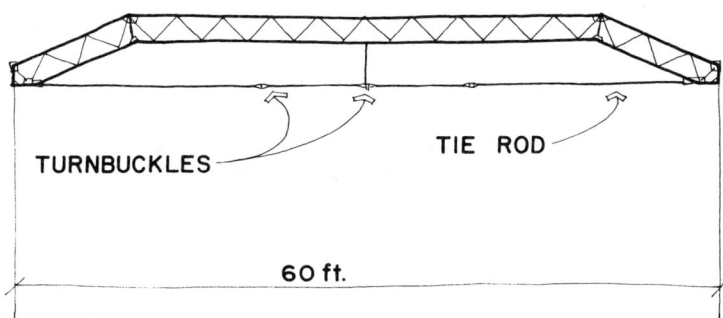

Figure 2-17. Prestressed steel truss supporting the roof structure of a factory building in Harlow, England (early 1950's). The 60 ft span Warren truss, tightened by turnbuckles, and the tie bars prestress the whole truss.

108 SIMPLIFIED TRUSS DESIGN

Prestressing the upper chord of a cantilevered truss would obviously reduce the tension in the upper chord and the compression in the lower chord. An example can be seen in the roof trusses of the United Airlines hangar at O'Hare International Airport in Chicago (Figure 2-18). The prestressed steel trusses, 40 ft on center, having a cantilevered span of 140 ft, were prestressed along the top chord. A series of prestressing rods, applied at different points on the top chord, increased the prestressing force as the point of application approached the fixed end support. The efficiency of prestressing was proven by the 20% weight reduction achieved in comparison to conventional design.

Another example of prestressed steel trusses is in the roof structure of the Melsbrock Airport building in Brussels (1954). Spanning 248 ft 10 in. the trusses (13 ft 9 in. deep at midspan) were prestressed with four cables, resulting in substantial weight reduction (see Figure 2-19).

Figure 2-18. Cantilevered steel trusses, supporting the roof of the United Air Lines hangar at O'Hare International Airport in Chicago. Engineer: Paul Rogers. Spaced 40 ft on center, the 140 ft cantilevered trusses were prestressed along the upper chord with several rods whose number increased toward the support. Savings amount to 20% weight reduction.

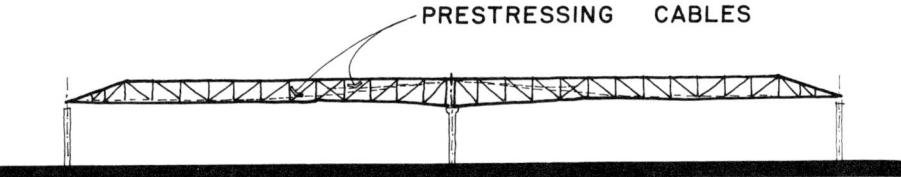

Figure 2-19. Prestressed steel trusses supporting the roof structure in the Melsbrock Airport building in Brussels (1954). Structural engineer: Gustave Magnel. Spanning 248 ft 10 in., the trusses are 13 ft 9 in. deep at midspan. The prestressing is achieved with four cables per truss, with stresses in the cables up to 105,000 psi.

CONCRETE TRUSSES

In contrast to timber and steel trusses, widely covered in technical literature, concrete trusses are not adequately presented. Especially in countries where steel and wood are economical, concrete trusses have not been popular because they could not compete economically with conventional steel and timber types. As we have seen in the previous chapter, concrete trusses have been very popular in Italy. Moreover, they are also popular in other European countries and are gaining in popularity in the United States as well.

Three-hinged arches made of trussed components and trussed two-hinged arches with a horizontal connecting tie, both made of reinforced concrete, are currently available as standard production in Italy for long-span industrial buildings (see Figure 2-20). Examples of concrete trusses are also found in England (see Plate 2-1), and examples of concrete trusses in the United States include those made of concrete masonry units built experimentally in 1957 by Texas Stressed Concrete Corporation, in San Antonio, Texas (designed by Eric C. Molke, a structural engineer). The two prototype trusses (see Plates 2-2a and 2-2b) were six-panel Bowstrings. The clear span was 100 ft

110 SIMPLIFIED TRUSS DESIGN

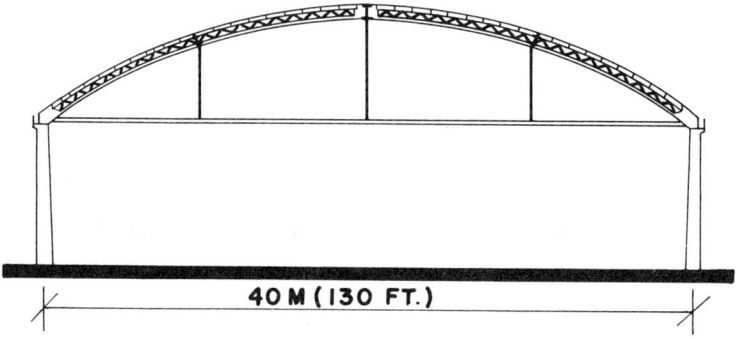

Figure 2-20. Reinforced concrete trussed arch. Standard production available and on the market in Italy (Courtesy S.p.A. Ing. A. Ghira, Rome).

Plate 2-1. Examples of prestressed concrete trusses combined with a grid system under construction. Transair Hangar at Gatwick Airport, London, 1959; project by London Ferro Concrete Company, Ltd. (Courtesy of Portland Cement Association.)

for each truss. The heights were 14 ft and 12 ft, respectively, with a spacing of 20 ft, *c.* to *c.* for both. The design live load capacity was 40 psf. All members were made of lightweight concrete block units, and diagonal truss members were omitted. The top chord was slightly prestressed, and the bottom chord was also prestressed to dissipate tension produced by dead loads.

a.

b.

Plate 2-2a, b. Experimental prototype trusses built out of lightweight concrete block units, by Texas Stressed Concrete Corporation, San Antonio, Texas, 1957: six-panel Bowstring, 100 ft span. (Courtesy of Portland Cement Association.)

Present standard production of concrete trusses in the United States includes trusses up to 80 ft in span, with heights up to 36 in. Current construction techniques include precast diagonals incorporated when the rest of the truss is cast. Casting is sometimes done with the form lying flat on a horizontal plane, but some other processes place the form on a vertical plane. Prestressing of the top and bottom chords is also an economical technique widely used in standard production. For roof construction, see Plate 2-3.

A SELECTIVE EVALUATION OF TRUSSES 113

Plate 2-3. Concrete trusses with lengths up to 80 ft. Standard product available in the United States.

TERMINOLOGY

The reader may find the definitions below helpful in understanding and analyzing trusses.

Axial forces: tension (+) and compression (−) generated in the truss members by the loads applied at the joints.

Bottom chord, or lower chord: perimetric members in the lower part of the truss.

Chords: exterior members of the truss.

Joints: points of intersection of members.

Secondary stresses: shear and bending; such stresses are occasionally added to the main stresses (tension and compression) due to departure from the major definitions of trusses. Typical causes of secondary stresses are application of loads along the member axis, and fixed end connections instead of frictionless hinges at joints.

Struts: compression members.

Ties: tension members.

Top chord, or upper chord: perimetric members in the upper part of the truss.

TYPES OF TRUSSES

Note that in our present study of truss types and their structural characteristics, we assume that the loads applied are only vertical; for simplicity, we neglect any horizontal effects, which are usually produced by wind loads. Note also in the following discussion there may be confusion concerning the exact nomenclature for different types of trusses. Moreover, the types described below are among the most popular for roofs only. Bridge trusses, usually characterized by very long spans, are somewhat different; the most popular types for bridges are illustrated in Appendix A.

Simple Triangle

The simplest form of truss is made of one triangle, with two members in compression and one in tension. Its main drawback is the unsupported length of compression members, which limits the span.

116 SIMPLIFIED TRUSS DESIGN

King Post

Subdivision of the previous truss with a king post does not reduce the length of the compression members (except for case b in Figure 2-21). The effects of the center posts on the axial forces, however, vary according to the three cases shown below.

1. If the bottom chord is in line with the supports, the post is not stressed, and the axial forces in the other members are unchanged (see Figure 2-21a).

2. If the lower chord sags under the central post, the post itself is in compression, and the forces in the other members are reduced (see Figure 2-21d).

3. If the lower chord is chambered (upward), then the center post is in tension, and the axial forces in the other members are increased (see Figure 2-21c).

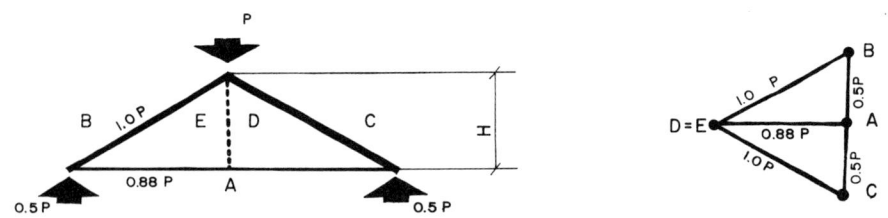

Figure 2-21. The consequences of adding a central post to a simple triangular truss can be understood by comparing the Cremona diagrams for the various cases illustrated herein.
    a. For the simple king post truss, the post does not carry any load nor does it vary the forces in the other members.

A SELECTIVE EVALUATION OF TRUSSES 117

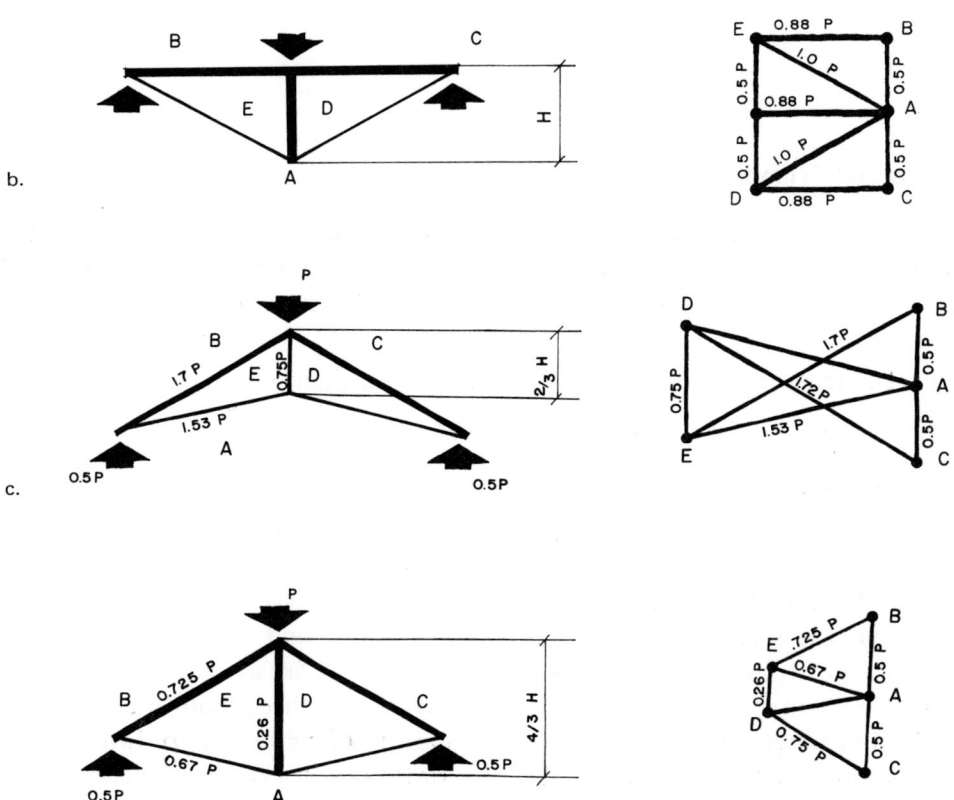

b. In the reversed king post truss, the central post is compressed and, unlike the previous case, it is strictly indispensable.
c. A shorter central post is in tension and it increases the forces in the top and bottom chords.
d. A longer central post is a strut that reduces the forces in the top and bottom chords.

Reinforced Triangles

If we follow the usual definition of a truss, stating that all members at any given joint are hinged, these trusses, and others similar to them, would be unstable. The trusses in question include the following four.

1. The truss reinforced with collar beams (see Figure 2-22a).
2. The queen post (see Figure 2-22b).
3. The Italian or Palladian (see Figure 2-22c).
4. Many other trusses typical of the Middle Ages.

In reality, such trusses can be considered statically determinate if we assume that the real truss members are only those around the perimeter, forming the major triangle. The others inside the triangle are assumed to be added only to stiffen the perimetric members. Only the three main members are considered hinged at the three corners, and their lengths are not interrupted by any intermediate hinge. In other words, the interior stiffeners do not break the continuity of the main members.

A SELECTIVE EVALUATION OF TRUSSES 119

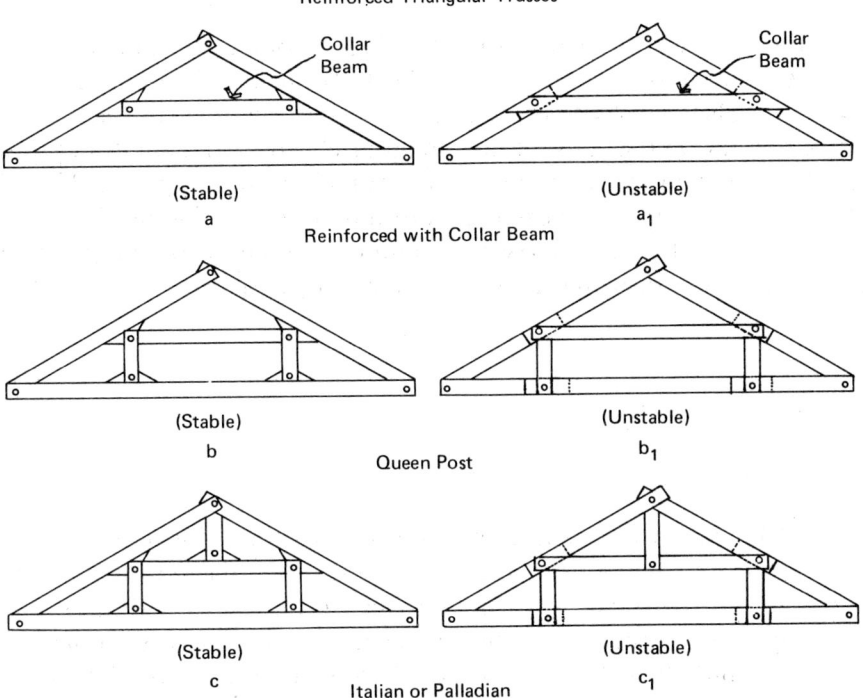

Figure 2-22. Trusses that could be interpreted as stable and statically determinate, because the hinges do not interrupt the perimetric members are shown in (a), (b), and (c). Unstable structures are given in (a_1), (b_1), and (c_1).

Scissors or German Truss

The scissors—or German—truss consists of four compression members in the top chord, two tension members in the bottom chord, a vertical tension member at midspan, and two other compression members. These last two members are the extensions of the bottom chord, and their function is to reduce the length of the members in the top chord (see Figure 2-23a). Two variations of the simple scissors truss are illustrated in Figures 2-23b and c. The objective of their geometry is to reduce the length of the members of the top chord, thus allowing the truss to have larger spans.

Palladian

The real Palladian, which some erroneously call the king post, consists of four compression members in the top chord, two tension members in the bottom chord, a vertical tension member at midspan, and two diagonal members in compression. In certain aspects, this truss is similar to the scissors truss, but they differ in terms of axial forces. In the Palladian, the forces in the top and bottom chord are much less than those in the scissors truss, because of the horizontality of the bottom chord in the Palladian (see Figure 2-24).

A SELECTIVE EVALUATION OF TRUSSES 121

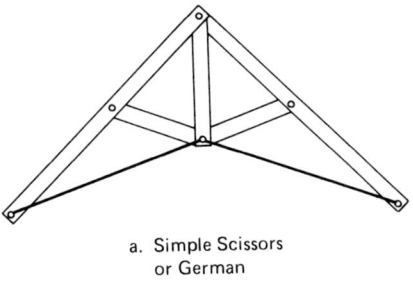

a. Simple Scissors or German

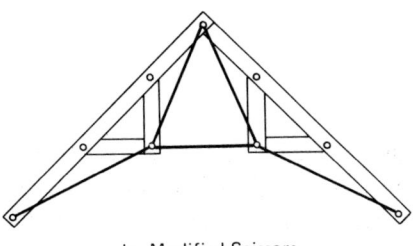

b. Modified Scissors

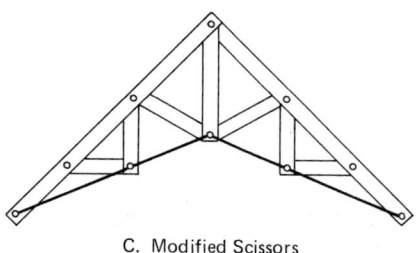

C. Modified Scissors

Figure 2-23. The scissors or German truss: (a) simple; (b) and (c) modified.

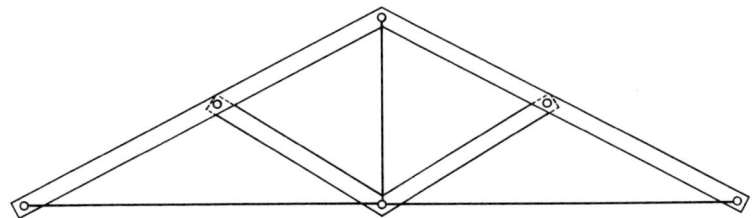

Figure 2-24. The Palladian, typical of the Renaissance.

Hammer-Beam

The hammer-beam truss, typical of the Middle Ages, except for some individual variations can be described as a triangular truss with top compression and bottom tension chords, both pitched and parallel to each other. Struts are vertical; ties, horizontal (see Figure 2-25).

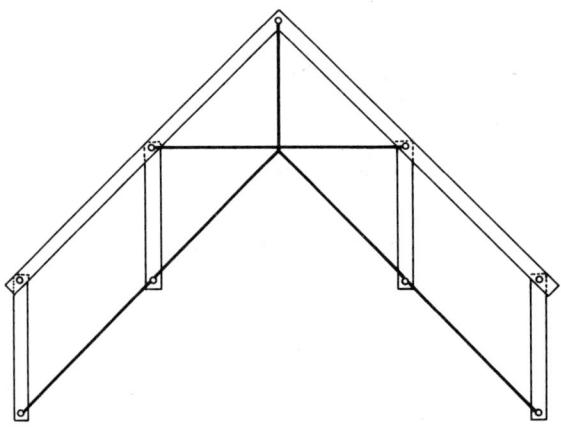

Figure 2-25. The hammer-beam, typical of the Middle Ages.

Polonceau or Fink

The Polonceau, called also the Fink, is a triangular truss characterized by diagonal members in compression, perpendicular to the top chord, and parallel to each other. Their length is deliberately kept short in comparison with other members. This truss can be simple, with a horizontal bottom chord or a chambered bottom chord; the truss can also be multi-paneled with a horizontal bottom chord, or multi-paneled with a chambered bottom chord (see Plate 2-4).

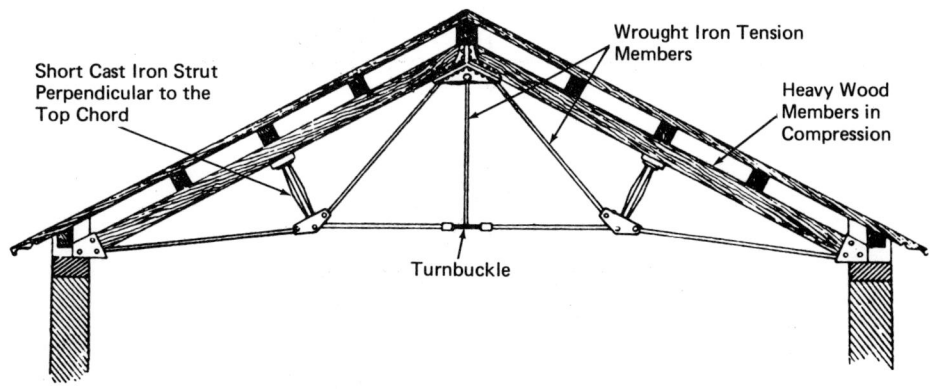

Plate 2-4. The Polonceau truss.

Fan

The fan truss is a triangular truss which can be simple or multi-paneled. The simple fan is characterized by the fact that, when half the truss is observed, all the diagonal members, two struts and one tie, radiate as a fan from one point on the lower chord toward the upper chord. These members subdivide the upper chord into several short members (see Figure 2-26a). The multi-paneled fan, although more complex than the above, employs the same principles of design as does the simple fan (see Figure 2-26b).

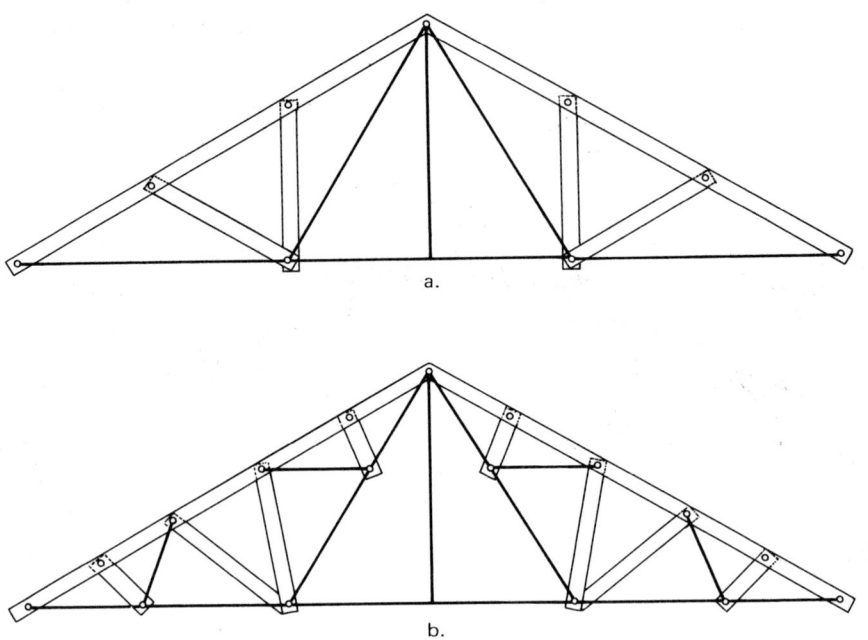

Figure 2-26. The fan: (a) simple and (b) multi-paneled.

Howe

The triangular Howe is characterized by the fact that tension members between the two chords are all vertical, while the diagonal compression members are not necessarily parallel to each other (see Figure 2-27a). The flat Howe is characterized by the fact that the tension members between the two chords are all vertical, while the inclined compression members are not necessarily parallel to each other (see Figure 2-27b).

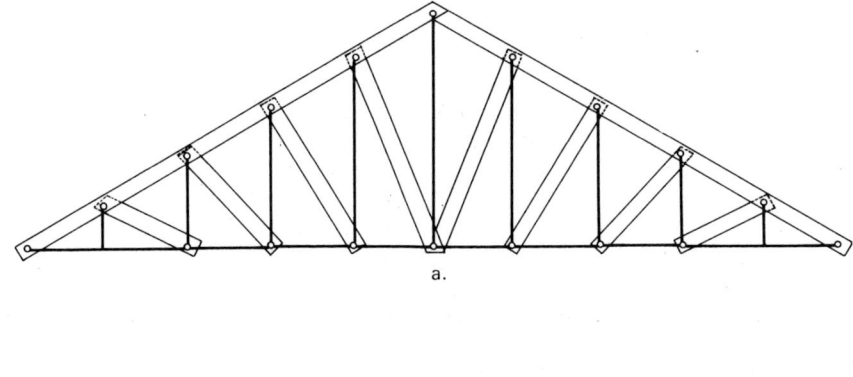

a.

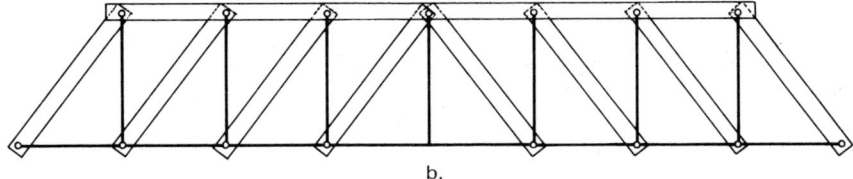

b.

Figure 2-27. The Howe truss: (a) pitched and (b) flat.

Pratt

The triangular Pratt is characterized by the fact that between the two chords, the compression members are all vertical, while the diagonal members are in tension and are not necessarily parallel to each other (see Figure 2-28a). The flat Pratt is characterized by the fact that between the two chords, the compression members are all vertical, and the diagonal members, all parallel to one another, are in tension (see Figure 2-28b).

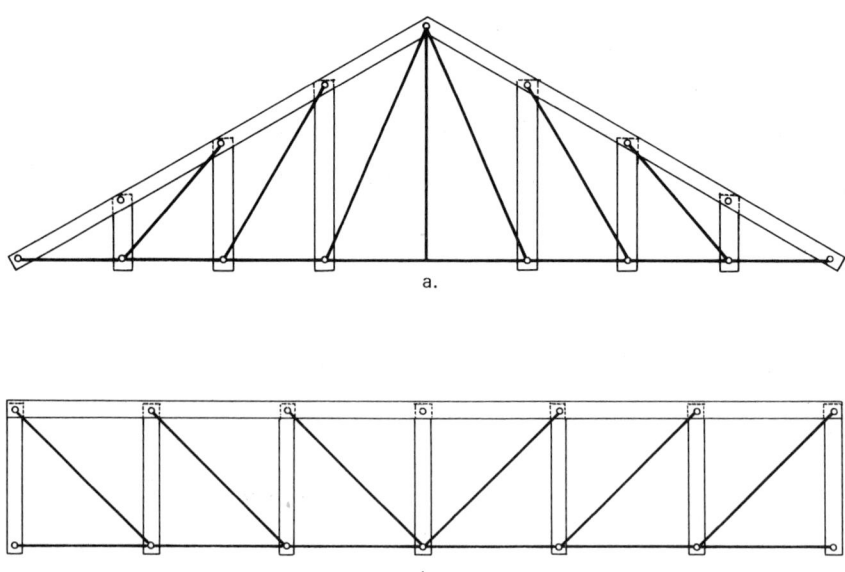

Figure 2-28. The Pratt truss: (a) triangular and (b) flat.

Belgian

The Belgian is a triangular truss, characterized by the fact that between the two chords, the diagonal compression members are all perpendicular to the top chord. The diagonal members, in tension, are not necessarily parallel to each other (see Figure 2-29). Although they both have diagonal compression members perpendicular to the top chord, the Belgian is distinguished from the Polonceau or Fink by the fact that in the Belgian, the diagonal compression members extend from the top to the bottom chord, while in the Polonceau or Fink, they may stop between the two chords.

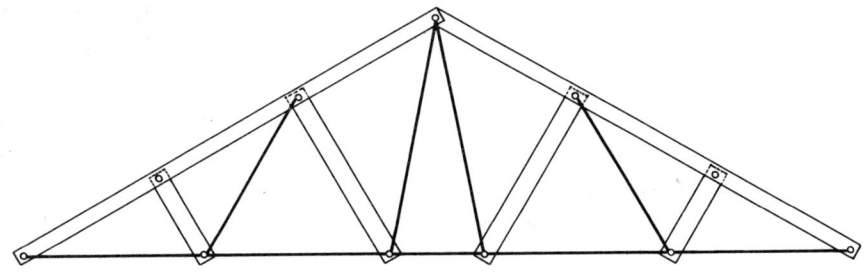

Figure 2-29. The Belgian truss.

Warren

The Warren is a flat truss characterized by all members of the top and bottom chords being of equal lengths; all diagonal members, whether in compression or in tension, have equal lengths; in each half of the truss, the diagonal compression members are parallel to each other; and the diagonal tension members are also parallel to each other (see Figure 2-30). Note that the original Warren was subdivided into equilateral triangles; the common types, however, may have isosceles triangles.

TRIDIMENSIONAL TRUSSING

All the trusses we have discussed, and will be discussing, share the common quality of being planar; i.e., all their members and all the applied forces and reactions lie in the same plane. Under the above heading, however, other types of trussed structures not having this characteristic will be mentioned briefly.

Trusses may lie, for instance, on a continuous surface which is not a plane, and may be stressed with forces which have perpendicular and tangential components to this surface. One example can be found in the Production Hall at the Instituto Tecnico de Constructiccion y del Cemento, Costillares, Spain, by Eduardo Torroja, engineer. The roof trusses lie on a cylindrical surface which spans, in the manner of a barrel shell, from one support to the other. Thus, a cross-section through the truss would show an arc whose angle at the center is 90° (see Figure 2-31).

A SELECTIVE EVALUATION OF TRUSSES 129

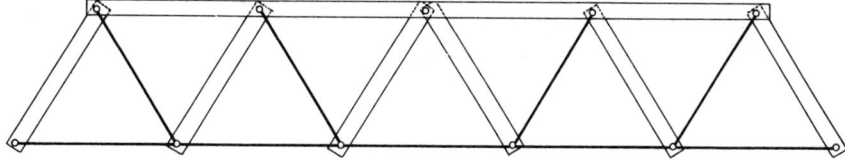

Figure 2-30. The Warren truss.

Figure 2-31. A non-planar truss that lies on a cylindrical surface whose cross section is an arc of a circle with a central angle of 90°. Acting like a barrel shell, the truss carries a metal skin of zinc sheeting. Several trusses of this type constitute the roof structure for the Production Hall of the Instituto Tecnico de Constructiccion y del Cemento, Costillares, in Spain. Engineer: Eduardo Torroja.

130 SIMPLIFIED TRUSS DESIGN

A truss can also consist of members contained in more than one plane. Some members may lie all in one plane while others lie on other planes, and still others may lie on the intersection of these planes. Figure 2-32 shows a simple tridimensional truss containing three planes; thus, a cross-section

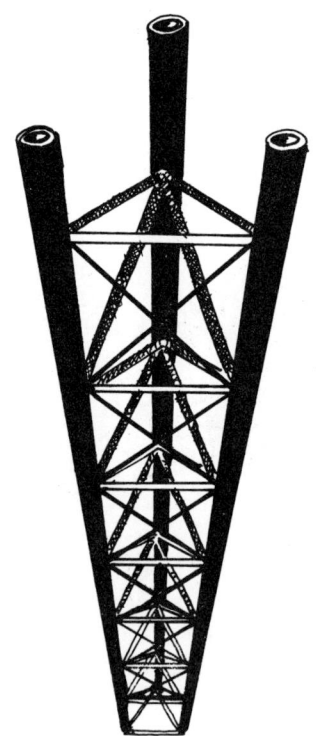

Figure 2-32. Stepping up from bidimensional planar structures to three-dimensional forms in space, the truss acquires lightness and elegance with its visual transparency and unobstructing body. In the hands of sensitive designers, the truss can become an exciting sculptural composition of lines in space. Notice the effectiveness of using variable bar size in the articulation of the truss vocabulary.

through the truss would show a triangle. Extremely efficient and simple to build even with primitive tools and materials, this tridimensional truss has very practical significance because of its lateral stability and integrity.

Space frames are also a form of tridimensional truss system which has gained an important position on the list of modern structural alternatives in recent years. Horizontally in roof and floor systems, as well as vertically in walls and columns, the space frame has gained such an acceptance in present architectural language that it sometimes is the dominating note in building design (see Figure 2-33).

For a better understanding of the structural behavior of space frames, consider a series of parallel-chord trusses placed side by side to support a floor. If a concentrated load is applied to one truss, the others would not share any part. However, if diagonal members, connecting the top chord of

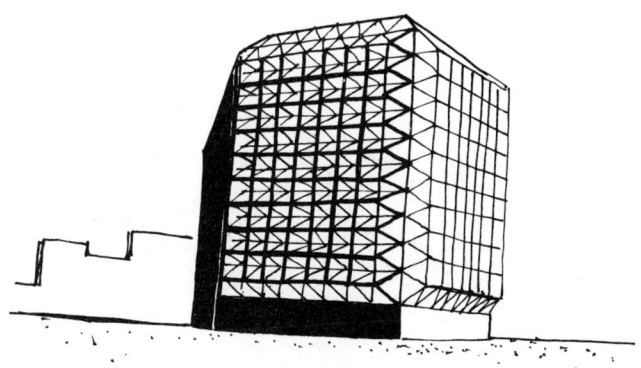

Figure 2-33. Architectural expression of tridimensional truss systems has become so popular at present that it is now a standard pre-engineered system commercially available. Galvanized tubular shapes are the most popular. The example shown in this figure is the Wilson Commons Campus Center at the University of Rochester, New York.

132 SIMPLIFIED TRUSS DESIGN

one truss with the bottom chords of the adjacent trusses, are installed, the load would be distributed among the trusses by the diagonal members. In other words, a concentrated load is distributed first through all the members converging at the joint where the load is applied and then will spread a considerable distance in all directions, eventually reaching the supports. The relationship between trusses and space frames is similar to that which exists between one-way and two-way slabs. An indication of its efficiency is given by the comparison between the depth required by a space frame ($\frac{1}{20}$ to $\frac{1}{30}$ of its span) and that required by a series of parallel trusses ($\frac{1}{10}$ of the span) to carry the same load.

The geometry of space frames is characterized by the two features below.

1. A space frame is an assembly of two parallel plane grids, equal or not, connected by diagonal web members.

2. A space frame is an assembly of modular tridimensional units formed by the edges of a tetrahedron, a square pyramid, or other polyhedra (see Figure 2-34).

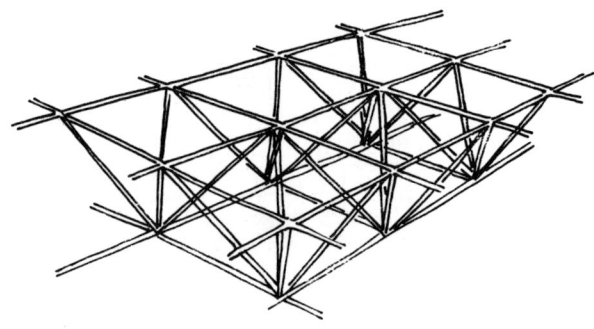

Figure 2-34. Typical space frame with top and bottom grids and interconnecting diagonals.

Domical space frames are also tridimensional trussed structures consisting of hinged rectilinear members, which are the edges of triangles or other regular polygons obtained by subdividing a spherical dome. Triangles can sometimes be grouped together to form larger hexagons, as in geodesic domes. More precisely, notice that the joints of the dome are on the surface of a sphere while the members intersect the sphere. In reality, the domical space frame is a polyhedron inscribed within the sphere. The chosen pattern in which the sphere is subdivided determines the type of dome. Famous types include the Schwedler dome (see Figure 2-35), the Zeiss-Dywidag dome (see Figure 2-36), and the geodesic domes by R. B. Fuller (see Figure 2-37).

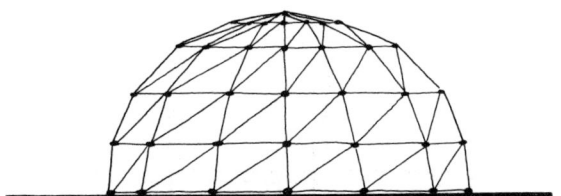

Figure 2-35. The Schwedler trussed dome with its nodal points generated by the intersection of equidistant meridians and parallels traced over a reference hemisphere.

Figure 2-36. The Zeiss-Dywidag trussed dome.

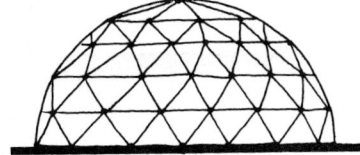

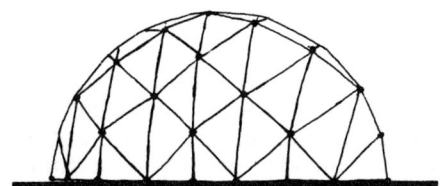

Figure 2-37. Example of popular "geodesic" domes by Buckminster Fuller.

134 SIMPLIFIED TRUSS DESIGN

Tensegrity structures, so called by R. B. Fuller in his work, are also tridimensional trussed assemblies that minimize the number of compression members in comparison to that of tension members, with consequential efficiency gains (see Figures 2-38 and 2-39). The elegance and feeling of purity projected by the tensegrity structure have entered the art field. An example of sculptural work based on these concepts is illustrated in Figure 2-40.

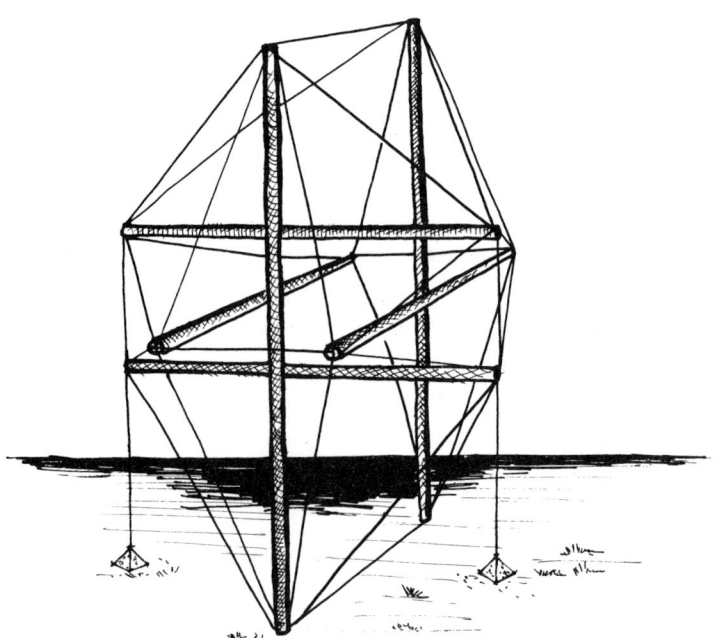

Figure 2-38. Tridimensional application of truss systems that includes only tension and compression members. A six-strut "tensegrity" structure, as R. B. Fuller calls it, consists of a continuous arrangement of tension members and a discontinuity of the struts that never come in contact with each other. With no redundant members, the structure is stable.

A SELECTIVE EVALUATION OF TRUSSES   135

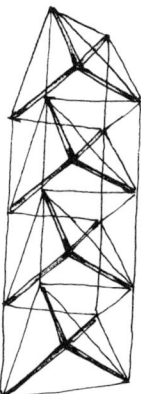

Figure 2-39. "Tensegrity column": a space truss structure that R.B. Fuller describes as a stack of "center-of-gravity radial tube tetrahedra." A stable structure with no redundant member that minimizes the number of compression members for structural efficiency.

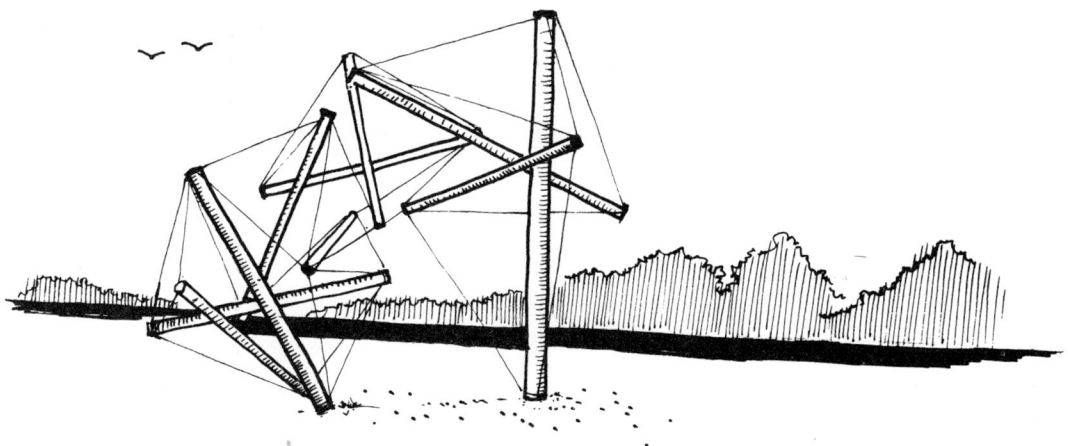

Figure 2-40. Made popular by Fuller's work, "tensegrity" structures have brought the fundamental principles of tridimensional trusses into the realm of sculpture. Characterized by a continuous succession of tensile members, this particular type of "tensegrity" includes also the condition that compression members are never in contact with each other. Entitled "Audry I", the sculpture shown above is by Kenneth Snelson (1967), the creator also of a similar piece entitled "Cantilever" (1967).

STRUCTURES OF EXPEDIENCY

No other structural system is as flexible as trussing for ease and rapidity of construction. Even in emergency situations, the system can be employed without special tools, equipment, and materials. Consider, for instance, that short tree branches or bamboos tied at the end with vines can be used to build trussed structures in a wilderness. Light, short members can form light components, which in turn can be joined to form the total structure, using only the muscles of man. Likewise, planar and space trusses, three-dimensional columns, trussed domes, space frames, etc., can all be constructed for emergency or temporary structures with such primitive materials and methods. Although this aspect may not have common daily applications in engineering or architectural practices and is usually overlooked in literature, it remains a significant characteristic.

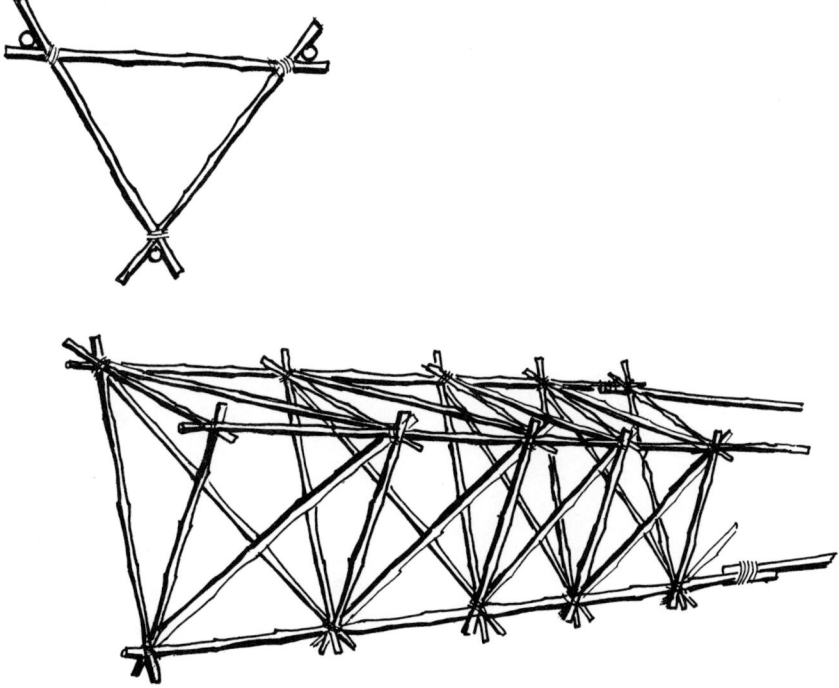

Figure 2-41. In the middle of a wilderness, away from any aid or product of the industrial era, a solid, rational, and elegant structure can be built with small bamboos tied with vines, by using three-dimensional truss systems. Roofs, columns, and walls could all be built in this simple and efficient way. Indeed, spatial trusses have important characteristics which often go unnoticed in literature.

3
Coefficients for Simplified Truss Analysis

If we select a truss over another system in a particular design, do we know what type of truss to select? To answer this, a designer needs a way to analyze that is more efficient and less complicated than conferring with a structural engineer or refreshing his memory of past course work. This chapter provides the designer with the medium he needs and, through the use of the following tables, the reader can quickly and easily analyze the most popular types of trusses. The designer who wishes to stray from these types can find methods for analyzing almost any type of truss problem by computer analysis in Chapter 4.

USE OF TABLES

The following tables spare the reader any algebraic or graphical analysis in determining the axial forces for the members of some of the most popular trusses. The tables also allow for the quick determination of the length of each member and the angle between members. Since the axial forces and the length of all members can be quickly computed for a given truss, it is possible to go back and change the truss configuration, its number of panels, and its height, to compute again axial forces and lengths until satisfactory results are found. Note that with the span kept constant, increasing the depth reduces the axial forces, and increasing the number of panels reduces the member length, which is of primary importance, particularly for compression members.

The different truss types included in the tables vary for the configuration and the number of panels. We distinguish them according to sloping, or

140 SIMPLIFIED TRUSS DESIGN

pitched types (Belgian, Polonceau (Fink), Fan, Howe, and Pratt) and flat types; i.e., with parallel chords (Howe, Pratt, modified Pratt, Warren).

Determination of Axial Forces

To determine axial forces in each member of the truss induced by the loads P on the top chord, multiply the load P by the factor K_1 in the tables. Similarly, the axial forces in each member of the truss induced by the loads p on the lower chord can be computed by multiplying p by the coefficient K_2 in the tables.

Notice that the positive sign (+) for the coefficients indicates a tension force in the member, and that a negative sign (−) indicates a compression force. Note also that loads P and p, acting independently of one another, must be applied at the joints only, in accordance to the figures illustrating the tables.

For the most common values of L/H, the coefficients K_1 and K_2 have been calculated in the tables. For any other value of L/H, the reader can find K_1 and K_2 from the general formulas given in the tables.

Determination of Length

Knowing the span L and height H of a given truss selected from the tables, the length for any member can be computed by multiplying the height H by the factor K obtained from the tables. Notice that K is a function of L/H. The tables give general formulas for the computation of K. For the most common values of L/H, the values of K have already been computed and inserted in the tables.

Example: Assume a span $L = 60$ ft and a height $H = 6$ ft for a truss loaded as in the following manner.

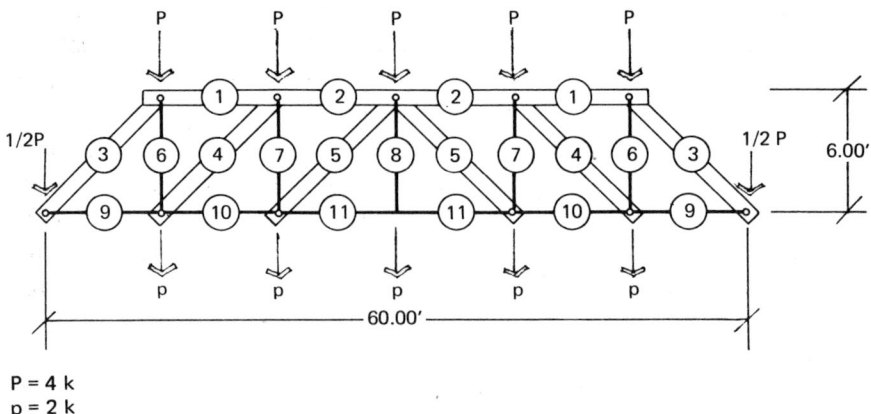

P = 4 k
p = 2 k

Figure 3-1.

Loading #1: (P = 4 kips) total load = 6 × P = 24 kips

Loading #2: (p = 2 kips) total load = 5 × p = 10 kips

From the tables, a six-panel flat Howe has been selected (see Figure 3-1). Compute the length of member #3, as indicated in the illustration in the table; also compute the axial force in member #3.

For $n = L/H = \dfrac{60}{6} = 10$

length of member #3 = $H \times K$ = 6 × 1.943651 = 11.66 ft

axial force for member #3, under Loading #1:

$P \times K_1$ = 4 × (−4.86) = −19.44K (compression)

axial force for member #3, under Loading #2:

$p \times K_2$ = 2 × (−4.86) = −9.72 K (compression).

TABLES FOR DETERMINING AXIAL FORCE COEFFICIENTS, LENGTH COEFFICIENTS, AND ANGLES

Table 3-1. Triangular Belgian

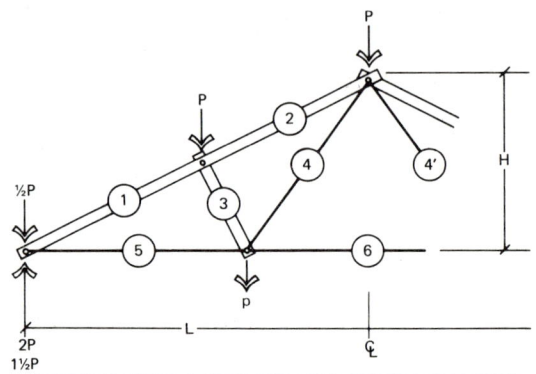

Pitch = H/L = $1/6\sqrt{3}$
$n = L/H = 2\sqrt{3} = 3.4642$
Roof Slope = 30° 00′

AXIAL FORCE COEFFICIENTS		
Member	K_1	K_2
1	−3.00	−2.00
2	−2.50	−2.00
3	−0.87	0
4	0.87	1.15
5	2.60	1.73
6	1.73	1.15

LENGTH COEFFICIENTS	
Member	K
1, 2	1.000000
3	0.577350
4, 5, 6	1.154701

ANGLES (IN DEGREES) BETWEEN MEMBERS	
Member	Deg
1-5, 2-4	30
3-4, 3-5, 4-5, 4-4	60
1-3, 2-3	90

Table 3-2. Triangular Belgian

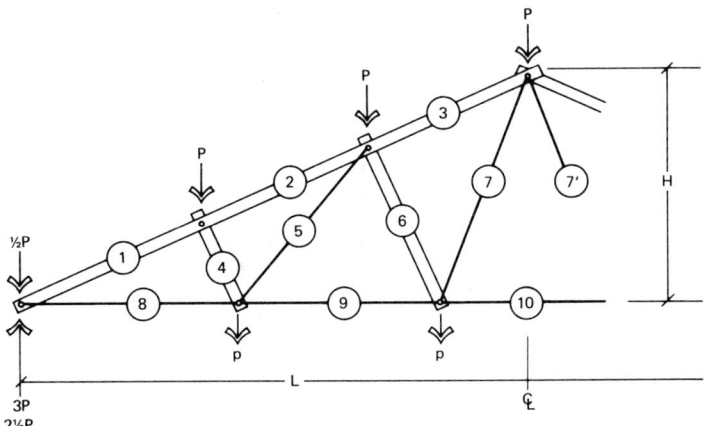

Pitch = $H/L = 1/10\sqrt{5}$
$n = L/H = 2\sqrt{5} = 4.4722$
Roof Slope = 24° 07'

AXIAL FORCE COEFFICIENTS

Member	K_1	K_2
1	−6.12	−4.90
2	−5.72	−4.90
3	−4.29	−3.67
4	−0.91	0
5	−1.37	−0.55
6	1.12	1.34
7	1.37	1.64
8	5.59	4.47
9	4.47	3.58
10	3.35	2.68

LENGTH COEFFICIENTS

Member	K
1, 2, 3,	0.816497
4	0.365148
5	0.730296
7	1.095445
6, 8, 9, 10	0.894427

144 SIMPLIFIED TRUSS DESIGN

Table 3-2. *(Continued)*

ANGLES (IN DEGREES) BETWEEN MEMBERS

Member	Deg.
1-8, 2-6 3-7 4-6, 4-8, 5-6, 5-9, 7-10	24 42 66
5-7, 6-9, 7-7 1-4, 2-4, 3-5	48 90

Table 3-3. Triangular Belgian

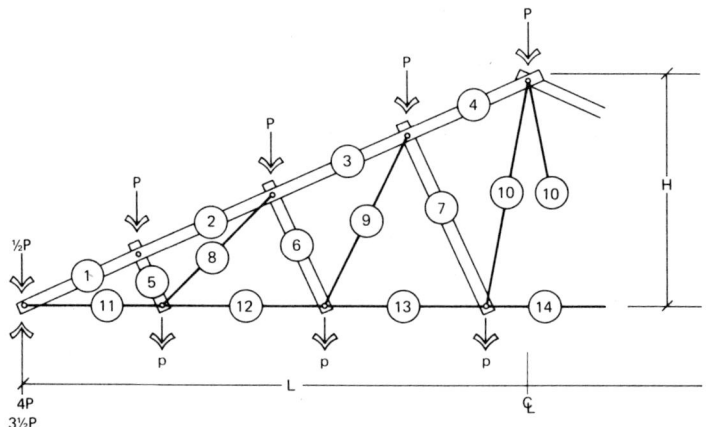

Pitch = $H/L = 1/14 \sqrt{7}$
$n = L/H = 2\sqrt{7} = 5.2915$
Roof Slope = $20° 42'$

AXIAL FORCE COEFFICIENTS		
Member	K_1	K_2
1	-9.90	-8.48
2	-9.55	-8.48
3	-7.95	-7.07
4	-6.36	-5.66
5	-0.94	0
6	-1.40	-0.53
7	-1.87	-1.07
8	1.32	1.51
9	1.55	1.77
10	1.87	2.14

Table 3-3. (*Continued*)

AXIAL FORCE COEFFICIENTS		
Member	K_1	K_2
11	9.26	7.94
12	7.94	6.80
13	6.61	5.67
14	5.29	4.54

LENGTH COEFFICIENTS	
Member	K
1, 2, 3, 4	0.707107
5	0.267261
6	0.534522
7	0.801784
8, 11, 12, 13, 14	0.755929
9	0.886405
10	1.069044

ANGLES (IN DEGREES) BETWEEN MEMBERS	
Member	Deg.
1-11, 2-8	21
3-9	37
4-10	49
5-8, 5-11, 6-8, 6-12, 7-13	69
6-9, 7-9, 7-10	53
	41
8-12	42
9-13	58
10-14	70
1-5, 2-5, 3-6, 4-7	90
10-10	40

146 SIMPLIFIED TRUSS DESIGN

Table 3-4. Triangular Belgian

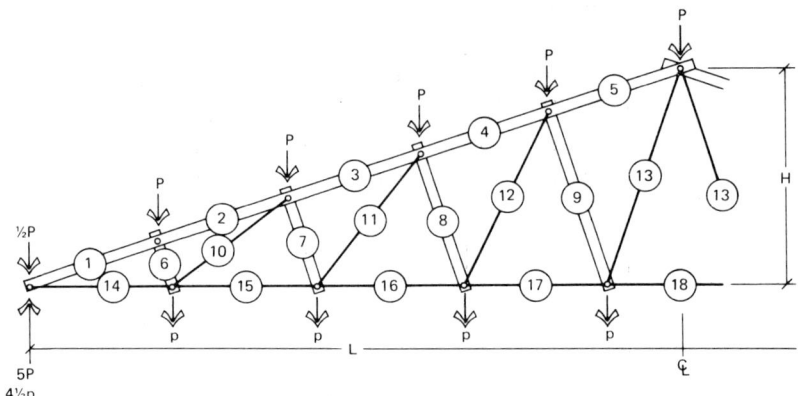

Pitch = H/L = 1/6 $n = L/H = 6$ Roof Slope = 18° 26′		
AXIAL FORCE COEFFICIENTS		
Member	K_1	K_2
1	−14.23	−12.65
2	−13.91	−12.65
3	−12.17	−11.07
4	−10.44	−9.49
5	−8.70	−7.91
6	−0.95	0
7	−1.42	−0.53
8	−1.90	−1.05
9	−2.37	−1.58
10	1.50	1.67
11	1.71	1.90
12	2.01	2.24
13	2.37	2.64
14	13.50	12.00
15	12.00	10.67
16	10.50	9.33
17	9.00	8.00
18	7.50	6.67

Table 3-4. (*Continued*)

LENGTH COEFFICIENTS	
Member	K
1, 2, 3, 4, 5, 8,	0.632456
6	0.210819
7	0.421637
9	0.843274
11	0.760117
12	0.894427
13	1.054092
10, 14, 15, 16, 17, 18	0.666667

ANGLES (IN DEGREES) BETWEEN MEMBERS	
Members	Deg
1-14, 2-10	18
3-11	34
4-12, 8-12, 9-12	45
5-13	53
6-10, 6-14, 7-10, 7-15, 8-16, 9-17	72
7-11, 8-11	56
9-13	37
10-15	36
11-16	52
12-17	63
13-18	71
1-6, 2-6, 3-7, 4-8, 5-9	90
13-13	38

148 SIMPLIFIED TRUSS DESIGN

Table 3-5. Triangular Belgian

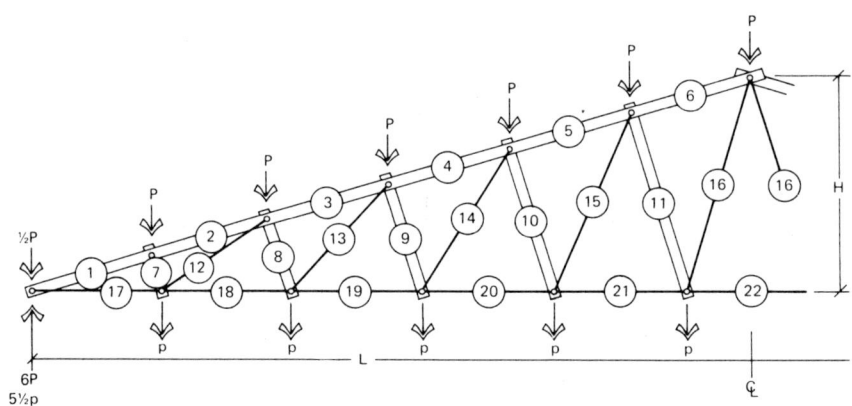

Pitch = $H/L = 1/22\sqrt{11}$ $n = L/H = 2\sqrt{11} = 6.6332$ Roof Slope = 16° 47'		
AXIAL FORCE COEFFICIENTS		
Member	K_1	K_2
1	-19.06	-17.32
2	-18.76	-17.32
3	-16.89	-15.59
4	-15.01	-13.86
5	-13.13	-12.12
6	-11.26	-10.39
7	-0.96	0
8	-1.44	-0.52
9	-1.91	-1.04
10	-2.39	-1.57
11	-2.87	-2.09
12	1.66	1.81
13	1.85	2.02
14	2.14	2.34
15	2.49	2.71
16	2.87	3.13
17	18.24	16.58
18	16.58	15.08
19	14.92	13.57
20	13.27	12.06
21	11.61	10.55
22	9.95	9.05

Table 3-5. (*Continued*)

LENGTH COEFFICIENTS	
Member	K
1, 2, 3, 4, 5, 6	0.577350
7	0.174078
8	0.348155
9	0.522233
10	0.696311
11	0.870388
12, 17, 18, 19, 20, 21, 22	0.603023
13	0.674200
14	0.778499
15	0.904534
16	1.044466
ANGLES (IN DEGREES) BETWEEN MEMBERS	
Member	Deg
1-17, 2-12	17
3-13	31
4-14	42
5-15	50
6-16	56
7-12, 7-17, 8-12, 8-18, 9-19, 10-20, 11-21, 16-22	73
8-13, 9-13, 14-20	59
9-14, 10-14, 13-19	48
10-15, 11-15	40
11-16, 12-18, 16-16	34
15-21	67
1-7, 2-7, 3-8, 4-9, 5-10, 6-11	90

150 SIMPLIFIED TRUSS DESIGN

Table 3-6. Triangular Belgian

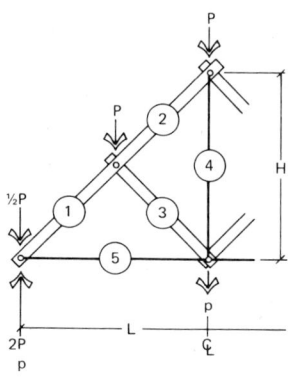

Pitch = H/L = 1/2	
$n = L/H = 2$	
Roof Slope = 45°	

AXIAL FORCE COEFFICIENTS		
Member	K_1	K_2
1	−2.12	−0.71
2	−1.41	−0.71
3	−0.71	0
4	1.00	1.00
5	1.50	0.50

LENGTH COEFFICIENTS	
Member	K
1, 2, 3	0.707107
4, 5	1.000000

ANGLES (IN DEGREES) BETWEEN MEMBERS	
Member	Deg
1-5, 2-4, 3-4, 3-5	45
1-3, 2-3	90

Table 3-7. Triangular Belgian

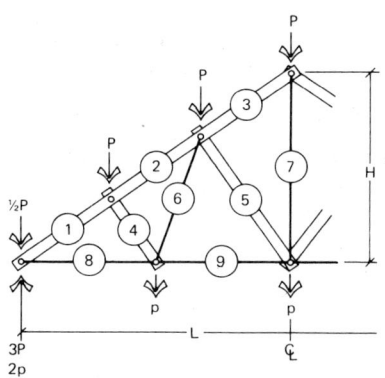

Pitch = $H/L = 1/4\sqrt{2}$
$n = L/H = 2\sqrt{2} = 2.8284$
Roof Slope = $35° 16'$

AXIAL FORCE COEFFICIENTS

Member	K_1	K_2
1	−4.33	−2.60
2	−3.75	−2.60
3	−2.60	−1.75
4	−0.82	0
5	−1.22	−0.61
6	0.71	1.06
7	2.00	2.00
8	3.54	2.12
9	2.83	1.77

LENGTH COEFFICIENTS

Member	K
1, 2, 3	0.577344
4	0.408248
5	0.816497
6, 8, 9	0.707107
7	1.000000

152   SIMPLIFIED TRUSS DESIGN

Table 3-7. *(Continued)*

ANGLES (IN DEGREES) BETWEEN MEMBERS	
Member	Deg
1-8, 2-6, 5-7	35
3-7, 4-6, 4-8, 5-6, 5-9	55
6-9	70
1-4, 2-4, 3-5	90

Table 3-8. Triangular Belgian

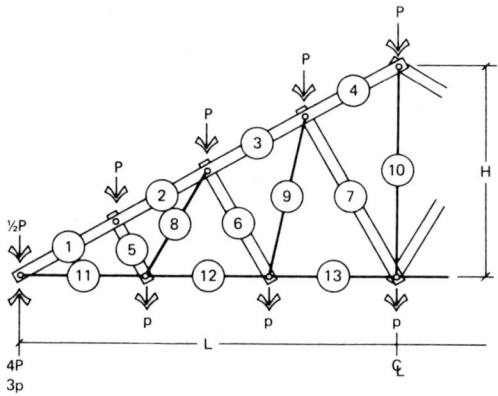

Pitch = $H/L = 1/6\sqrt{3}$		
$n = L/H = 2\sqrt{3} = 3.4642$		
Roof Slope = $30°$		
AXIAL FORCE COEFFICIENTS		
Member	K_1	K_2
1	−7.00	−5.00
2	−6.50	−5.00
3	−5.25	−4.00
4	−4.00	−3.00
5	−0.87	0
6	−1.30	−0.57
7	−1.73	−1.15
8	0.87	1.15
9	1.15	1.53

Table 3-8. (Continued)

AXIAL FORCE COEFFICIENTS		
Member	K_1	K_2
10	3.00	3.00
11	6.06	4.33
12	5.20	3.75
13	4.33	3.18

LENGTH COEFFICIENTS	
Member	K
1, 2, 3, 4	0.500000
5	0.288675
6, 8, 11, 12, 13	0.577350
7	0.866025
9	0.763783
10	1.000000

ANGLES (IN DEGREES) BETWEEN MEMBERS	
Member	Deg
1-11, 2-8, 7-10	30
3-9	49
4-10, 5-8, 5-11, 6-8, 6-12, 7-13, 8-12	60
6-9, 7-9	41
9-13	79
1-5, 2-5, 3-6, 4-7	90

154 SIMPLIFIED TRUSS DESIGN

Table 3-9. Triangular Belgian

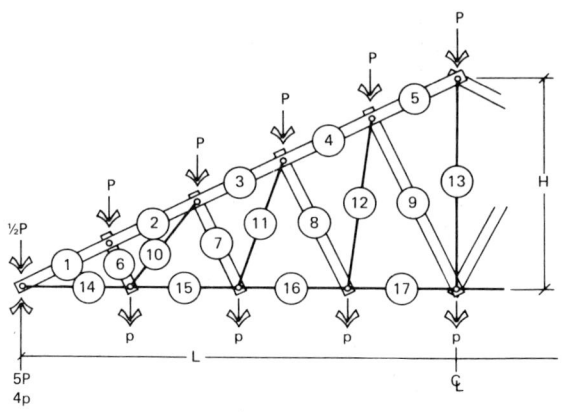

Pitch = H/L = 1/4 $n = L/H = 4$ Roof Slope = 26° 34′		
AXIAL FORCE COEFFICIENTS		
Member	K_1	K_2
1	−10.06	−7.83
2	−9.62	−7.83
3	−8.27	−6.71
4	−6.95	−5.59
5	−5.59	−4.47
6	−0.89	0
7	−1.34	−0.56
8	−1.79	−1.12
9	−2.24	−1.68
10	1.00	1.25
11	1.26	1.58
12	1.61	2.02
13	4.00	4.00
14	9.00	7.00
15	8.00	6.25
16	7.00	5.50
17	6.00	4.75

Table 3-9. (*Continued*)

LENGTH COEFFICIENTS	
Member	K
1, 2, 3, 4, 5, 7	0.447214
6	0.223607
8	0.670820
9	0.894427
10, 14, 15, 16, 17	0.500000
11	0.632456
12	0.806226
13	1.000000

ANGLES (IN DEGREES) BETWEEN MEMBERS	
Member	Deg
1-14, 2-10, 9-13	27
3-11, 7-11, 8-11	45
4-12	56
5-13, 6-10, 6-14, 7-10, 7-15, 8-16, 9-17	63
8-12, 9-12	34
10-15	54
11-16	72
12-17	83
1-6, 2-6, 3-7, 4-8, 5-9	90

156 SIMPLIFIED TRUSS DESIGN

Table 3-10. Triangular Belgian

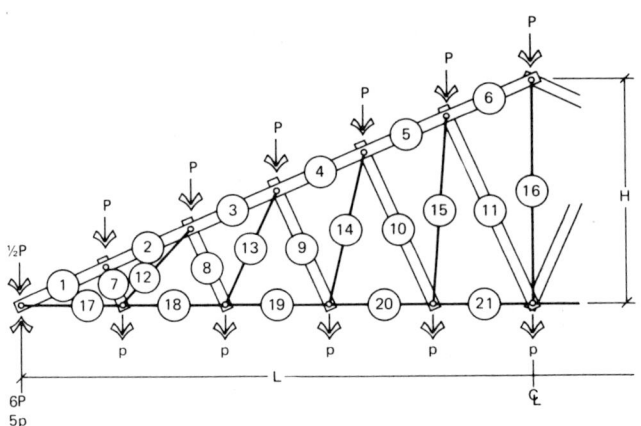

Pitch = $H/L = 1/10\sqrt{5}$
$n = L/H = 2\sqrt{5} = 4.4722$
Roof Slope = $24°\ 07'$

Member	K_1	K_2
1	−13.47	−11.02
2	−13.06	−11.02
3	−11.64	−9.80
4	−10.21	−8.57
5	−8.78	−7.35
6	−7.35	−6.12
7	−0.91	0
8	−1.37	−0.55
9	−1.83	−1.10
10	−2.28	−1.64
11	−2.74	−2.19
12	1.12	1.34
13	1.37	1.64
14	1.71	2.05
15	2.09	2.51
16	5.00	5.00
17	12.30	10.06
18	11.16	9.17
19	10.06	8.27
20	8.94	7.38
21	7.83	6.48

Table 3-10. (*Continued*)

LENGTH COEFFICIENTS	
Member	K
1, 2, 3, 4, 5, 6	0.408248
7	0.182574
8	0.365149
9, 13	0.547723
10	0.730297
11	0.912871
12, 17, 18, 19, 20, 21	0.447214
14	0.683130
15	0.836660
16	1.000000
ANGLES (IN DEGREES) BETWEEN MEMBERS	
Member	Deg
1-17, 2-12, 11-16	24
3-13	42
4-14	53
5-15	61
6-16, 7-12, 7-17, 8-12, 8-18, 9-19, 10-20, 11-21, 13-19	66
8-13, 9-13, 12-18	48
9-14, 10-14	37
10-15, 11-15	29
14-20	77
15-21	85
1-7, 2-7, 3-8, 4-9, 5-10, 6-11	90

Table 3-11. Polonceau (or Fink) 4 Panels at Top

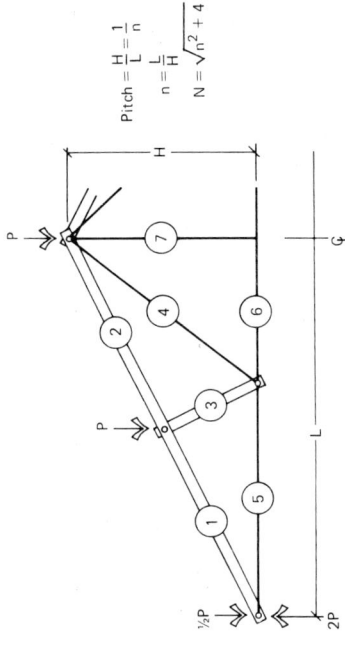

Pitch = $\frac{H}{L} = \frac{1}{n}$
$n = \frac{L}{H}$
$N = \sqrt{n^2 + 4}$

Member	VALUES OF n							General Formulas
	2	3	$2\sqrt{3}$	4	5	6	7	
	K_1	K_1	K_1	K_1	K_1	K_1	K_1	K_1
	AXIAL FORCE COEFFICIENTS							
1	-2.12	-2.70	-3.00	-3.35	-4.04	-4.74	-5.46	-3/4 N
2	-1.41	-2.15	-2.50	-2.91	-3.67	-4.43	-5.19	-1/N (3/4 n^2 + 1)
3	-0.71	-0.83	-0.87	-0.89	-0.93	-0.95	-0.96	-n/N
4	0.50	0.75	0.87	1.00	1.25	1.50	1.75	1/4 n
5	1.50	2.25	2.60	3.00	3.75	4.50	5.25	3/4 n
6	1.00	1.50	1.73	2.00	2.50	3.00	3.50	1/2 n
7	0	0	0	0	0	0	0	0

LENGTH COEFFICIENTS

	K	K	K	K	K	K	K	K	K
1, 2	0.707107	0.901388	1.000000	1.118034	1.346291	1.581139	1.820027	1/4 N
3	0.707107	0.600925	0.577367	0.559017	0.538516	0.527046	0.520008	1/2 N/n
4, 5	1.000000	1.083333	1.154734	1.250000	1.450000	1.666667	1.892857	1/4 N²/n
6	0.000000	0.416667	0.577250	0.750000	1.050000	1.333333	1.607143	1/4 (n − 4/n)
7	1.000000	1.000000	1.000000	1.000000	1.000000	1.000000	1.000000	1

ANGLES (IN DEGREES) BETWEEN MEMBERS

1-5, 2-4	45	34	30	27	22	18	16	2/n = tan a
4-6	90	68	60	54	44	36	32	2 − a
3-4, 3-5	45	56	60	63	68	72	74	90 − a
4-7	0	22	30	36	46	54	58	90 − 2a
1-3, 2-3, 6-7	90	90	90	90	90	90	90	90

Table 3-12. Polonceau (or Fink) 6 Panels at Top

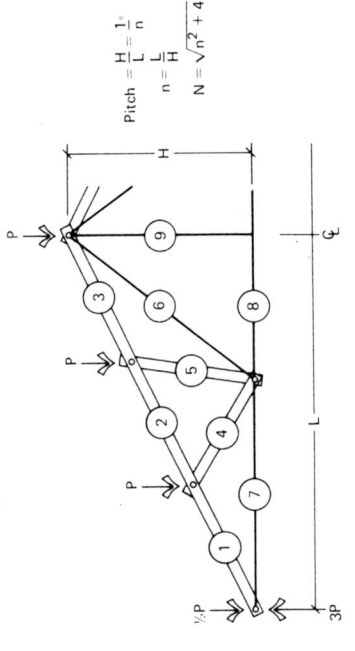

$\text{Pitch} = \dfrac{H}{L} = \dfrac{1}{n}$

$n = \dfrac{L}{H}$

$N = \sqrt{n^2 + 4}$

Member	\multicolumn{7}{c	}{VALUES OF n}	General Formulas
	2	3	$2\sqrt{3}$	4	5	6	7	
	K_1	K_1	K_1	K_1	K_1	K_1	K_1	K_1
	\multicolumn{7}{c	}{AXIAL FORCE COEFFICIENTS}	
1	−3.54	−4.51	−5.00	−5.59	−6.73	−7.91	−9.10	$-\dfrac{5/4\,N}{13N^2-16}$
2	−2.59	−3.54	−4.00	−4.55	−5.59	−6.64	−7.70	$-\dfrac{12N}{5n^2+4}$
3	−2.12	−3.40	−4.00	−4.70	−5.99	−7.27	−8.55	$-\dfrac{4N}{n\sqrt{n^2+36}}$
4	−0.75	−0.93	−1.00	−1.07	−1.21	−1.34	−1.48	$-\dfrac{6N}{n\sqrt{n^2+36}}$
5	−0.75	−0.93	−1.00	−1.07	−1.21	−1.34	−1.48	$-\dfrac{6N}{6N}$

	1.00	1.50	1.73	2.00	2.50	3.00	3.50	
6	1.00	1.50	1.73	2.00	2.50	3.00	3.50	$1/2\,n$
7	2.50	3.75	4.33	5.00	6.25	7.50	8.75	$5/4\,n$
8	1.50	2.25	2.60	3.00	3.75	4.50	5.25	$3/4\,n$
9	0	0	0	0	0	0	0	0

LENGTH COEFFICIENTS

	K	K	K	K	K	K	K	K
1, 2, 3	0.471405	0.600925	0.666667	0.745356	0.897527	1.054093	1.213352	$1/6\,N$
4, 5	0.745356	0.671855	0.666686	0.671855	0.700991	0.745356	0.799039	$1/12\,N\left(\dfrac{\sqrt{n^2+36}}{n}\right)$
6, 7	1.000000	1.083333	1.154734	1.250000	1.450000	1.666667	1.892857	$1/4\,N^2/n$
8	0.000000	0.416667	0.577250	0.750000	1.050000	1.333333	1.607150	$1/4\,(n-4/n)$
9	1.000000	1.000000	1.000000	1.000000	1.000000	1.000000	1.000000	1

ANGLES (IN DEGREES) BETWEEN MEMBERS

1-7, 3-6	45	34	30	27	22	18	16	$2/n = \tan a$
2-4, 2-5	72	63	60	56	50	45	41	$6/n = \tan b$
6-8	90	68	60	54	44	36	32	$2a$
6-9	0	22	30	36	46	54	58	$90 - 2a$
4-7, 5-6	27	29	30	29	28	27	25	$b - a$
1-4, 3-5	108	117	120	124	130	135	139	$180 - b$
4-5	36	54	60	68	80	90	98	$180 - 2b$
8-9	90	90	90	90	90	90	90	90

Table 3-13. Polonceau (or Fink) 8 Panels at Top

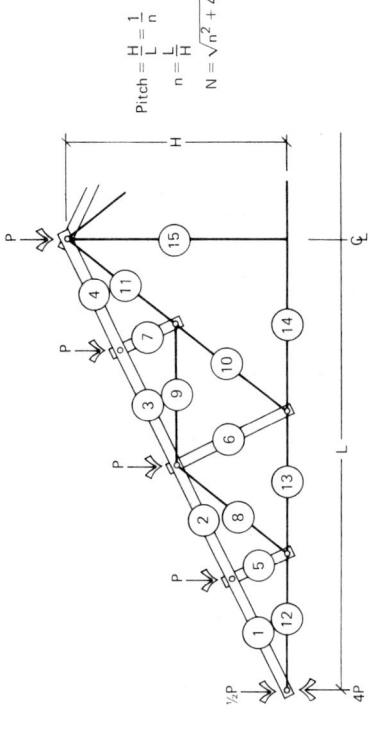

Pitch = $\frac{H}{L} = \frac{1}{n}$
$n = \frac{L}{H}$
$N = \sqrt{n^2 + 4}$

Member	VALUES OF n						General Formulas
	2	3	$2\sqrt{3}$	4	5	6	7	
	K_1	K_1	K_1	K_1	K_1	K_1	K_1	K_1
	AXIAL FORCE COEFFICIENTS							
1	−4.95	−6.31	−7.00	−7.83	−9.42	−11.07	−12.74	−7/4 N
2	−4.24	−5.76	−6.50	−7.38	−9.05	−10.75	−12.47	−(7/4 N − 2/N)
3	−3.54	−5.20	−6.00	−6.93	−8.68	−10.43	−12.19	−(7/4 N − 4/N)
4	−2.83	−4.65	−5.50	−6.48	−8.31	−10.12	−11.92	−(7/4 N − 6/N)
5	−0.71	−0.83	−0.87	−0.89	−0.93	−0.95	−0.96	−n/N
6	−1.41	−1.66	−1.73	−1.79	−1.86	−1.90	−1.92	−2 n/N
7	−0.71	−0.85	−0.87	−0.89	−0.93	−0.95	−0.96	−n/N
8	0.50	0.75	0.87	1.00	1.25	1.50	1.75	1/4 n
9	0.50	0.75	0.87	1.00	1.25	1.50	1.75	1/4 n
10	1.00	1.50	1.73	2.00	2.50	3.00	3.50	1/2 n

	1.50	2.25	2.60	3.00	3.75	4.50	5.25	
11	1.50	2.25	2.60	3.00	3.75	4.50	5.25	$3/4\,n$
12	3.50	5.25	6.06	7.00	8.75	10.50	12.25	$7/4\,n$
13	3.00	4.50	5.20	6.00	7.50	9.00	10.50	$3/2\,n$
14	2.00	3.00	3.46	4.00	5.00	6.00	7.00	n
15	0	0	0	0	0	0	0	0

LENGTH COEFFICIENTS

	K	K	K	K	K	K	K	K
1, 2, 3, 4	0.353553	0.450694	0.500000	0.559017	0.673146	0.790569	0.910014	$1/8\,N$
5, 7	0.353553	0.300463	0.288684	0.279509	0.269258	0.263523	0.260004	$1/4\,N/n$
6	0.707107	0.600925	0.577367	0.559017	0.538516	0.527046	0.520008	$1/2\,N/n$
8, 9, 10, 11, 12, 13	0.500000	0.541667	0.577367	0.625000	0.725000	0.833333	0.946429	$1/8\,N^2/n$
14	0.000000	0.416667	0.577250	0.750000	1.050000	1.333333	1.607150	$1/4\,(n - 4/n)$
15	1.000000	1.000000	1.000000	1.000000	1.000000	1.000000	1.000000	1

ANGLES (IN DEGREES) BETWEEN MEMBERS

1-12, 2-8, 3-9, 4-11	45	34	30	27	22	18	16	$2/n = \tan a$
5-8, 6-12, 6-8, 6-9, 6-10, 6-13, 7-9, 7-11	45	56	60	63	68	72	74	$90 - a$
8-13, 9-10	90	68	60	54	44	36	32	$2a$
10-14, 11-15	0	22	30	36	46	54	58	$90 - 2a$
1-5, 2-5, 3-7, 4-7, 14-15	90	90	90	90	90	90	90	90

Table 3-14. Polonceau (or Fink) 10 Panels at Top

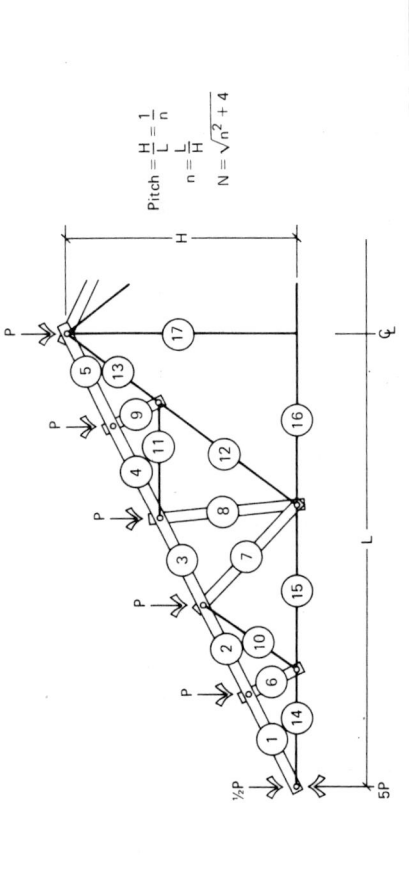

Pitch = $H/L = 1/n$
$n = L/H$
$N = \sqrt{n^2 + 4}$

Member	\multicolumn{7}{c}{VALUES OF n — AXIAL FORCE COEFFICIENTS}	General Formulas
	2	3	$2\sqrt{3}$	4	5	6	7	
	K_1	K_1	K_1	K_1	K_1	K_1	K_1	K_1
1	−6.36	−8.11	−9.00	−10.06	−12.12	−14.25	−16.38	$-9/4\, N$
2	−5.66	−7.55	−8.50	−9.62	−11.75	−13.91	−16.11	$-1/4\, \dfrac{(9n^2 + 28)}{N}$
3	−4.37	−6.00	−6.80	−7.74	−9.52	−11.31	−13.14	$-1/20\, \dfrac{(37n^2 + 100)}{N}$
4	−4.24	−6.44	−7.50	−8.72	−11.00	−13.28	−15.56	$-3/4\, \dfrac{(3n^2 + 4)}{N}$
5	−3.54	−5.89	−7.00	−8.27	−10.63	−12.97	−15.28	$-1/4\, \dfrac{(9n^2 + 4)}{N}$
6	−0.71	−0.83	−0.87	−0.89	−0.93	−0.95	−0.96	$-n/N$
7	−1.08	−1.31	−1.38	−1.44	−1.56	−1.66	−1.76	$-3/20\, n/N\sqrt{n^2 + 100}$
8	−1.08	−1.31	−1.38	−1.44	−1.56	−1.66	−1.76	$-3/20\, n/N\sqrt{n^2 + 100}$
9	−0.71	−0.83	−0.87	−0.89	−0.93	−0.95	−0.96	$-n/N$

Rows	K	K	K	K	K	K	K	K
10	0.50	0.75	0.87	1.00	1.25	1.50	1.75	$1/4\,n$
11	0.50	0.75	0.87	1.00	1.25	1.50	1.75	$1/4\,n$
12	1.50	2.25	2.60	3.00	3.75	4.50	5.25	$3/4\,n$
13	2.00	3.00	3.46	4.00	5.00	6.00	7.00	n
14	4.50	6.75	7.79	9.00	11.25	13.50	15.75	$9/4\,n$
15	4.00	6.00	6.92	8.00	10.00	12.00	14.00	$2n$
16	2.50	3.75	4.34	5.00	6.25	7.50	8.75	$5/4\,n$
17	0	0	0	0	0	0	0	0

LENGTH COEFFICIENTS

Rows	K	K	K	K	K	K	K	K
1, 2, 3, 4, 5	0.282843	0.360555	0.400000	0.447214	0.538516	0.632456	0.728011	$1/10\,N$
6, 9	0.282843	0.240370	0.230947	0.223607	0.215407	0.210819	0.208003	$1/5\,N/n$
7, 8	0.721110	0.627384	0.611028	0.602080	0.602080	0.614636	0.634750	$1/20\,N/n\,\sqrt{n^2+100}$
10, 11, 13, 14	0.400000	0.433333	0.461894	0.500000	0.580000	0.666667	0.757143	$1/10\,N^2/n$
12, 15	0.600000	0.650000	0.692841	0.750000	0.870000	1.000000	1.135715	$3/20\,N^2/n$
16	0.000000	0.416667	0.575000	0.750000	1.050000	1.333333	1.607150	$1/4\,(n-4/n)$
17	1.000000	1.000000	1.000000	1.000000	1.000000	1.000000	1.000000	1

ANGLES (IN DEGREES) BETWEEN MEMBERS

Rows	K	K	K	K	K	K	K	K
1-14, 2-10, 4-11, 5-13	45	34	30	27	22	18	16	$2/n = \tan a$
3-7, 3-8, 6-10, 6-14, 9-11, 9-13	79	73	71	68	63	59	55	$10/n = \tan b$
10-15, 11-12, 12-16	45	56	60	63	68	72	74	$90 - a$
	90	68	60	54	44	36	32	$2a$
7-15, 8-12	34	39	41	41	41	41	39	$b - a$
7-10, 8-11	56	73	79	85	95	103	109	$180 - a - b$
13-17	0	22	30	36	46	54	58	$90 - 2a$
7-8	22	34	38	44	54	62	70	$180 - 2b$
1-6, 2-6, 4-9, 5-9, 16-17	90	90	90	90	90	90	90	90

Table 3-15. Triangular Fan 12 Panels at Top

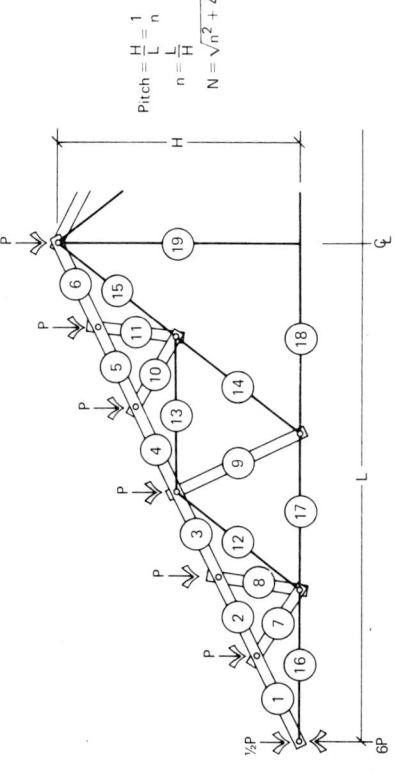

Pitch = $\frac{H}{L} = 1$
$n = \frac{L}{H}$
$N = \sqrt{n^2 + 4}$

Member	VALUES OF n							General Formulas
	2	3	$2\sqrt{3}$	4	5	6	7	
	K_1	K_1	K_1	K_1	K_1	K_1	K_1	K_1
	AXIAL FORCE COEFFICIENTS							
1	-7.78	-9.92	-11.00	-12.30	-14.81	-17.39	-20.02	$-11/4\, N$
2	-6.84	-8.94	-10.00	-11.25	-13.66	-16.13	-18.62	$-1/12\, \dfrac{(31N^2 - 16)}{N}$
3	-6.36	-8.81	-10.00	-11.40	-14.07	-16.76	-19.47	$-1/12\, \dfrac{(33N^2 - 48)}{N}$
4	-5.66	-8.25	-9.50	-10.96	-13.70	-16.44	-19.20	$-1/12\, \dfrac{(33N^2 - 72)}{N}$

	K	K	K	K	K	K	K	K
5	−4.71	−7.28	−8.50	−9.91	−12.55	−15.18	−17.80	$-1/12\,\dfrac{(31N^2-88)}{N}$
6	−4.24	−7.14	−8.50	−10.06	−12.95	−15.93	−18.65	$-1/12\,\dfrac{(33N^2-120)}{N}$
7	−0.75	−0.93	−1.00	−1.07	−1.21	−1.34	−1.48	$-1/6\,n/N\,\sqrt{n^2+36}$
8	−0.75	−0.93	−1.00	−1.07	−1.21	−1.34	−1.48	$-1/6\,n/N\,\sqrt{n^2+36}$
9	−2.12	−2.50	−2.60	−2.68	−2.79	−2.85	−2.88	$-3\,n/N$
10	−0.75	−0.93	−1.00	−1.07	−1.21	−1.34	−1.48	$-1/6\,n/N\,\sqrt{n^2+36}$
11	−0.75	−0.93	−1.00	−1.07	−1.21	−1.34	−1.48	$-1/6\,n/N\,\sqrt{n^2+36}$
12	1.00	1.50	1.73	2.00	2.50	3.00	3.50	$1/2\,n$
13	1.00	1.50	1.73	2.00	2.50	3.00	3.50	$1/2\,n$
14	1.50	2.25	2.60	3.00	3.75	4.50	5.25	$3/4\,n$
15	2.50	3.75	4.33	5.00	6.25	7.50	8.75	$5/4\,n$
16	5.50	8.25	9.53	11.00	13.75	16.50	19.25	$11/4\,n$
17	4.50	6.75	7.79	9.00	11.25	13.50	15.75	$9/4\,n$
18	3.00	4.50	5.20	6.00	7.50	9.00	10.50	$3/2\,n$
19	0	0	0	0	0	0	0	0

LENGTH COEFFICIENTS

	K	K	K	K	K	K	K	K
1, 2, 3, 4, 5, 6	0.235702	0.300463	0.333333	0.372678	0.448764	0.527046	0.606676	$1/12\,N$
7, 8, 10, 11	0.372678	0.335927	0.333343	0.335927	0.350496	0.372678	0.399520	$1/24\,N\,\dfrac{\sqrt{n^2+36}}{n}$
9	0.707107	0.600925	0.577367	0.559017	0.538516	0.527046	0.520008	$1/2\,N/n$
12, 13, 14, 15, 16, 17	0.500000	0.541667	0.577367	0.625000	0.725000	0.833333	0.946429	$1/8\,N^2/n$
18	0.000000	0.416667	0.577250	0.750000	1.050000	1.333333	1.607150	$1/4\,(n-4/n)$
19	1.000000	1.000000	1.000000	1.000000	1.000000	1.000000	1.000000	1

Table 3-15. (Continued)

	ANGLES (IN DEGREES) BETWEEN MEMBERS						
1-16, 3-12, 4-13, 6-15	45	34	30	27	22	18	16	$2/n = \tan a$
2-7, 2-8, 5-10, 5-11	72	63	60	56	50	45	41	$6/n = \tan b$
12-17, 13-14, 14-18	90	68	60	54	44	36	32	$2a$
9-12, 9-13, 9-14, 9-17	45	56	60	63	68	72	74	$90 - a$
15-19	0	22	30	36	46	54	58	$90 - 2a$
7-16, 8-12, 10-13, 11-15	27	29	30	29	28	27	25	$b - a$
1-7, 3-8, 4-10, 6-11	108	117	120	124	130	135	139	$180 - b$
7-8, 10-11	36	54	60	68	80	90	98	$180 - 2b$
18-19	90	90	90	90	90	90	90	90

Table 3-16. Triangular Howe 4 Panels at Top & Bottom

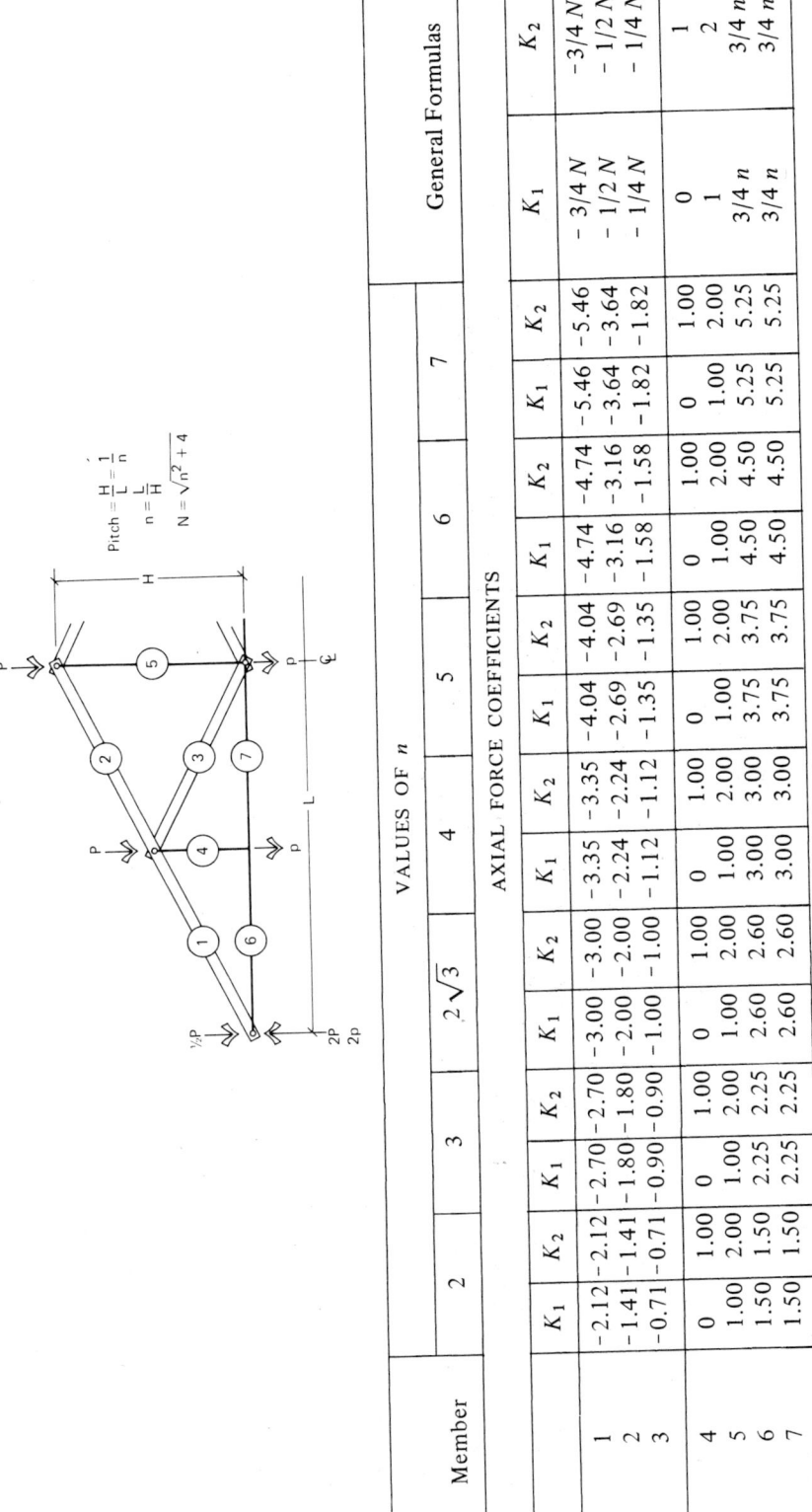

Pitch = $\frac{H}{L} = \frac{1}{n}$
$n = \frac{L}{H}$
$N = \sqrt{n^2 + 4}$

Member	VALUES OF n												General Formulas	
	2		3		$2\sqrt{3}$		4		5		6		7		K_1	K_2
	K_1	K_2	K_1	K_2	K_1	K_2	K_1	K_2	K_1	K_2	K_1	K_2	K_1	K_2		
AXIAL FORCE COEFFICIENTS																
1	-2.12	-2.12	-2.70	-2.70	-3.00	-3.00	-3.35	-3.35	-4.04	-4.04	-4.74	-4.74	-5.46	-5.46	-3/4 N	-3/4 N
2	-1.41	-1.41	-1.80	-1.80	-2.00	-2.00	-2.24	-2.24	-2.69	-2.69	-3.16	-3.16	-3.64	-3.64	-1/2 N	-1/2 N
3	-0.71	-0.71	-0.90	-0.90	-1.00	-1.00	-1.12	-1.12	-1.35	-1.35	-1.58	-1.58	-1.82	-1.82	-1/4 N	-1/4 N
4	0	1.00	0	1.00	0	1.00	0	1.00	0	1.00	0	1.00	0	1.00	0	1
5	1.00	2.00	1.00	2.00	1.00	2.00	1.00	2.00	1.00	2.00	1.00	2.00	1.00	2.00	1	2
6	1.50	1.50	2.25	2.25	2.60	2.60	3.00	3.00	3.75	3.75	4.50	4.50	5.25	5.25	3/4 n	3/4 n
7	1.50	1.50	2.25	2.25	2.60	2.60	3.00	3.00	3.75	3.75	4.50	4.50	5.25	5.25	3/4 n	3/4 n

Table 3-16. (Continued)

Member	VALUES OF n							General Formulas
	2	3	$2\sqrt{3}$	4	5	6	7	
	K	K	K	K	K	K	K	K
	LENGTH COEFFICIENTS							
1, 2, 3	0.707107	0.901388	1.000000	1.118034	1.346291	1.581139	1.820027	$1/4\,N$
4	0.500000	0.500000	0.500000	0.500000	0.500000	0.500000	0.500000	$1/2$
5	1.000000	1.000000	1.000000	1.000000	1.000000	1.000000	1.000000	1
6, 7	0.500000	0.750000	0.866000	1.000000	1.250000	1.500000	1.750000	$1/4\,n$
	ANGLES (IN DEGREES) BETWEEN MEMBERS							
1-6, 3-7	45	34	30	27	22	16	16	$2/n = \tan a$
1-4, 2-5, 3-4, 3-5	45	56	60	63	68	72	74	$90 - a$
2-3	90	68	60	54	44	36	32	$2a$
4-6, 4-7	90	90	90	90	90	90	90	90

Table 3-17. Triangular Howe 6 Panels at Top & Bottom

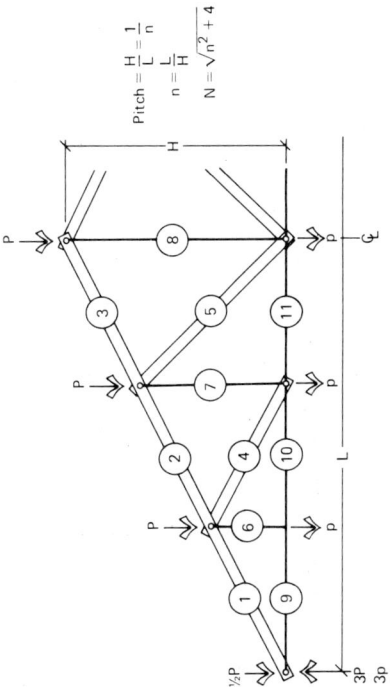

Pitch = $\frac{H}{L} = \frac{1}{n}$
$n = \frac{L}{H}$
$N = \sqrt{n^2 + 4}$

Member	VALUES OF n												General Formulas	
	2		3		$2\sqrt{3}$		4		5		6		7			
	K_1	K_2	K_1	K_2	K_1	K_2	K_1	K_2	K_1	K_2	K_1	K_2	K_1	K_1	K_2
	AXIAL FORCE COEFFICIENTS														
1	-3.54	-3.54	-4.51	-4.51	-5.00	-5.00	-5.59	-5.59	-6.73	-6.73	-7.91	-7.91	-9.10	-5/4 N	-5/4 N
2	-2.83	-2.83	-3.61	-3.61	-4.00	-4.00	-4.47	-4.47	-5.39	-5.39	-6.32	-6.32	-7.28	-N	-N
3	-2.12	-2.12	-2.70	-2.70	-3.00	-3.00	-3.35	-3.35	-4.04	-4.04	-4.74	-4.74	-5.46	-3/4 N	-3/4 N
4	-0.71	-0.71	-0.90	-0.90	-1.00	-1.00	-1.12	-1.12	-1.35	-1.35	-1.58	-1.58	-1.82	-1/4 N	-1/4 N
5	-1.12	-1.12	-1.25	-1.25	-1.32	-1.32	-1.41	-1.41	-1.60	-1.60	-1.80	-1.80	-2.02	$-1/4\sqrt{n^2+16}$	$-1/4\sqrt{n^2+16}$
6	0	1.00	0	1.00	0	1.00	0	1.00	0	1.00	0	1.00	1.00	0	1
7	0.50	1.50	0.50	1.50	0.50	1.50	0.50	1.50	0.50	1.50	0.50	1.50	1.50	1/2	3/2
8	2.00	3.00	2.00	3.00	2.00	3.00	2.00	3.00	2.00	3.00	2.00	3.00	3.00	2	3
9	2.50	2.50	3.75	3.75	4.33	4.33	5.00	5.00	6.25	6.25	7.50	7.50	8.75	5/4 n	5/4 n

Table 3-17. (Continued)

Member	VALUES OF n													General Formulas	
	2		3		$2\sqrt{3}$		4		5		6		7		
	K_1	K_2	K_1	K_2	K_1	K_2	K_1	K_2	K_1	K_2	K_1	K_2	K_1	K_2	K_1	K_2
AXIAL FORCE COEFFICIENTS																
10	2.50	2.50	3.75	3.75	4.33	4.33	5.00	5.00	6.25	6.25	7.50	7.50	8.75	8.75	$5/4\,n$	$5/4\,n$
11	2.00	2.00	3.00	3.00	3.46	3.46	4.00	4.00	5.00	5.00	6.00	6.00	7.00	7.00	n	n

Member	2	3	$2\sqrt{3}$	4	5	6	7	General Formulas
	K	K	K	K	K	K	K	K
LENGTH COEFFICIENTS								
1, 2, 3, 4	0.471405	0.600925	0.666667	0.745356	0.897527	1.054093	1.213352	$1/6\,N$
5	0.745356	0.833333	0.881917	0.942809	1.067187	1.201850	1.343710	$1/6\sqrt{n^2+16}$
6	0.333333	0.333333	0.333333	0.333333	0.333333	0.333333	0.333333	$1/3$
7	0.666667	0.666667	0.666667	0.666667	0.666667	0.666667	0.666667	$2/3$
8	1.000000	1.000000	1.000000	1.000000	1.000000	1.000000	1.000000	1
9, 10, 11	0.333333	0.500000	0.577350	0.666667	0.833333	1.000000	1.166667	$1/6\,n$
ANGLES (IN DEGREES) BETWEEN MEMBERS								
1-9, 4-10	45	34	30	27	22	18	16	$2/n = \tan a$
5-11	63	53	49	45	39	34	30	$4/n = \tan b$
1-6, 2-7, 3-8, 4-6, 4-7	45	56	60	63	68	72	74	$90 - a$
5-7, 5-8	27	37	41	45	51	56	60	$90 - b$
2-4	90	68	60	54	44	36	32	$2a$
3-5	108	87	79	72	61	52	46	$a + b$
6-9, 6-10, 7-11	90	90	90	90	90	90	90	90

Table 3-18. Triangular Howe 8 Panels at Top & Bottom

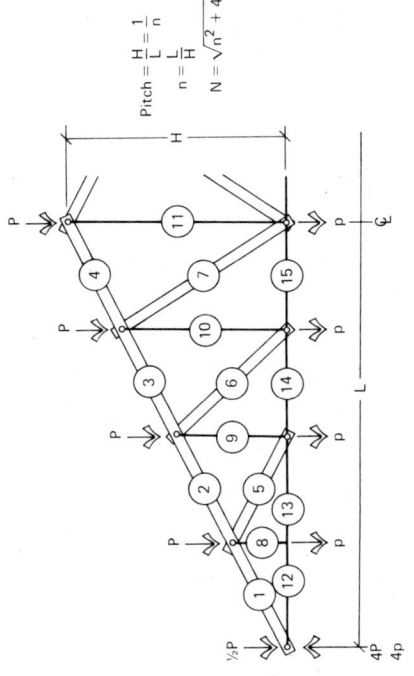

Member	2		3		$2\sqrt{3}$		4		5		6		7		General Formulas	
	K_1	K_2	K_1	K_2	K_1	K_2	K_1	K_2	K_1	K_2	K_1	K_2	K_1	K_2	K_1	K_2
	VALUES OF n															
	AXIAL FORCE COEFFICIENTS															
1	-4.95	-4.95	-6.31	-6.31	-7.00	-7.00	-7.83	-7.83	-9.42	-9.42	-11.07	-11.07	-12.74	-12.74	-7/4 N	-7/4 N
2	-4.24	-4.24	-5.41	-5.41	-6.00	-6.00	-6.71	-6.71	-8.08	-8.08	-9.49	-9.49	-10.92	-10.92	-3/2 N	-3/2 N
3	-3.54	-3.54	-4.51	-4.51	-5.00	-5.00	-5.59	-5.59	-6.73	-6.73	-7.91	-7.91	-9.10	-9.10	-5/4 N	-5/4 N
4	-2.83	-2.83	-3.61	-3.61	-4.00	-4.00	-4.47	-4.47	-5.39	-5.39	-6.32	-6.32	-7.28	-7.28	-N	-N
5	-0.71	-0.71	-0.90	-0.90	-1.00	-1.00	-1.12	-1.12	-1.35	-1.35	-1.58	-1.58	-1.82	-1.82	-1/4 N	-1/4 N

Pitch = $\frac{H}{L} = \frac{1}{n}$
$n = \frac{L}{H}$
$N = \sqrt{n^2 + 4}$

Table 3-18. (Continued)

Member	VALUES OF n														General Formulas	
	2		3		$2\sqrt{3}$		4		5		6		7		K_1	K_2
	K_1	K_2	K_1	K_2	K_1	K_2	K_1	K_2	K_1	K_2	K_1	K_2	K_1	K_2		
	AXIAL FORCE COEFFICIENTS															
6	-1.12	-1.12	-1.25	-1.25	-1.32	-1.32	-1.41	-1.41	-1.60	-1.60	-1.80	-1.80	-2.02	-2.02	$-1/4\sqrt{n^2+16}$	$-1/4\sqrt{n^2+16}$
7	-1.58	-1.58	-1.68	-1.68	-1.73	-1.73	-1.80	-1.80	-1.95	-1.95	-2.12	-2.12	-2.30	-2.30	$-1/4\sqrt{n^2+36}$	$-1/4\sqrt{n^2+36}$
8	0	1.00	0	1.00	0	1.00	0	1.00	0	1.00	0	1.00	0	1.00	0	1
9	0.50	1.50	0.50	1.50	0.50	1.50	0.50	1.50	0.50	1.50	0.50	1.50	0.50	1.50	1/2	3/2
10	1.00	2.00	1.00	2.00	1.00	2.00	1.00	2.00	1.00	2.00	1.00	2.00	1.00	2.00	1	2
11	3.00	4.00	4.00	4.00	4.00	4.00	3.00	4.00	3.00	4.00	3.00	4.00	3.00	4.00	3	4
12	3.50	3.50	5.25	5.25	6.06	6.06	7.00	7.00	8.75	8.75	10.50	10.50	12.25	12.25	$7/4\,n$	$7/4\,n$
13	3.50	3.50	5.25	5.25	6.06	6.06	7.00	7.00	8.75	8.75	10.50	10.50	12.25	12.25	$7/4\,n$	$7/4\,n$
14	3.00	3.00	4.50	4.50	5.20	5.20	6.00	6.00	7.50	7.50	9.00	9.00	10.50	10.50	$3/2\,n$	$3/2\,n$
15	2.50	2.50	3.75	3.75	4.33	4.33	5.00	5.00	6.25	6.25	7.50	7.50	8.75	8.75	$5/4\,n$	$5/4\,n$

Member	LENGTH COEFFICIENTS							
	2	3	$2\sqrt{3}$	4	5	6	7	K
	K	K	K	K	K	K	K	
1, 2, 3, 4, 5	0.353553	0.450694	0.500000	0.559017	0.673146	0.790569	0.910014	$1/8\,N$
6	0.559017	0.625000	0.661438	0.707107	0.800391	0.901388	1.007782	$1/8\sqrt{n^2+36}$
7	0.790569	0.838525	0.866025	0.901388	0.976281	1.060660	1.152443	$1/8\sqrt{n^2+36}$
8	0.250000	0.250000	0.250000	0.250000	0.250000	0.250000	0.250000	1/4
9	0.500000	0.500000	0.500000	0.500000	0.500000	0.500000	0.500000	1/2
10	0.750000	0.750000	0.750000	0.750000	0.750000	0.750000	0.750000	3/4
11	1.000000	1.000000	1.000000	1.000000	1.000000	1.000000	1.000000	1
12, 13, 14, 15	0.250000	0.375000	0.433000	0.500000	0.625000	0.750000	0.875000	$1/8\,n$

ANGLES (IN DEGREES) BETWEEN MEMBERS							
1-12, 5-13	45	34	30	27	22	18	16	$2/n = \tan a$
6-14	63	53	49	45	39	34	30	$4/n = \tan b$
7-15	72	63	60	56	50	45	41	$6/n = \tan c$
1-8, 2-9, 3-10, 4-11, 5-8, 5-9	45	56	60	63	68	72	74	$90 - a$
6-9, 6-10	27	37	41	45	51	56	60	$90 - b$
7-10, 7-11	18	27	30	34	40	45	49	$90 - c$
2-5	90	68	60	54	44	36	32	$2a$
3-6	108	87	79	72	61	52	46	$a + b$
4-7	117	97	90	83	72	63	57	$a + c$
6-12, 8-13, 9-14, 10-15	90	90	90	90	90	90	90	90

Table 3-19. Triangular Howe 10 Panels at Top & Bottom

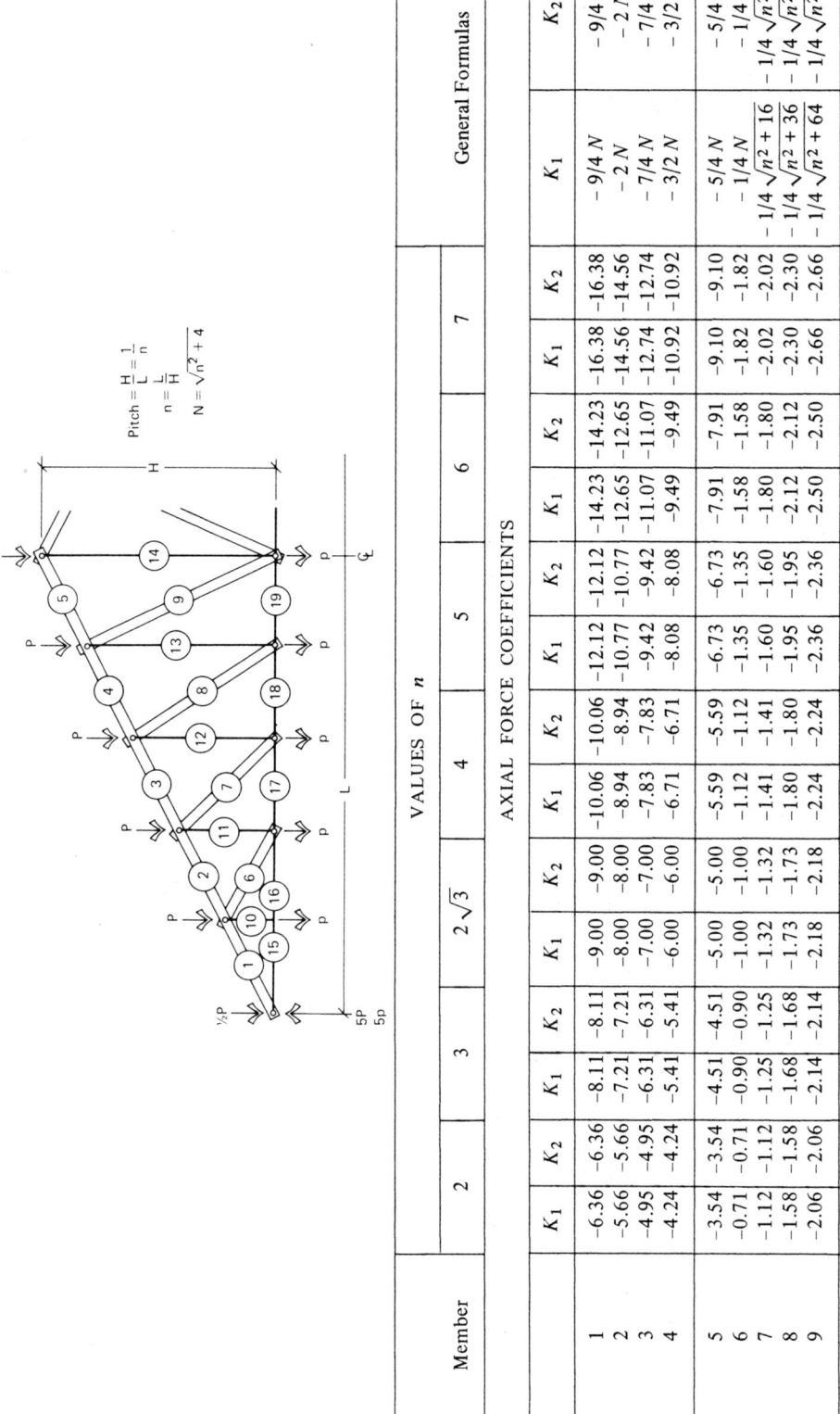

Member	2		3		$2\sqrt{3}$		4		5		6		7		General Formulas	
	K_1	K_2	K_1	K_2	K_1	K_2	K_1	K_2	K_1	K_2	K_1	K_2	K_1	K_2	K_1	K_2
						VALUES OF n										
					AXIAL FORCE COEFFICIENTS											
1	−6.36	−6.36	−8.11	−8.11	−9.00	−9.00	−10.06	−10.06	−12.12	−12.12	−14.23	−14.23	−16.38	−16.38	$-9/4\,N$	$-9/4\,N$
2	−5.66	−5.66	−7.21	−7.21	−8.00	−8.00	−8.94	−8.94	−10.77	−10.77	−12.65	−12.65	−14.56	−14.56	$-2\,N$	$-2\,N$
3	−4.95	−4.95	−6.31	−6.31	−7.00	−7.00	−7.83	−7.83	−9.42	−9.42	−11.07	−11.07	−12.74	−12.74	$-7/4\,N$	$-7/4\,N$
4	−4.24	−4.24	−5.41	−5.41	−6.00	−6.00	−6.71	−6.71	−8.08	−8.08	−9.49	−9.49	−10.92	−10.92	$-3/2\,N$	$-3/2\,N$
5	−3.54	−3.54	−4.51	−4.51	−5.00	−5.00	−5.59	−5.59	−6.73	−6.73	−7.91	−7.91	−9.10	−9.10	$-5/4\,N$	$-5/4\,N$
6	−0.71	−0.71	−0.90	−0.90	−1.00	−1.00	−1.12	−1.12	−1.35	−1.35	−1.58	−1.58	−1.82	−1.82	$-1/4\,N$	$-1/4\,N$
7	−1.12	−1.12	−1.25	−1.25	−1.32	−1.32	−1.41	−1.41	−1.60	−1.60	−1.80	−1.80	−2.02	−2.02	$-1/4\sqrt{n^2+16}$	$-1/4\sqrt{n^2+16}$
8	−1.58	−1.58	−1.68	−1.68	−1.73	−1.73	−1.80	−1.80	−1.95	−1.95	−2.12	−2.12	−2.30	−2.30	$-1/4\sqrt{n^2+36}$	$-1/4\sqrt{n^2+36}$
9	−2.06	−2.06	−2.14	−2.14	−2.18	−2.18	−2.24	−2.24	−2.36	−2.36	−2.50	−2.50	−2.66	−2.66	$-1/4\sqrt{n^2+64}$	$-1/4\sqrt{n^2+64}$

Pitch $= \dfrac{H}{L} = \dfrac{1}{n}$
$n = \dfrac{L}{H}$
$N = \sqrt{n^2+4}$

LENGTH COEFFICIENTS

Member																		
10	0	1.00	0	1.00	0	1.00	0	1.00	0	1.00	0	1.00	0	1.00	0	1.00	0	1
11	0.50	1.50	0.50	1.50	0.50	1.50	0.50	1.50	0.50	1.50	0.50	1.50	0.50	1.50	0.50	1.50	1/2	3/2
12	1.00	2.00	1.00	2.00	1.00	2.00	1.00	2.00	1.00	2.00	1.00	2.00	1.00	2.00	1.00	2.00	1	2
13	1.50	2.50	1.50	2.50	1.50	2.50	1.50	2.50	1.50	2.50	1.50	2.50	1.50	2.50	1.50	2.50	3/2	5/2
14	4.00	5.00	4.00	5.00	4.00	5.00	4.00	5.00	4.00	5.00	4.00	5.00	4.00	5.00	4.00	5.00	4	5
15	4.50	4.50	6.75	6.75	7.79	7.79	9.00	9.00	11.25	11.25	13.50	13.50	15.75	15.75	15.75	15.75	$9/4\,n$	$9/4\,n$
16	4.50	4.50	6.75	6.75	7.79	7.79	9.00	9.00	11.25	11.25	13.50	13.50	15.75	15.75	15.75	15.75	$9/4\,n$	$9/4\,n$
17	4.00	4.00	6.00	6.00	6.93	6.93	8.00	8.00	10.00	10.00	12.00	12.00	14.00	14.00	14.00	14.00	$2n$	$2n$
18	3.50	3.50	5.25	5.25	6.06	6.06	7.00	7.00	8.75	8.75	10.50	10.50	12.25	12.25	12.25	12.25	$7/4\,n$	$7/4\,n$
19	3.00	3.00	4.50	4.50	5.20	5.20	6.00	6.00	7.50	7.50	9.00	9.00	10.50	10.50	10.50	10.50	$3/2\,n$	$3/2\,n$

Member	K	K	K	K	K	K	K	K	K
1, 2, 3, 4, 5, 6	0.282843	0.360555	0.400000	0.447214	0.538516	0.632456	0.728011		$1/10\,N$
7	0.447214	0.500000	0.529150	0.565685	0.640312	0.721110	0.806226		$1/10\,\sqrt{n^2+16}$
8	0.632456	0.670820	0.692820	0.721110	0.781025	0.848528	0.921954		$1/10\,\sqrt{n^2+36}$
9	0.824621	0.854400	0.871780	0.894427	0.943398	1.000000	1.063015		$1/10\,\sqrt{n^2+64}$
10	0.200000	0.200000	0.200000	0.200000	0.200000	0.200000	0.200000		1/5
11	0.400000	0.400000	0.400000	0.400000	0.400000	0.400000	0.400000		2/5
12	0.600000	0.600000	0.600000	0.600000	0.600000	0.600000	0.600000		3/5
13	0.800000	0.800000	0.800000	0.800000	0.800000	0.800000	0.800000		4/5
14	1.000000	1.000000	1.000000	1.000000	1.000000	1.000000	1.000000		1
15, 16, 17, 18, 19	0.200000	0.300000	0.346400	0.400000	0.500000	0.600000	0.700000		$1/10\,n$

ANGLES (IN DEGREES) BETWEEN MEMBERS

Member								
1-15, 6-16	45	34	30	27	22	18	16	$2/n = \tan a$
7-17	63	53	49	45	39	34	30	$4/n = \tan b$
8-18	72	63	60	56	50	45	41	$6/n = \tan c$
9-19	76	69	67	63	58	53	49	$8/n = \tan d$

Table 3-19. (Continued)

Member	VALUES OF n / ANGLES (IN DEGREES) BETWEEN MEMBERS							General Formulas
	2	3	$2\sqrt{3}$	4	5	6	7	
	K	K	K	K	K	K	K	K
1-10, 2-11, 3-12, 4-13, 5-14,								
6-10, 6-11	45	56	60	63	68	72	74	$90-a$
7-11, 7-12	27	37	41	45	51	56	60	$90-b$
8-12, 8-13	18	27	30	34	40	45	49	$90-c$
9-13, 9-14	14	21	23	27	32	37	41	$90-d$
2-6	90	68	60	54	44	36	32	$2a$
3-7	108	87	79	72	61	52	46	$a+b$
4-8	117	97	90	83	72	63	57	$a+c$
5-9	121	108	97	90	80	71	65	$a+d$
10-15, 10-16, 11-17, 12-18, 13-19	90	90	90	90	90	90	90	90

Table 3-20. Triangular Howe 12 Panels at Top & Bottom

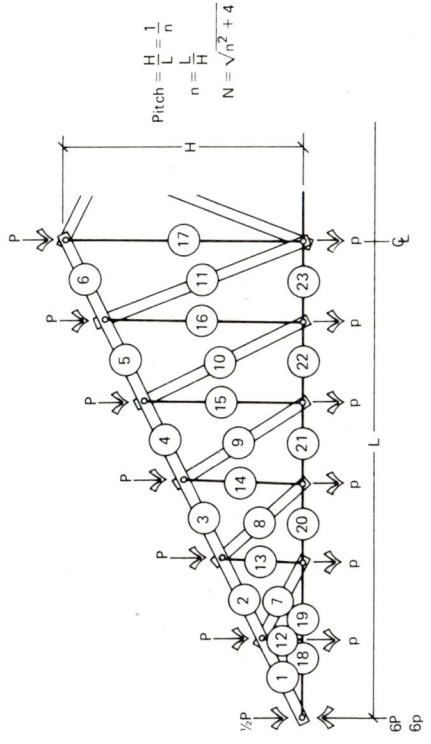

Pitch = $\frac{H}{L} = \frac{1}{n}$
$n = \frac{L}{H}$
$N = \sqrt{n^2 + 4}$

Member	\multicolumn{2}{c}{2}		\multicolumn{2}{c}{3}		\multicolumn{2}{c}{$2\sqrt{3}$}		\multicolumn{2}{c}{4}		\multicolumn{2}{c}{5}		\multicolumn{2}{c}{6}		\multicolumn{2}{c}{7}		\multicolumn{2}{c}{General Formulas}
	K_1	K_2	K_1	K_2	K_1	K_2	K_1	K_2	K_1	K_2	K_1	K_2	K_1	K_2	K_1	K_2
1	−7.78	−7.78	−9.92	−9.92	−11.00	−11.00	−12.30	−12.30	−14.81	−14.81	−17.39	−17.39	−20.02	−20.02	−11/4 N	−11/4 N
2	−7.07	−7.07	−9.01	−9.01	−10.00	−10.00	−11.18	−11.18	−13.46	−13.46	−15.81	−15.81	−18.20	−18.20	−5/2 N	−5/2 N
3	−6.36	−6.36	−8.11	−8.11	−9.00	−9.00	−10.06	−10.06	−12.12	−12.12	−14.23	−14.23	−16.38	−16.38	−9/4 N	−9/4 N
4	−5.66	−5.66	−7.21	−7.21	−8.00	−8.00	−8.94	−8.94	−10.77	−10.77	−12.65	−12.65	−14.56	−14.56	−2 N	−2 N
5	−4.95	−4.95	−6.31	−6.31	−7.00	−7.00	−7.83	−7.83	−9.42	−9.42	−11.07	−11.07	−12.74	−12.74	−7/4 N	−7/4 N
6	−4.24	−4.24	−5.41	−5.41	−6.00	−6.00	−6.71	−6.71	−8.08	−8.08	−9.49	−9.49	−10.92	−10.92	−3/2 N	−3/2 N
7	−0.71	−0.71	−0.90	−0.90	−1.00	−1.00	−1.12	−1.12	−1.35	−1.35	−1.58	−1.58	−1.82	−1.82	−1/4 N	−1/4 N
8	−1.12	−1.12	−1.25	−1.25	−1.32	−1.32	−1.41	−1.41	−1.60	−1.60	−1.80	−1.80	−2.02	−2.02	−1/4 $\sqrt{n^2+16}$	−1/4 $\sqrt{n^2+16}$

VALUES OF n — AXIAL FORCE COEFFICIENT

Table 3-20. (Continued)

Member	VALUES OF n												General Formulas	
	2		3		$2\sqrt{3}$		4		5		6		7	
	K_1	K_2	K_1	K_2	K_1	K_2	K_1	K_2	K_1	K_2	K_1	K_2	K_1	K_2

AXIAL FORCE COEFFICIENT

Member	K_1	K_2	K_1	K_2	K_1	K_2	K_1	K_2	K_1	K_2	K_1	K_2	K_1	K_2	K_1	K_2
9	−1.58	−1.58	−1.68	−1.68	−1.73	−1.73	−1.80	−1.80	−1.95	−1.95	−2.12	−2.12	−2.30	−2.30	$-1/4\sqrt{n^2+36}$	$-1/4\sqrt{n^2+36}$
10	−2.06	−2.06	−2.14	−2.14	−2.18	−2.18	−2.24	−2.24	−2.36	−2.36	−2.50	−2.50	−2.66	−2.66	$-1/4\sqrt{n^2+64}$	$-1/4\sqrt{n^2+64}$
11	−2.55	−2.55	−2.61	−2.61	−2.65	−2.65	−2.69	−2.69	−2.80	−2.80	−2.92	−2.92	−3.05	−3.05	$-1/4\sqrt{n^2+100}$	$-1/4\sqrt{n^2+100}$
12	0	1.00	0	1.00	0	1.00	0	1.00	0	1.00	0	1.00	0	1.00	0	1
13	0.50	1.50	0.50	1.50	0.50	1.50	0.50	1.50	0.50	1.50	0.50	1.50	0.50	1.50	1/2	3/2
14	1.00	2.00	1.00	2.00	1.00	2.00	1.00	2.00	1.00	2.00	1.00	2.00	1.00	2.00	1	2
15	1.50	2.50	1.50	2.50	1.50	2.50	1.50	2.50	1.50	2.50	1.50	2.50	1.50	2.50	3/2	5/2
16	2.00	3.00	2.00	3.00	2.00	3.00	2.00	3.00	2.00	3.00	2.00	3.00	2.00	3.00	2	3
17	5.00	6.00	5.00	6.00	5.00	6.00	5.00	6.00	5.00	6.00	5.00	6.00	5.00	6.00	5	6
18	5.50	5.50	8.25	8.25	9.53	9.53	11.00	11.00	13.75	13.75	16.50	16.50	19.25	19.25	11/4 n	11/4 n
19	5.50	5.50	8.25	8.25	9.53	9.53	11.00	11.00	13.75	13.75	16.50	16.50	19.25	19.25	11/4 n	11/4 n
20	5.00	5.00	7.50	7.50	8.66	8.66	10.00	10.00	12.50	12.50	15.00	15.00	17.50	17.50	5/2 n	5/2 n
21	4.50	4.50	6.75	6.75	7.79	7.79	9.00	9.00	11.25	11.25	13.50	13.50	15.75	15.75	9/4 n	9/4 n
22	4.00	4.00	6.00	6.00	6.93	6.93	8.00	8.00	10.00	10.00	12.00	12.00	14.00	14.00	2 n	2 n
23	3.50	3.50	5.25	5.25	6.06	6.06	7.00	7.00	8.75	8.75	10.50	10.50	12.25	12.25	7/4 n	7/4 n

LENGTH COEFFICIENTS

Member	K (n=2)	K (n=3)	K (n=$2\sqrt{3}$)	K (n=4)	K (n=5)	K (n=6)	K (n=7)	K
1, 2, 3, 4, 5, 6, 7	0.235702	0.300463	0.333333	0.372678	0.448764	0.527046	0.606676	$1/12\,N$
8	0.372678	0.416667	0.440959	0.471405	0.533594	0.600925	0.671855	$1/12\sqrt{n^2+16}$
9	0.527046	0.559017	0.577350	0.600925	0.650854	0.707107	0.768295	$1/12\sqrt{n^2+36}$
10	0.687184	0.712000	0.726483	0.745356	0.786165	0.833333	0.885845	$1/12\sqrt{n^2+64}$

Members								
11	0.849837	0.870026	0.881917	0.897527	0.931695	0.971825	1.017213	$1/12 \sqrt{n^2 + 100}$
12	0.166667	0.166667	0.166667	0.166667	0.166667	0.166667	0.166667	1/6
13	0.333333	0.333333	0.333333	0.333333	0.333333	0.333333	0.333333	1/3
14	0.500000	0.500000	0.500000	0.500000	0.500000	0.500000	0.500000	1/2
15	0.666667	0.666667	0.666667	0.666667	0.666667	0.666667	0.666667	2/3
16	0.833333	0.833333	0.833333	0.833333	0.833333	0.833333	0.833333	5/6
17	1.000000	1.000000	1.000000	1.000000	1.000000	1.000000	1.000000	1
18, 19, 20, 21, 22, 23	0.166667	0.250000	0.288667	0.333333	0.416667	0.500000	0.583333	$1/12\, n$

ANGLES (IN DEGREES) BETWEEN MEMBERS

Members								
1-18, 7-19	45	34	30	27	22	18	16	$2/n = \tan a$
8-20	63	53	49	45	39	34	30	$4/n = \tan b$
9-21	72	63	60	56	50	45	41	$6/n = \tan c$
10-22	76	69	67	63	58	53	49	$8/n = \tan d$
11-23	79	73	71	68	63	59	55	$10/n = \tan e$
1-12, 2-13, 3-14, 4-15, 5-16, 6-17, 7-13	45	56	60	63	68	72	74	$90 - a$
8-13, 8-14	27	37	41	45	51	56	60	$90 - b$
9-14, 9-15	18	27	30	34	40	45	49	$90 - c$
10-15, 10-16	14	21	23	27	32	37	41	$90 - d$
11-16, 11-17	11	17	19	22	27	31	35	$90 - e$
2-7	90	68	60	54	44	36	32	$2a$
3-8	108	87	79	72	61	52	46	$a + b$
4-9	117	97	90	83	72	63	57	$a + c$
5-10	121	103	97	90	80	71	65	$a + d$
6-11	124	107	101	95	85	77	71	$a + e$
12-18, 12-19, 13-20, 14-21, 14-22, 16-23	90	90	90	90	90	90	90	90

Table 3-21. Triangular Pratt 4 Panels at Top & Bottom

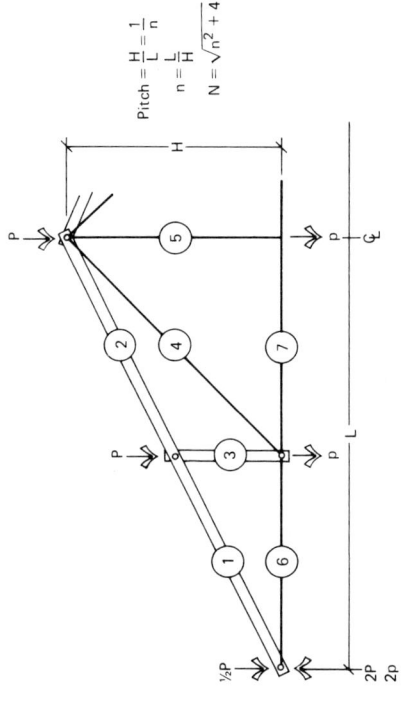

Member	\multicolumn{2}{c}{2}		3		$2\sqrt{3}$		4		5		6		7		\multicolumn{2}{c}{General Formulas}
	K_1	K_2	K_1	K_2	K_1	K_2	K_1	K_2	K_1	K_2	K_1	K_2	K_1	K_2	K_1	K_2
1	−2.12	−2.12	−2.70	−2.70	−3.00	−3.00	−3.35	−3.35	−4.04	−4.04	−4.74	−4.74	−5.46	−5.46	−3/4 N	−3/4 N
2	−2.12	−2.12	−2.70	−2.70	−3.00	−3.00	−3.35	−3.35	−4.04	−4.04	−4.74	−4.74	−5.46	−5.46	−3/4 N	−3/4 N
3	−1.00	0	−1.00	0	−1.00	0	−1.00	0	−1.00	0	−1.00	0	−1.00	0	−1	0
4	1.12	1.12	1.25	1.25	1.32	1.32	1.41	1.41	1.60	1.60	1.80	1.80	2.02	2.02	$1/4\sqrt{n^2+16}$	$1/4\sqrt{n^2+16}$
5	0	1.00	0	1.00	0	1.00	0	1.00	0	1.00	0	1.00	0	1.00	0	1
6	1.50	1.50	2.25	2.25	2.60	2.60	3.00	3.00	3.75	3.75	4.50	4.50	5.25	5.25	3/4 n	3/4 n
7	1.00	1.00	1.50	1.50	1.73	1.73	2.00	2.00	2.50	2.50	3.00	3.00	3.50	3.50	1/2 n	1/2 n

VALUES OF n — AXIAL FORCE COEFFICIENTS

Pitch $= \dfrac{H}{L} = \dfrac{1}{n}$
$n = \dfrac{L}{H}$
$N = \sqrt{n^2 + 4}$

LENGTH COEFFICIENTS

	K	K	K	K	K	K	K	K
1, 2	0.707107	0.901388	1.000000	1.118034	1.346291	1.581139	1.820027	$1/4\,N$
3	0.500000	0.500000	0.500000	0.500000	0.500000	0.500000	0.500000	$1/2$
4	1.118034	1.250000	1.322876	1.414214	1.600781	1.802776	2.015564	$1/4\sqrt{n^2+16}$
5	1.000000	1.000000	1.000000	1.000000	1.000000	1.000000	1.000000	1
6, 7	0.500000	0.750000	0.866000	1.000000	1.250000	1.500000	1.750000	$1/4\,n$

ANGLES (IN DEGREES) BETWEEN MEMBERS

1-6	45	34	30	27	22	18	16	$2/n = \tan a$
4-7	63	53	49	45	39	34	30	$4/n = \tan b$
1-3	45	56	60	63	68	72	74	$90 - a$
3-4, 4-5	27	37	41	45	51	56	60	$90 - b$
2-4	18	19	19	18	17	16	14	$b - a$
2-3	135	124	120	117	112	108	106	$90 + a$
3-6, 5-7	90	90	90	90	90	90	90	90

Table 3-22. Triangular Pratt 6 Panels at Top & Bottom

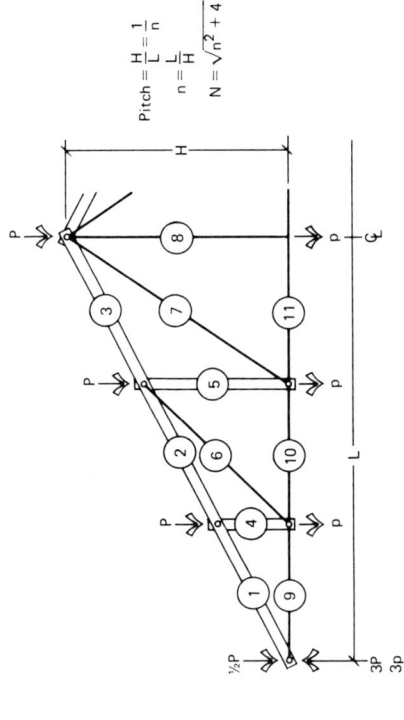

$$\text{Pitch} = \frac{H}{L} = \frac{1}{n}$$
$$n = \frac{L}{H}$$
$$N = \sqrt{n^2 + 4}$$

Member	VALUES OF n												General Formulas	
	2		3		$2\sqrt{3}$		4		5		6		7		K_1	K_2
	K_1	K_2	K_1	K_2	K_1	K_2	K_1	K_2	K_1	K_2	K_1	K_2	K_1	K_2	K_1	K_2
	AXIAL FORCE COEFFICIENTS															
1	-3.54	-3.54	-4.51	-4.51	-5.00	-5.00	-5.59	-5.59	-6.73	-6.73	-7.91	-7.91	-9.10	-9.10	-5/4 N	-5/4 N
2	-3.54	-3.54	-4.51	-4.51	-5.00	-5.00	-5.59	-5.59	-6.73	-6.73	-7.91	-7.91	-9.10	-9.10	-5/4 N	-5/4 N
3	-2.83	-2.83	-3.61	-3.61	-4.00	-4.00	-4.47	-4.47	-5.39	-5.39	-6.32	-6.32	-7.28	-7.28	-N	-N
4	-1.00	0	-1.00	0	-1.00	0	-1.00	0	-1.00	0	-1.00	0	-1.00	0	-1	0
5	-1.50	-0.50	-1.50	-0.50	-1.50	-0.50	-1.50	-0.50	-1.50	-0.50	-1.50	-0.50	-1.50	-0.50	-3/2	-1/2
6	1.12	1.12	1.25	1.25	1.32	1.32	1.41	1.41	1.60	1.60	1.80	1.80	2.02	2.02	$1/4\sqrt{n^2+16}$	$1/4\sqrt{n^2+16}$
7	1.58	1.58	1.68	1.68	1.73	1.73	1.80	1.80	1.95	1.95	2.12	2.12	2.30	2.30	$1/4\sqrt{n^2+36}$	$1/4\sqrt{n^2+36}$

8	0	1.00	0	1.00	0	1.00	0	1.00	0	1.00	0	1.00	0	1.00	0	1	
9	2.50	2.50	3.75	3.75	4.33	4.33	5.00	5.00	6.25	6.25	7.50	7.50	8.75	8.75	5/4 n	5/4 n
10	2.00	2.00	3.00	3.00	3.46	3.46	4.00	4.00	5.00	5.00	6.00	6.00	7.00	7.00	n	n
11	1.50	1.50	2.25	2.25	2.60	2.60	3.00	3.00	3.75	3.75	4.50	4.50	5.25	5.25	3/4 n	3/4 n

LENGTH COEFFICIENTS

	K	K	K	K	K	K	K	K
1, 2, 3	0.471405	0.600925	0.666075	0.745356	0.897527	1.054093	1.213352	$1/6\,N$
4	0.333333	0.333333	0.333333	0.333333	0.333333	0.333333	0.333333	$1/3$
5	0.666667	0.666667	0.666667	0.666667	0.666667	0.666667	0.666667	$2/3$
6	0.745356	0.833333	0.881917	0.942809	1.067187	1.201850	1.343710	$1/6\sqrt{n^2+16}$
7	1.054093	1.118034	1.154701	1.201850	1.301708	1.414214	1.536591	$1/6\sqrt{n^2+36}$
8	1.000000	1.000000	1.000000	1.000000	1.000000	1.000000	1.000000	1
9, 10, 11	0.333333	0.500000	0.577333	0.666667	0.833333	1.000000	1.166667	$1/6\,n$

ANGLES (IN DEGREES) BETWEEN MEMBERS

1-9	45	34	30	27	22	18	16	$2/n = \tan a$
6-10	63	53	49	45	39	34	30	$4/n = \tan b$
7-11	72	63	60	56	50	45	41	$6/n = \tan c$
1-4	45	56	60	63	68	72	74	$90 - a$
4-6, 5-6	27	37	41	45	51	56	60	$90 - b$
5-7, 7-8	18	27	30	34	40	45	49	$90 - c$
2-5	18	19	19	18	17	16	14	$b - a$
3-7	27	29	30	29	28	27	25	$c - a$
2-4, 3-5	135	124	120	117	112	108	106	$90 + a$
1-9, 5-10, 8-11	90	90	90	90	90	90	90	90

Table 3-23. Triangular Pratt 8 Panels at Top & Bottom

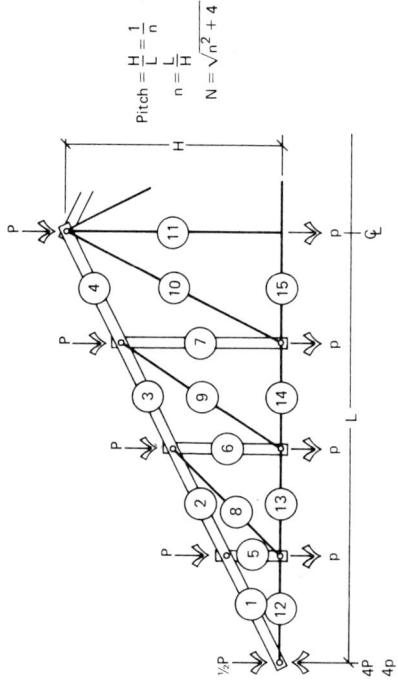

Pitch = $\frac{H}{L} = \frac{1}{n}$
$n = \frac{L}{H}$
$N = \sqrt{n^2 + 4}$

Member	\multicolumn{14}{c}{VALUES OF n}	General Formulas	
	2		3		$2\sqrt{3}$		4		5		6		7			
	K_1	K_2	K_1	K_2	K_1	K_2	K_1	K_2	K_1	K_2	K_1	K_2	K_1	K_2	K_1	K_2
	\multicolumn{14}{c}{AXIAL FORCE COEFFICIENTS}		
1	-4.95	-4.95	-6.31	-6.31	-7.00	-7.00	-7.83	-7.83	-9.42	-9.42	-11.07	-11.07	-12.74	-12.74	-7/4 N	-7/4 N
2	-4.95	-4.95	-6.31	-6.31	-7.00	-7.00	-7.83	-7.83	-9.42	-9.42	-11.07	-11.07	-12.74	-12.74	-7/4 N	-7/4 N
3	-4.24	-4.24	-5.41	-5.41	-6.00	-6.00	-6.71	-6.71	-8.08	-8.08	-9.49	-9.49	-10.92	-10.92	-3/2 N	-3/2 N
4	-3.54	-3.54	-4.51	-4.51	-5.00	-5.00	-5.59	-5.59	-6.73	-6.73	-7.91	-7.91	-9.10	-9.10	-5/4 N	-5/4 N
5	-1.00	0	-1.00	0	-1.00	0	-1.00	0	-1.00	0	-1.00	0	-1.00	0	-1	0
6	-1.50	-0.50	-1.50	-0.50	-1.50	-0.50	-1.50	-0.50	-1.50	-0.50	-1.50	-0.50	-1.50	-0.50	-3/2	-1/2
7	-2.00	-1.00	-2.00	-1.00	-2.00	-1.00	-2.00	-1.00	-2.00	-1.00	-2.00	-1.00	-2.00	-1.00	-2	-1
8	1.12	1.12	1.25	1.25	1.32	1.32	1.41	1.41	1.60	1.60	1.80	1.80	2.02	2.02	$1/4\sqrt{n^2+16}$	$1/4\sqrt{n^2+16}$
9	1.58	1.58	1.68	1.68	1.73	1.73	1.80	1.80	1.95	1.95	2.12	2.12	2.30	2.30	$1/4\sqrt{n^2+36}$	$1/4\sqrt{n^2+36}$
10	2.06	2.06	2.14	2.14	2.18	2.18	2.24	2.24	2.36	2.36	2.50	2.50	2.66	2.66	$1/4\sqrt{n^2+64}$	$1/4\sqrt{n^2+64}$

LENGTH COEFFICIENTS

Member	K		K		K		K		K		K		K		K		K	
11	0	1.00	0	1.00	0	1.00	0	1.00	0	1.00	0	1.00	0	1.00	0	1		
12	3.50	3.50	5.25	5.25	6.06	6.06	7.00	7.00	8.75	8.75	10.50	10.50	12.25	12.25	7/4 n	7/4 n		
13	3.00	3.00	4.50	4.50	5.20	5.20	6.00	6.00	7.50	7.50	9.00	9.00	10.50	10.50	3/2 n	3/2 n		
14	2.50	2.50	3.75	3.75	4.33	4.33	5.00	5.00	6.25	6.25	7.50	7.50	8.75	8.75	5/4 n	5/4 n		
15	2.00	2.00	3.00	3.00	3.46	3.46	4.00	4.00	5.00	5.00	6.00	6.00	7.00	7.00	n	n		

Member	K	K	K	K	K	K	K	K	K
1, 2, 3, 4	0.353553	0.450694	0.500000	0.559017	0.673146	0.790569	0.910014	1/8 N	
5	0.250000	0.250000	0.250000	0.250000	0.250000	0.250000	0.250000	1/4	
6	0.500000	0.500000	0.500000	0.500000	0.500000	0.500000	0.500000	1/2	
7	0.750000	0.750000	0.750000	0.750000	0.750000	0.750000	0.750000	3/4	
8	0.559017	0.625000	0.661438	0.707107	0.800391	0.901388	1.007782	$1/8\sqrt{n^2+16}$	
9	0.790569	0.838525	0.866025	0.901388	0.976281	1.060660	1.152443	$1/8\sqrt{n^2+36}$	
10	1.030776	1.068000	1.089725	1.118034	1.179248	1.250000	1.328768	$1/8\sqrt{n^2+64}$	
11	1.000000	1.000000	1.000000	1.000000	1.000000	1.000000	1.000000	1	
12, 13, 14, 15	0.250000	0.375000	0.433000	0.500000	0.625000	0.750000	0.875000	$1/8\,n$	

ANGLES (IN DEGREES) BETWEEN MEMBERS

Members									
1-12	45	34	30	27	22	18	16	$2/n = \tan a$	
8-13	63	53	49	45	39	34	30	$4/n = \tan b$	
9-14	72	63	60	56	50	45	41	$6/n = \tan c$	
10-15	76	69	67	63	58	53	49	$8/n = \tan d$	
1-5	45	56	60	63	68	72	74	$90 - a$	
5-8, 6-8	27	37	41	45	51	56	60	$90 - b$	
6-9, 7-9	18	27	30	34	40	45	49	$90 - c$	
7-10, 10-11	14	21	23	27	32	37	41	$90 - d$	
2-8	18	19	19	18	17	16	14	$b - a$	
3-9	27	29	30	29	28	27	25	$c - a$	
4-10	31	35	37	36	36	35	33	$d - a$	
2-5; 3-6, 4-7	135	124	120	117	112	108	106	$90 + a$	
5-12, 6-13, 7-14, 11-15	90	90	90	90	90	90	90	90	

Table 3-24. Triangular Pratt 10 Panels at Top & Bottom

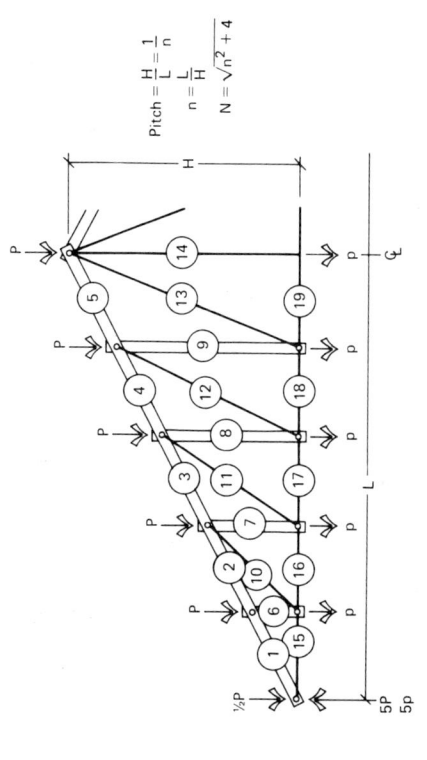

Pitch = $\frac{H}{L} = \frac{1}{n}$
$n = \frac{L}{H}$
$N = \sqrt{n^2 + 4}$

Member	VALUES OF n												General Formulas	
	2		3		$2\sqrt{3}$		4		5		6		7	
	K_1	K_2	K_1	K_2	K_1	K_2	K_1	K_2	K_1	K_2	K_1	K_2	K_1	K_2
	AXIAL FORCE COEFFICIENTS													
1	−6.36	−6.36	−8.11	−8.11	−9.00	−9.00	−10.06	−10.06	−12.12	−12.12	−14.23	−14.23	−16.38	−16.38	−9/4 N	−9/4 N
2	−6.36	−6.36	−8.11	−8.11	−9.00	−9.00	−10.06	−10.06	−12.12	−12.12	−14.23	−14.23	−16.38	−16.38	−9/4 N	−9/4 N
3	−5.66	−5.66	−7.21	−7.21	−8.00	−8.00	−8.94	−8.94	−10.77	−10.77	−12.65	−12.65	−14.56	−14.56	−2 N	−2 N
4	−4.95	−4.95	−6.31	−6.31	−7.00	−7.00	−7.83	−7.83	−9.42	−9.42	−11.07	−11.07	−12.74	−12.74	−7/4 N	−7/4 N
5	−4.24	−4.24	−5.41	−5.41	−6.00	−6.00	−6.71	−6.71	−8.08	−8.08	−9.49	−9.49	−10.92	−10.92	−3/2 N	−3/2 N
6	−1.00	0	−1.00	0	−1.00	0	−1.00	0	−1.00	0	−1.00	0	−1.00	0	−1	0
7	−1.50	−0.50	−1.50	−0.50	−1.50	−0.50	−1.50	−0.50	−1.50	−0.50	−1.50	−0.50	−1.50	−0.50	−3/2	−1/2
8	−2.00	−1.00	−2.00	−1.00	−2.00	−1.00	−2.00	−1.00	−2.00	−1.00	−2.00	−1.00	−2.00	−1.00	−2	−1
9	−2.50	−1.50	−2.50	−1.50	−2.50	−1.50	−2.50	−1.50	−2.50	−1.50	−2.50	−1.50	−2.50	−1.50	−5/2	−3/2

10	1.12	1.12	1.25	1.25	1.32	1.32	1.41	1.41	1.60	1.60	1.80	1.80	2.02	2.02	$1/4\sqrt{n^2+16}$	$1/4\sqrt{n^2+16}$
11	1.58	1.58	1.68	1.68	1.73	1.73	1.80	1.80	1.95	1.95	2.12	2.12	2.30	2.30	$1/4\sqrt{n^2+36}$	$1/4\sqrt{n^2+36}$
12	2.06	2.06	2.14	2.14	2.18	2.18	2.24	2.24	2.36	2.36	2.50	2.50	2.66	2.66	$1/4\sqrt{n^2+64}$	$1/4\sqrt{n^2+64}$
13	2.55	2.55	2.61	2.61	2.65	2.65	2.69	2.69	2.80	2.80	2.92	2.92	3.05	3.05	$1/4\sqrt{n^2+100}$	$1/4\sqrt{n^2+100}$
14	1.00	1.00	0	0	1.00	1.00	0	0	1.00	1.00	1.00	1.00	0	0	1	1
15	4.50	4.50	6.75	6.75	7.79	7.79	9.00	9.00	11.25	11.25	13.50	13.50	15.75	15.75	$9/4\,n$	$9/4\,n$
16	4.00	4.00	6.00	6.00	6.93	6.93	8.00	8.00	10.00	10.00	12.00	12.00	14.00	14.00	$2\,n$	$2\,n$
17	3.50	3.50	5.25	5.25	6.06	6.06	7.00	7.00	8.75	8.75	10.50	10.50	12.25	12.25	$7/4\,n$	$7/4\,n$
18	3.00	3.00	4.50	4.50	5.20	5.20	6.00	6.00	7.50	7.50	9.00	9.00	10.50	10.50	$3/2\,n$	$3/2\,n$
19	2.50	2.50	3.75	3.75	4.33	4.33	5.00	5.00	6.25	6.25	7.50	7.50	8.75	8.75	$5/4\,n$	$5/4\,n$

LENGTH COEFFICIENTS

	K	K	K	K	K	K	K	K
1, 2, 3, 4, 5	0.282843	0.360555	0.400000	0.447214	0.538516	0.632456	0.728011	$1/10\,N$
6	0.200000	0.200000	0.200000	0.200000	0.200000	0.200000	0.200000	1/5
7	0.400000	0.400000	0.400000	0.400000	0.400000	0.400000	0.400000	2/5
8	0.600000	0.600000	0.600000	0.600000	0.600000	0.600000	0.600000	3/5
9	0.800000	0.800000	0.800000	0.800000	0.800000	0.800000	0.800000	4/5
10	0.447214	0.500000	0.529150	0.565685	0.640312	0.721110	0.806226	$1/10\sqrt{n^2+16}$
11	0.632456	0.670820	0.692820	0.721110	0.781025	0.846528	0.921954	$1/10\sqrt{n^2+36}$
12	0.824621	0.854400	0.871780	0.894427	0.943398	1.000000	1.063016	$1/10\sqrt{n^2+64}$
13	1.019804	1.044031	1.058301	1.077033	1.118034	1.166190	1.220656	$1/10\sqrt{n^2+100}$
14	1.000000	1.000000	1.000000	1.000000	1.000000	1.000000	1.000000	1
15, 16, 17, 18, 19	0.200000	0.300000	0.346400	0.400000	0.500000	0.600000	0.700000	$1/10\,n$

ANGLES (IN DEGREES) BETWEEN MEMBERS

1-15	45	34	30	27	22	18	16	$2/n = \tan a$
10-16	63	53	49	45	39	34	30	$4/n = \tan b$
11-17	72	63	60	56	50	45	41	$6/n = \tan c$
12-18	76	69	67	63	58	53	49	$8/n = \tan d$

Table 3-24. (Continued)

Member	VALUES OF n							General Formulas
	2	3	$2\sqrt{3}$	4	5	6	7	
	ANGLES (IN DEGREES) BETWEEN MEMBERS							
	K	K	K	K	K	K	K	K
13-19	79	73	71	68	64	59	55	$10/n = \tan e$
1-6	45	56	60	63	68	72	74	$90 - a$
6-10, 7-10	27	37	41	45	51	56	60	$90 - b$
7-11, 8-11	18	27	30	34	40	45	49	$90 - c$
8-12, 9-12	14	21	23	27	32	37	41	$90 - d$
9-13, 13-14	11	17	19	22	26	31	35	$90 - e$
2-10	18	19	19	18	17	16	14	$b - a$
3-11	27	29	30	29	28	27	25	$c - a$
4-12	31	35	37	36	36	35	33	$d - a$
5-13	34	39	41	41	42	41	39	$e - a$
2-8, 3-7, 4-8 5-9	135	124	120	117	112	108	106	$90 + a$
6-15, 7-16, 8-17 9-18, 14-19	90	90	90	90	90	90	90	90

Table 3-25. Triangular Pratt 12 Panels at Top & Bottom

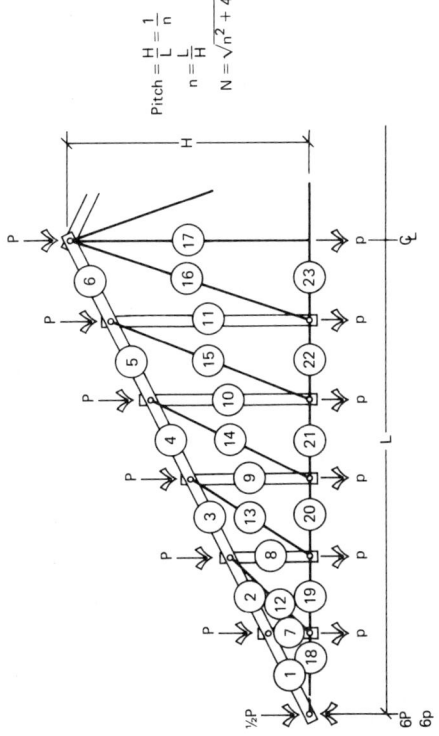

Pitch = $\frac{H}{L} = \frac{1}{n}$
$n = \frac{L}{H}$
$N = \sqrt{n^2 + 4}$

Member	VALUES OF n														General Formulas	
	2		3		$2\sqrt{3}$		4		5		6		7			
	K_1	K_2	K_1	K_2	K_1	K_2	K_1	K_2	K_1	K_2	K_1	K_2	K_1	K_2	K_1	K_2
	AXIAL FORCE COEFFICIENTS															
1	−7.78	−7.78	−9.92	−9.92	−11.00	−11.00	−12.30	−12.30	−14.81	−14.81	−17.39	−17.39	−20.02	−20.02	−11/4 N	−11/4 N
2	−7.78	−7.78	−9.92	−9.92	−11.00	−11.00	−12.30	−12.30	−14.81	−14.81	−17.39	−17.39	−20.02	−20.02	−11/4 N	−11/4 N
3	−7.07	−7.07	−9.01	−9.01	−10.00	−10.00	−11.18	−11.18	−13.46	−13.46	−15.81	−15.81	−18.20	−18.20	−5/2 N	−5/2 N
4	−6.36	−6.36	−8.11	−8.11	−9.00	−9.00	−10.06	−10.06	−12.12	−12.12	−14.23	−14.23	−16.38	−16.38	−9/4 N	−9/4 N
5	−5.66	−5.66	−7.21	−7.21	−8.00	−8.00	−8.94	−8.94	−10.77	−10.77	−12.65	−12.65	−14.56	−14.56	−2 N	−2 N
6	−4.95	−4.95	−6.31	−6.31	−7.00	−7.00	−7.83	−7.83	−9.42	−9.42	−11.07	−11.07	−12.74	−12.74	−7/4 N	−7/4 N
7	−1.00	0	−1.00	0	−1.00	0	−1.00	0	−1.00	0	−1.00	0	−1.00	0	−1	0
8	−1.50	−0.50	−1.50	−0.50	−1.50	−0.50	−1.50	−0.50	−1.50	−0.50	−1.50	−0.50	−1.50	−0.50	−3/2	−1/2

Table 3-25. (Continued)

Member	VALUES OF n														General Formulas	
	2		3		$2\sqrt{3}$		4		5		6		7			
	K_1	K_2	K_1	K_2	K_1	K_2	K_1	K_2	K_1	K_2	K_1	K_2	K_1	K_2	K_1	K_2
AXIAL FORCE COEFFICIENTS																
9	−2.00	−1.00	−2.00	−1.00	−2.00	−1.00	−2.00	−1.00	−2.00	−1.00	−2.00	−1.00	−2.00	−1.00	−2	−1
10	−2.50	−1.50	−2.50	−1.50	−2.50	−1.50	−2.50	−1.50	−2.50	−1.50	−2.50	−1.50	−2.50	−1.50	−5/2	−3/2
11	−3.00	−2.00	−3.00	−2.00	−3.00	−2.00	−3.00	−2.00	−3.00	−2.00	−3.00	−2.00	−3.00	−2.00	−3	−2
12	1.12	1.12	1.25	1.25	1.32	1.32	1.41	1.41	1.60	1.60	1.80	1.80	2.02	2.02	$1/4\sqrt{n^2+16}$	$1/4\sqrt{n^2+16}$
13	1.58	1.58	1.68	1.68	1.73	1.73	1.80	1.80	1.95	1.95	2.12	2.12	2.30	2.30	$1/4\sqrt{n^2+36}$	$1/4\sqrt{n^2+36}$
14	2.06	2.06	2.14	2.14	2.18	2.18	2.24	2.24	2.36	2.36	2.50	2.50	2.66	2.66	$1/4\sqrt{n^2+64}$	$1/4\sqrt{n^2+64}$
15	2.55	2.55	2.61	2.61	2.65	2.65	2.69	2.69	2.80	2.80	2.92	2.92	3.05	3.05	$1/4\sqrt{n^2+100}$	$1/4\sqrt{n^2+100}$
16	3.04	3.04	3.09	3.09	3.12	3.12	3.16	3.16	3.25	3.25	3.35	3.35	3.47	3.47	$1/4\sqrt{n^2+144}$	$1/4\sqrt{n^2+144}$
17	0	1.00	0	1.00	0	1.00	0	1.00	0	1.00	0	1.00	0	1.00	0	1
18	5.50	5.50	8.25	8.25	9.52	9.52	11.00	11.00	13.75	13.75	16.50	16.50	19.25	19.25	$11/4\,n$	$11/4\,n$
19	5.00	5.00	7.50	7.50	8.66	8.66	10.00	10.00	12.50	12.50	15.00	15.00	17.50	17.50	$5/2\,n$	$5/2\,n$
20	4.50	4.50	6.75	6.75	7.79	7.79	9.00	9.00	11.25	11.25	13.50	13.50	15.75	15.75	$9/4\,n$	$9/4\,n$
21	4.00	4.00	6.00	6.00	6.93	6.93	8.00	8.00	10.00	10.00	12.00	12.00	14.00	14.00	$2\,n$	$2\,n$
22	3.50	3.50	5.25	5.25	6.06	6.06	7.00	7.00	8.75	8.75	10.50	10.50	12.25	12.25	$7/4\,n$	$7/4\,n$
23	3.00	3.00	4.50	4.50	5.20	5.20	6.00	6.00	7.50	7.50	9.00	9.00	10.50	10.50	$3/2\,n$	$3/2\,n$

Member	K (n=2)	K (n=3)	K (n=$2\sqrt{3}$)	K (n=4)	K (n=5)	K (n=6)	K (n=7)	K
LENGTH COEFFICIENTS								
1, 2, 3, 4, 5, 6,	0.235702	0.300463	0.333333	0.372678	0.448764	0.527046	0.606676	$1/12\,N$
7	0.166667	0.166667	0.166667	0.166667	0.166667	0.166667	0.166667	1/6
8	0.333333	0.333333	0.333333	0.333333	0.333333	0.333333	0.333333	1/3
9	0.500000	0.500000	0.500000	0.500000	0.500000	0.500000	0.500000	1/2
10	0.666667	0.666667	0.666667	0.666667	0.666667	0.666667	0.666667	2/3
11	0.833333	0.833333	0.833333	0.833333	0.833333	0.833333	0.833333	5/6

	12	0.372678	0.416667	0.440959	0.471405	0.533594	0.600925	0.671855	$1/12 \sqrt{n^2 + 16}$
	13	0.527046	0.559017	0.577350	0.600925	0.650854	0.707107	0.768295	$1/12 \sqrt{n^2 + 36}$
	14	0.687184	0.712000	0.726483	0.745356	0.786165	0.833333	0.885845	$1/12 \sqrt{n^2 + 64}$
	15	0.849837	0.870026	0.881917	0.897527	0.931695	0.971825	1.017213	$1/12 \sqrt{n^2 + 100}$
	16	1.013794	1.030776	1.040833	1.054093	1.083333	1.118034	1.157704	$1/12 \sqrt{n^2 + 144}$
	17	1.000000	1.000000	1.000000	1.000000	1.000000	1.000000	1.000000	1
	18, 19, 20, 21, 22, 23	0.166667	0.250000	0.288667	0.333333	0.416667	0.500000	0.583333	$1/12\, n$

ANGLES (IN DEGREES) BETWEEN MEMBERS

1-18, 2-12	45	34	30	27	22	18	16	$2/n = \tan a$
12-19	63	53	49	45	39	34	30	$4/n = \tan b$
13-20	72	63	60	56	50	45	41	$6/n = \tan c$
14-21	76	69	67	63	58	53	49	$8/n = \tan d$
15-22	79	73	71	68	63	59	55	$10/n = \tan e$
16-23	80	76	74	72	67	63	60	$12/n = \tan f$
1-7	45	56	60	63	68	72	74	$90 - a$
7-12, 8-12	27	37	41	45	51	56	60	$90 - b$
8-13, 9-13	18	27	30	34	40	45	49	$90 - c$
9-14, 10-14	14	21	23	27	32	37	41	$90 - d$
10-15, 11-15	11	17	19	22	27	31	35	$90 - e$
11-16, 16-17	10	14	16	18	23	27	30	$90 - f$
2-12	18	19	19	18	17	16	14	$b - a$
3-13	27	29	30	29	28	27	25	$c - a$
4-14	31	35	37	36	36	35	33	$d - a$
5-15	34	39	41	41	41	41	39	$e - a$
6-16	35	42	44	45	45	45	44	$f - a$
2-7, 3-8, 4-9, 5-10, 6-11	135	124	120	117	112	108	106	$90 + a$
7-18, 8-19, 9-20, 10-21, 11-22, 17-25	90	90	90	90	90	90	90	90

Table 3-26. Flat Howe 4 Panels at Top & Bottom

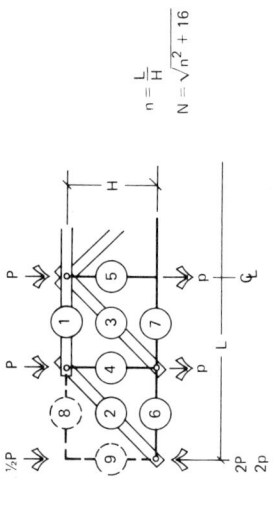

$n = \dfrac{L}{H}$
$N = \sqrt{n^2 + 16}$

Member	VALUES OF n													General Formulas	
	7		8		9		10		11		12		13			
	K_1	K_2	K_1	K_2	K_1	K_2	K_1	K_2	K_1	K_2	K_1	K_2	K_1	K_2	K_1	K_2
	AXIAL FORCE COEFFICIENTS															
1	−2.63	−2.63	−3.00	−3.00	−3.38	−3.38	−3.75	−3.75	−4.13	−4.13	−4.50	−4.50	−4.88	−4.88	−3/8 n	−3/8 n
2	−3.02	−3.02	−3.35	−3.35	−3.69	−3.69	−4.04	−4.04	−4.39	−4.39	−4.74	−4.74	−5.10	−5.10	−3/8 N	−3/8 N
3	−1.01	−1.01	−1.12	−1.12	−1.23	−1.23	−1.35	−1.35	−1.46	−1.46	−1.58	−1.58	−1.70	−1.70	−1/8 N	−1/8 N
4	0.50	1.50	0.50	1.50	0.50	1.50	0.50	1.50	0.50	1.50	0.50	1.50	0.50	1.50	1/2	3/2
5	0	1.00	0	1.00	0	1.00	0	1.00	0	1.00	0	1.00	0	1.00	0	1
6	2.63	2.63	3.00	3.00	3.38	3.38	3.75	3.75	4.13	4.13	4.50	4.50	4.88	4.88	3/8 n	3/8 n
7	3.50	3.50	4.00	4.00	4.50	4.50	5.00	5.00	5.50	5.50	6.00	6.00	6.50	6.50	1/2 n	1/2 n
8	0	0	0	0	0	0	0	0	0	0	0	0	0	0	0	0
9	−0.50	0	−0.50	0	−0.50	0	−0.50	0	−0.50	0	−0.50	0	−0.50	0	−1/2	0

LENGTH COEFFICIENTS

	K	K	K	K	K	K	K	K
Horizontal	1.750000	2.000000	2.250000	2.500000	2.750000	3.000000	3.250000	$1/4\,n$
Diagonal	2.015564	2.236068	2.462214	2.692582	2.926175	3.162278	3.400368	$1/4\,N$
Vertical	1.000000	1.000000	1.000000	1.000000	1.000000	1.000000	1.000000	1

ANGLES (IN DEGREES) BETWEEN MEMBERS

	K	K	K	K	K	K	K	K
Between diagonal and horizontal members	30	27	24	22	20	18	17	$4/n = \tan a$
Between diagonal and vertical members	60	63	66	68	70	72	73	$90 - a$
Between vertical and horizontal members	90	90	90	90	90	90	90	90

Table 3-27. Flat Howe 6 Panels at Top & Bottom

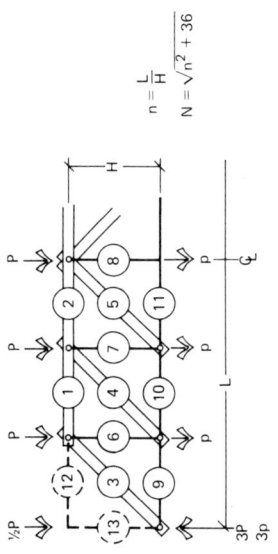

$n = \frac{L}{H}$
$N = \sqrt{n^2 + 36}$

AXIAL FORCE COEFFICIENTS

Member	7		8		9		10		11		12		13		General Formulas	
							VALUES OF n									
	K_1	K_2	K_1	K_2	K_1	K_2	K_1	K_2	K_1	K_2	K_1	K_2	K_1	K_2	K_1	K_2
1	-2.92	-2.92	-3.33	-3.33	-3.75	-3.75	-4.17	-4.17	-4.58	-4.58	-5.00	-5.00	-5.42	-5.42	$-5/12\,n$	$-5/12\,n$
2	-4.67	-4.67	-5.33	-5.33	-6.00	-6.00	-6.67	-6.67	-7.33	-7.33	-8.00	-8.00	-8.67	-8.67	$-2/3\,n$	$-2/3\,n$
3	-3.84	-3.84	-4.17	-4.17	-4.51	-4.51	-4.86	-4.86	-5.22	-5.22	-5.59	-5.59	-5.97	-5.97	$-5/12\,N$	$-5/12\,N$
4	-2.30	-2.30	-2.50	-2.50	-2.70	-2.70	-2.92	-2.92	-3.13	-3.13	-3.35	-3.35	-3.58	-3.58	$-1/4\,N$	$-1/4\,N$
5	-0.77	-0.77	-0.83	-0.83	-0.90	-0.90	-0.97	-0.97	-1.04	-1.04	-1.12	-1.12	-1.19	-1.19	$-1/12\,N$	$-1/12\,N$
6	1.50	2.50	1.50	2.50	1.50	2.50	1.50	2.50	1.50	2.50	1.50	2.50	1.50	2.50	3/2	5/2
7	0.50	1.50	0.50	1.50	0.50	1.50	0.50	1.50	0.50	1.50	0.50	1.50	0.50	1.50	1/2	3/2
8	0	1.00	0	1.00	0	1.00	0	1.00	0	1.00	0	1.00	0	1.00	0	1
9	2.92	2.92	3.33	3.33	3.75	3.75	4.17	4.17	4.58	4.58	5.00	5.00	5.42	5.42	$5/12\,n$	$5/12\,n$
10	4.67	4.67	5.33	5.33	6.00	6.00	6.67	6.67	7.33	7.33	8.00	8.00	8.67	8.67	$2/3\,n$	$2/3\,n$
11	5.25	5.25	6.00	6.00	6.75	6.75	7.50	7.50	8.25	8.25	9.00	9.00	9.75	9.75	$3/4\,n$	$3/4\,n$
12	0	0	0	0	0	0	0	0	0	0	0	0	0	0	0	0
13	-0.50	0	-0.50	0	-0.50	0	-0.50	0	-0.50	0	-0.50	0	-0.50	0	-1/2	0

LENGTH COEFFICIENTS

	K	K	K	K	K	K	K	K	K
Horizontal	1.166667	1.333333	1.500000	1.666667	1.833333	2.000000	2.166667	1/6 n
Diagonal	1.536591	1.666667	1.802776	1.943651	2.008327	2.236068	2.386304	1/6 N
Vertical	1.000000	1.000000	1.000000	1.000000	1.000000	1.000000	1.000000	1

ANGLES (IN DEGREES) BETWEEN MEMBERS

Between diagonal and horizontal members	41	37	34	31	29	27	25	$6/n = \tan a$
Between diagonal and vertical members	49	53	56	59	61	63	65	$90 - a$
Between vertical and horizontal members	90	90	90	90	90	90	90	90

Table 3-28. Flat Howe 8 Panels at Top & Bottom

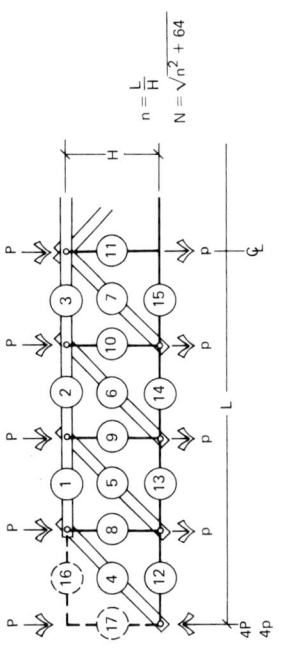

$n = \dfrac{L}{H}$

$N = \sqrt{n^2 + 64}$

Member	7		8		9		VALUES OF n 10		11		12		13		General Formulas K_1	K_2
	K_1	K_2	K_1	K_2	K_1	K_2	K_1	K_2	K_1	K_2	K_1	K_2	K_1	K_2		
	AXIAL FORCE COEFFICIENTS															
1	−3.06	−3.06	−3.50	−3.50	−3.94	−3.94	−4.38	−4.38	−4.81	−4.81	−5.25	−5.25	−5.69	−5.69	−7/16 n	−7/16 n
2	−5.25	−5.25	−6.00	−6.00	−6.75	−6.75	−7.50	−7.50	−8.25	−8.25	−9.00	−9.00	−9.75	−9.75	−3/4 n	−3/4 n
3	−6.56	−6.56	−7.50	−7.50	−8.44	−8.44	−9.38	−9.38	−10.31	−10.31	−11.25	−11.25	−12.19	−12.19	−15/16 n	−15/16 n
4	−4.65	−4.65	−4.95	−4.95	−5.27	−5.27	−5.60	−5.60	−5.95	−5.95	−6.31	−6.31	−6.68	−6.68	−7/16 N	−7/16 N
5	−3.32	−3.32	−3.54	−3.54	−3.76	−3.76	−4.00	−4.00	−4.25	−4.25	−4.51	−4.51	−4.77	−4.77	−5/16 N	−5/16 N
6	−1.99	−1.99	−2.12	−2.12	−2.26	−2.26	−2.40	−2.40	−2.55	−2.55	−2.70	−2.70	−2.86	−2.86	−3/16 N	−3/16 N
7	−0.66	−0.66	−0.71	−0.71	−0.75	−0.75	−0.80	−0.80	−0.85	−0.85	−0.90	−0.90	−0.95	−0.95	−1/16 N	−1/16 N
8	2.50	3.50	2.50	3.50	2.50	3.50	2.50	3.50	2.50	3.50	2.50	3.50	2.50	3.50	5/2	7/2
9	1.50	2.50	1.50	2.50	1.50	2.50	1.50	2.50	1.50	2.50	1.50	2.50	1.50	2.50	3/2	5/2
10	0.50	1.50	0.50	1.50	0.50	1.50	0.50	1.50	0.50	1.50	0.50	1.50	0.50	1.50	1/2	3/2
11	0	1.00	0	1.00	0	1.00	0	1.00	0	1.00	0	1.00	0	1.00	0	1
12	3.06	3.06	3.50	3.50	3.94	3.94	4.38	4.38	4.81	4.81	5.25	5.25	5.69	5.69	7/16 n	7/16 n

13	5.25	5.25	6.00	6.00	6.75	6.75	7.50	7.50	8.25	8.25	9.00	9.00	9.75	9.75	3/4 n	3/4 n
14	6.56	6.56	7.50	7.50	8.44	8.44	9.38	9.38	10.31	10.31	11.25	11.25	12.19	12.19	15/16 n	15/16 n
15	7.00	7.00	8.00	8.00	9.00	9.00	10.00	10.00	11.00	11.00	12.00	12.00	13.00	13.00	n	n
16	0	0	0	0	0	0	0	0	0	0	0	0	0	0	0	0
17	−0.50	0	−0.50	0	−0.50	0	−0.50	0	−0.50	0	−0.50	0	−0.50	0	−1/2	0

LENGTH COEFFICIENTS

	K	K	K	K	K	K	K	K	K
Horizontal	0.875000	1.000000	1.125000	1.250000	1.375000	1.500000	1.625000	1/8 n	
Diagonal	1.328768	1.414214	1.505199	1.600781	1.700184	1.802776	1.908042	1/8 N	
Vertical	1.000000	1.000000	1.000000	1.000000	1.000000	1.000000	1.000000	1	

ANGLES (IN DEGREES) BETWEEN MEMBERS

Between diagonal and horizontal members	49	45	42	39	36	34	32	$8/n = \tan a$	
Between diagonal and vertical members	41	45	46	51	54	56	56	$90 - a$	
Between vertical and horizontal members	90	90	90	90	90	90	90	90	

Table 3-29. Flat Howe 10 Panels at Top & Bottom

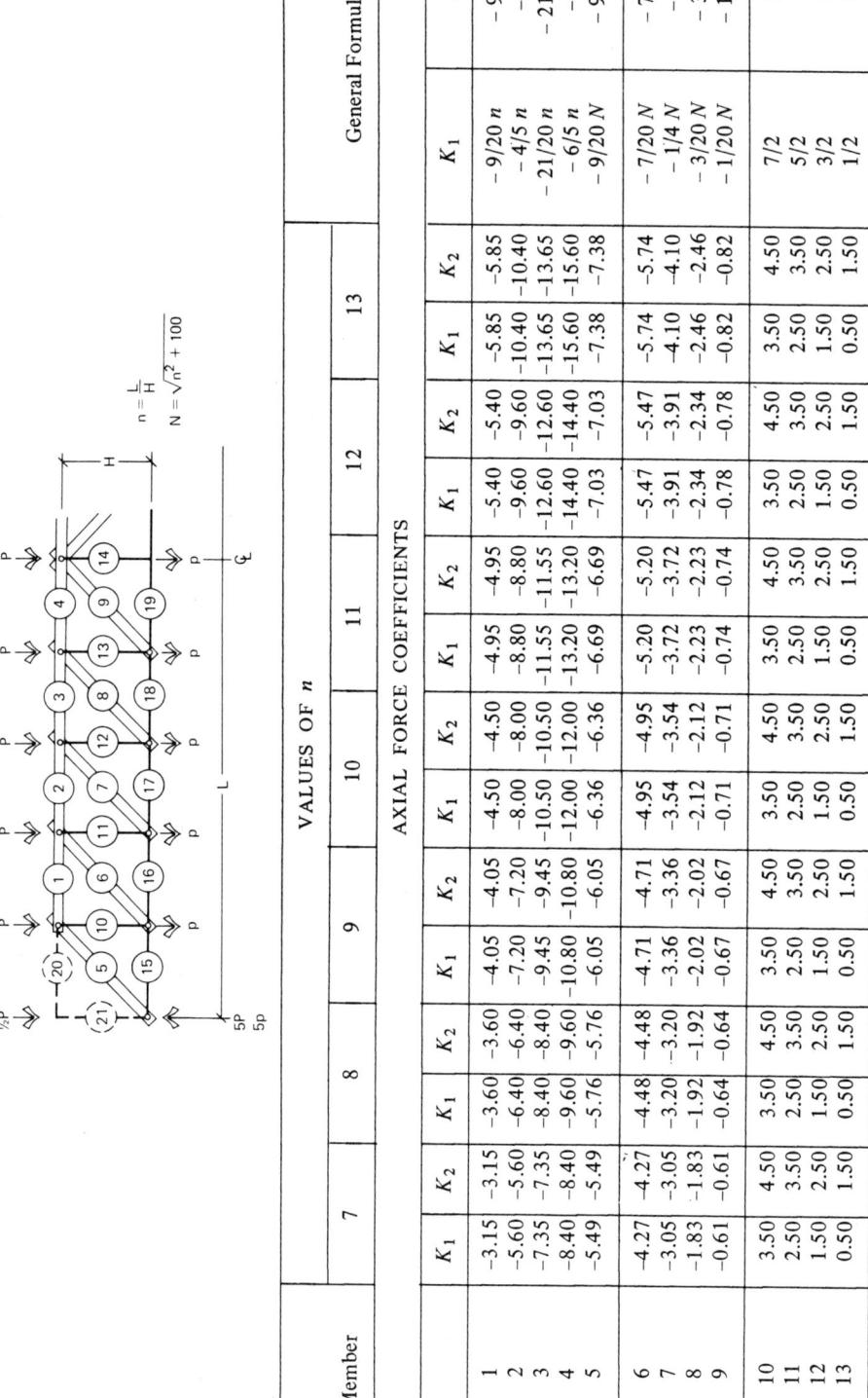

Member	7		8		9		10		11		12		13		General Formulas	
	\multicolumn{14}{c}{VALUES OF n}		
	\multicolumn{14}{c}{AXIAL FORCE COEFFICIENTS}		
	K_1	K_2	K_1	K_2	K_1	K_2	K_1	K_2	K_1	K_2	K_1	K_2	K_1	K_2	K_1	K_2
1	-3.15	-3.15	-3.60	-3.60	-4.05	-4.05	-4.50	-4.50	-4.95	-4.95	-5.40	-5.40	-5.85	-5.85	-9/20 n	-9/20 n
2	-5.60	-5.60	-6.40	-6.40	-7.20	-7.20	-8.00	-8.00	-8.80	-8.80	-9.60	-9.60	-10.40	-10.40	-4/5 n	-4/5 n
3	-7.35	-7.35	-8.40	-8.40	-9.45	-9.45	-10.50	-10.50	-11.55	-11.55	-12.60	-12.60	-13.65	-13.65	-21/20 n	-21/20 n
4	-8.40	-8.40	-9.60	-9.60	-10.80	-10.80	-12.00	-12.00	-13.20	-13.20	-14.40	-14.40	-15.60	-15.60	-6/5 n	-6/5 n
5	-5.49	-5.49	-5.76	-5.76	-6.05	-6.05	-6.36	-6.36	-6.69	-6.69	-7.03	-7.03	-7.38	-7.38	-9/20 N	-9/20 N
6	-4.27	-4.27	-4.48	-4.48	-4.71	-4.71	-4.95	-4.95	-5.20	-5.20	-5.47	-5.47	-5.74	-5.74	-7/20 N	-7/20 N
7	-3.05	-3.05	-3.20	-3.20	-3.36	-3.36	-3.54	-3.54	-3.72	-3.72	-3.91	-3.91	-4.10	-4.10	-1/4 N	-1/4 N
8	-1.83	-1.83	-1.92	-1.92	-2.02	-2.02	-2.12	-2.12	-2.23	-2.23	-2.34	-2.34	-2.46	-2.46	-3/20 N	-3/20 N
9	-0.61	-0.61	-0.64	-0.64	-0.67	-0.67	-0.71	-0.71	-0.74	-0.74	-0.78	-0.78	-0.82	-0.82	-1/20 N	-1/20 N
10	3.50	4.50	3.50	4.50	3.50	4.50	3.50	4.50	3.50	4.50	3.50	4.50	3.50	4.50	7/2	9/2
11	2.50	3.50	2.50	3.50	2.50	3.50	2.50	3.50	2.50	3.50	2.50	3.50	2.50	3.50	5/2	7/2
12	1.50	2.50	1.50	2.50	1.50	2.50	1.50	2.50	1.50	2.50	1.50	2.50	1.50	2.50	3/2	5/2
13	0.50	1.50	0.50	1.50	0.50	1.50	0.50	1.50	0.50	1.50	0.50	1.50	0.50	1.50	1/2	3/2

$n = \dfrac{L}{H}$

$N = \sqrt{n^2 + 100}$

14	0	1.00	0	1.00	0	1.00	0	1.00	0	1.00	0	1.00	0	1.00	1
15	3.15	3.15	3.60	3.60	4.05	4.05	4.50	4.50	4.95	4.95	5.40	5.40	5.85	5.85	9/20 n
16	5.60	5.60	6.40	6.40	7.20	7.20	8.00	8.00	8.80	8.80	9.60	9.60	10.40	10.40	4/5 n
17	7.35	7.35	8.40	8.40	9.45	9.45	10.50	10.50	11.55	11.55	12.60	12.60	13.65	13.65	21/20 n
18	8.40	8.40	9.60	9.60	10.80	10.80	12.00	12.00	13.20	13.20	14.40	14.40	15.60	15.60	6/5 n
19	8.75	8.75	10.00	10.00	11.25	11.25	12.50	12.50	13.75	13.75	15.00	15.00	16.25	16.25	5/4 n
20	0	0	0	0	0	0	0	0	0	0	0	0	0	0	0
21	−0.50	0	−0.50	0	−0.50	0	−0.50	0	−0.50	0	−0.50	0	−0.50	0	0

LENGTH COEFFICIENTS

	K	K	K	K	K	K	K	K
Horizontal	0.700000	0.800000	0.900000	1.000000	1.100000	1.200000	1.300000	1/10 n
Diagonal	1.220656	1.280625	1.345362	1.414214	1.486607	1.562050	1.640122	1/10 N
Vertical	1.000000	1.000000	1.000000	1.000000	1.000000	1.000000	1.000000	1

ANGLES (IN DEGREES) BETWEEN MEMBERS

Between diagonal and horizontal members	56	51	48	45	42	40	38	10/n = tan a
Between diagonal and vertical members	35	39	42	45	48	50	52	90 − a
Between vertical and horizontal members	90	90	90	90	90	90	90	90

Table 3-30. Flat Howe 12 Panels at Top & Bottom

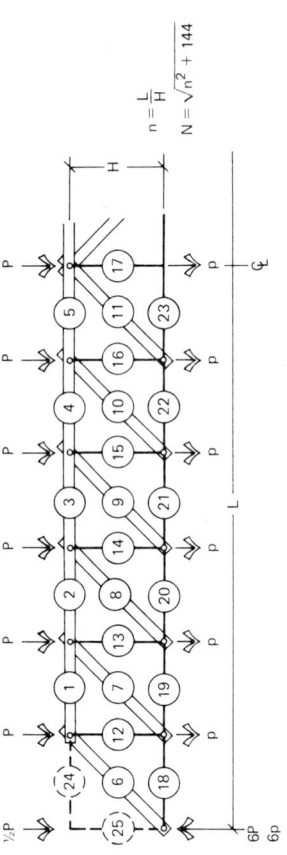

$n = \dfrac{L}{H}$

$N = \sqrt{n^2 + 144}$

VALUES OF n
AXIAL FORCE COEFFICIENTS

Member	7		8		9		10		11		12		13		General Formulas	
	K_1	K_2	K_1	K_2	K_1	K_2	K_1	K_2	K_1	K_2	K_1	K_2	K_1	K_2	K_1	K_2
1	−3.21	−3.21	−3.67	−3.67	−4.13	−4.13	−4.58	−4.58	−5.04	−5.04	−5.50	−5.50	−5.96	−5.96	−11/24 n	−11/24 n
2	−5.83	−5.83	−6.67	−6.67	−7.50	−7.50	−8.33	−8.33	−9.17	−9.17	−10.00	−10.00	−10.83	−10.83	−5/6 n	−5/6 n
3	−7.88	−7.88	−9.00	−9.00	−10.13	−10.13	−11.25	−11.25	−12.38	−12.38	−13.50	−13.50	−14.63	−14.63	−9/8 n	−9/8 n
4	−9.33	−9.33	−10.67	−10.67	−12.00	−12.00	−13.33	−13.33	−14.67	−14.67	−16.00	−16.00	−17.33	−17.33	−4/3 n	−4/3 n
5	−10.21	−10.21	−11.67	−11.67	−13.13	−13.13	−14.58	−14.58	−16.04	−16.04	−17.50	−17.50	−18.96	−18.96	−35/24 n	−35/24 n
6	−6.37	−6.37	−6.61	−6.61	−6.88	−6.88	−7.16	−7.16	−7.46	−7.46	−7.78	−7.78	−8.11	−8.11	−11/24 N	−11/24 N
7	−5.21	−5.21	−5.41	−5.41	−5.63	−5.63	−5.86	−5.86	−6.10	−6.10	−6.36	−6.36	−6.63	−6.63	−3/8 N	−3/8 N
8	−4.05	−4.05	−4.21	−4.21	−4.38	−4.38	−4.56	−4.56	−4.75	−4.75	−4.95	−4.95	−5.16	−5.16	−7/24 N	−7/24 N
9	−2.89	−2.89	−3.00	−3.00	−3.13	−3.13	−3.25	−3.25	−3.39	−3.39	−3.54	−3.54	−3.69	−3.69	−5/24 N	−5/24 N
10	−1.74	−1.74	−1.80	−1.80	−1.88	−1.88	−1.95	−1.95	−2.03	−2.03	−2.12	−2.12	−2.21	−2.21	−1/8 N	−1/8 N
11	−0.58	−0.58	−0.60	−0.60	−0.63	−0.63	−0.65	−0.65	−0.68	−0.68	−0.71	−0.71	−0.74	−0.74	−1/24 N	−1/24 N
12	4.50	5.50	4.50	5.50	4.50	5.50	4.50	5.50	4.50	5.50	4.50	5.50	4.50	5.50	9/2	11/2

13	3.50	4.50	3.50	4.50	3.50	4.50	3.50	4.50	3.50	4.50	3.50	4.50	7/2	9/2
14	2.50	3.50	2.50	3.50	2.50	3.50	2.50	3.50	2.50	3.50	2.50	3.50	5/2	7/2
15	1.50	2.50	1.50	2.50	1.50	2.50	1.50	2.50	1.50	2.50	1.50	2.50	3/2	5/2
16	0.50	1.50	0.50	1.50	0.50	1.50	0.50	1.50	0.50	1.50	0.50	1.50	1/2	3/2
17	0	1.00	0	1.00	0	1.00	0	1.00	0	1.00	0	1.00	0	1
18	3.21	3.67	3.67	4.13	4.13	4.58	4.58	5.04	5.04	5.50	5.50	5.96	5.96	11/24 n	11/24 n
19	5.83	6.67	6.67	7.50	7.50	8.33	8.33	9.17	9.17	10.00	10.00	10.83	10.83	5/6 n	5/6 n
20	7.88	9.00	9.00	10.13	10.13	11.25	11.25	12.38	12.38	13.50	13.50	14.63	14.63	9/8 n	9/8 n
21	9.33	10.67	10.67	12.00	12.00	13.33	13.33	14.67	14.67	16.00	16.00	17.33	17.33	4/3 n	4/3 n
22	10.21	11.67	11.67	13.13	13.13	14.58	14.58	16.04	16.04	17.50	17.50	18.96	18.96	35/24 n	35/24 n
23	10.50	12.00	12.00	13.50	13.50	15.00	15.00	16.50	16.50	18.00	18.00	19.50	19.50	3/2 n	3/2 n
24	0	0	0	0	0	0	0	0	0	0	0	0	0	0	0
25	−0.50	0	−0.50	0	−0.50	0	−0.50	0	−0.50	0	−0.50	0	−1/2	0

LENGTH COEFFICIENTS

	K	K	K	K	K	K	K	K
Horizontal	0.583333	0.666667	0.750000	0.833333	0.916667	1.000000	1.083333	1/12 n
Diagonal	1.157704	1.201850	1.250000	1.301708	1.356568	1.414214	1.474317	1/12 N
Vertical	1.000000	1.000000	1.000000	1.000000	1.000000	1.000000	1.000000	1

ANGLES (IN DEGREES) BETWEEN MEMBERS

Between diagonal and horizontal members	60	56	53	50	47	45	43	$12/n = \tan a$
Between diagonal and vertical members	30	34	37	40	43	45	47	$90 - a$
Between vertical and horizontal members	90	90	90	90	90	90	90	90

Table 3-31. Flat Pratt 4 Panels at Top & Bottom

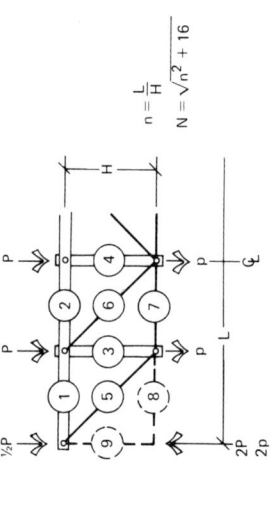

$n = \dfrac{L}{H}$

$N = \sqrt{n^2 + 16}$

VALUES OF n

AXIAL FORCE COEFFICIENTS

Member	7		8		9		10		11		12		13		General Formulas	
	K_1	K_2	K_1	K_2	K_1	K_2	K_1	K_2	K_1	K_2	K_1	K_2	K_1	K_2	K_1	K_2
1	-2.63	-2.63	-3.00	-3.00	-3.38	-3.38	-3.75	-3.75	-4.13	-4.13	-4.50	-4.50	-4.88	-4.88	$-3/8\,n$	$-3/8\,n$
2	-3.50	-3.50	-4.00	-4.00	-4.50	-4.50	-5.00	-5.00	-5.50	-5.50	-6.00	-6.00	-6.50	-6.50	$-1/2\,n$	$-1/2\,n$
3	-1.50	-0.50	-1.50	-0.50	-1.50	-0.50	-1.50	-0.50	-1.50	-0.50	-1.50	-0.50	-1.50	-0.50	$-3/2$	$-1/2$
4	-1.00	0	-1.00	0	-1.00	0	-1.00	0	-1.00	0	-1.00	0	-1.00	0	-1	0
5	3.02	3.02	3.35	3.35	3.69	3.69	4.04	4.04	4.39	4.39	4.74	4.74	5.10	5.10	$3/8\,N$	$3/8\,N$
6	1.01	1.01	1.12	1.12	1.23	1.23	1.35	1.35	1.46	1.46	1.58	1.58	1.70	1.70	$1/8\,N$	$1/8\,N$
7	2.63	2.63	3.00	3.00	3.38	3.38	3.75	3.75	4.13	4.13	4.50	4.50	4.88	4.88	$3/8\,N$	$3/8\,N$
8	0	0	0	0	0	0	0	0	0	0	0	0	0	0	0	0
9	-2.00	0	-2.00	0	-2.00	0	-2.00	0	-2.00	0	-2.00	0	-2.00	0	-2	0

LENGTH COEFFICIENTS

	K	K	K	K	K	K	K	K
Horizontal	1.750000	2.000000	2.250000	2.500000	2.750000	3.000000	3.250000	$1/4\,n$
Diagonal	2.015564	2.236068	2.462214	2.692582	2.926175	3.162278	3.400368	$1/4\,N$
Vertical	1.000000	1.000000	1.000000	1.000000	1.000000	1.000000	1.000000	1

	ANGLES (IN DEGREES) BETWEEN MEMBRES							
Between horizontal and diagonal members	30	27	24	22	20	18	17	$4/n = \tan a$
Between diagonal and vertical members	60	63	66	68	70	72	73	$90 - a$
Between vertical and hortizontal members	90	90	90	90	90	90	90	90

Table 3-32. Flat Pratt 6 Panels at Top & Bottom

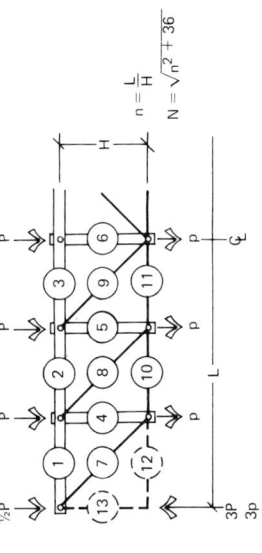

$n = \dfrac{L}{H}$

$N = \sqrt{n^2 + 36}$

Member	\multicolumn{14}{c}{VALUES OF n — AXIAL FORCE COEFFICIENTS}	General Formulas	
	7		8		9		10		11		12		13			
	K_1	K_2	K_1	K_2	K_1	K_2	K_1	K_2	K_1	K_2	K_1	K_2	K_1	K_2	K_1	K_2
1	−2.92	−2.92	−3.33	−3.33	−3.75	−3.75	−4.17	−4.17	−4.58	−4.58	−5.00	−5.00	−5.42	−5.42	−5/12 n	−5/12 n
2	−4.67	−4.67	−5.33	−5.33	−6.00	−6.00	−6.67	−6.67	−7.33	−7.33	−8.00	−8.00	−8.67	−8.67	−8/12 n	−8/12 n
3	−5.25	−5.25	−6.00	−6.00	−6.75	−6.75	−7.50	−7.50	−8.25	−8.25	−9.00	−9.00	−9.75	−9.75	−9/12 n	−9/12 n
4	−2.50	−1.50	−2.50	−1.50	−2.50	−1.50	−2.50	−1.50	−2.50	−1.50	−2.50	−1.50	−25.0	−1.50	−5/2	−3/2
5	−1.50	−0.50	−1.50	−0.50	−1.50	−0.50	−1.50	−0.50	−1.50	−0.50	−1.50	−0.50	−1.50	−0.50	−3/2	−1/2
6	−1.00	0	−1.00	0	−1.00	0	−1.00	0	−1.00	0	−1.00	0	−1.00	0	−1	0
7	3.84	3.84	4.17	4.17	4.51	4.51	4.86	4.86	5.22	5.22	5.59	5.59	5.97	5.97	5/12 N	5/12 N
8	2.30	2.30	2.50	2.50	2.70	2.70	2.92	2.92	3.13	3.13	3.35	3.35	3.58	3.58	3/12 N	3/12 N
9	0.77	0.77	0.83	0.83	0.90	0.90	0.97	0.97	1.04	1.04	1.12	1.12	1.19	1.19	1/12 N	1/12 N
10	2.92	2.92	3.33	3.33	3.75	3.75	4.17	4.17	4.58	4.58	5.00	5.00	5.42	5.42	5/12 n	5/12 n
11	4.67	4.67	5.33	5.33	6.00	6.00	6.67	6.67	7.33	7.33	8.00	8.00	8.67	8.67	8/12 n	8/12 n
12	0	0	0	0	0	0	0	0	0	0	0	0	0	0	0	0
13	−3.00	0	−3.00	0	−3.00	0	−3.00	0	−3.00	0	−3.00	0	−3.00	0	−3	0

LENGTH COEFFICIENTS

	K	K	K	K	K	K	K	K	K
Horizontal	1.166667	1.333333	1.500000	1.666667	1.833333	2.000000	2.166667	1/6 n
Diagonal	1.536591	1.666667	1.802776	1.943651	2.088327	2.236068	2.366304	1/6 N
Vertical	1.000000	1.000000	1.000000	1.000000	1.000000	1.000000	1.000000	1

ANGLES (IN DEGREES) BETWEEN MEMBERS

Between horizontal and diagonal members	41	37	34	31	29	27	25	6/n = tan a
Between diagonal and vertical members	49	53	56	59	61	63	65	90 − a
Between vertical and horizontal members	90	90	90	90	90	90	90	90

Table 3-33. Flat Pratt 8 Panels at Top & Bottom

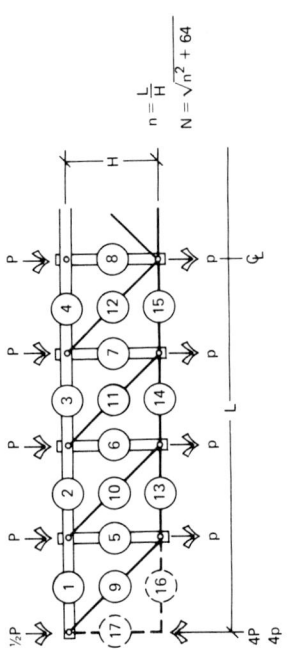

$n = \dfrac{L}{H}$

$N = \sqrt{n^2 + 64}$

Member	7		8		9		10		11		12		13		General Formulas	
	K_1	K_2	K_1	K_2	K_1	K_2	K_1	K_2	K_1	K_2	K_1	K_2	K_1	K_2	K_1	K_2
							VALUES OF n									
							AXIAL FORCE COEFFICIENTS									
1	−3.06	−3.06	−3.50	−3.50	−3.94	−3.94	−4.38	−4.38	−4.81	−4.81	−5.25	−5.25	−5.69	−5.69	−7/16 n	−7/16 n
2	−5.25	−5.25	−6.00	−6.00	−6.75	−6.75	−7.50	−7.50	−8.25	−8.25	−9.00	−9.00	−9.75	−9.75	−12/15 n	−12/15 n
3	−6.56	−6.56	−7.50	−7.50	−8.44	−8.44	−9.38	−9.38	−10.31	−10.31	−11.25	−11.25	−12.19	−12.19	−15/16 n	−15/16 n
4	−7.00	−7.00	−8.00	−8.00	−9.00	−9.00	−10.00	−10.00	−11.00	−11.00	−12.00	−12.00	−13.00	−13.00	−n	−n
5	−3.50	−2.50	−3.50	−2.50	−3.50	−2.50	−3.50	−2.50	−3.50	−2.50	−3.50	−2.50	−3.50	−2.50	−7/2	−5/2
6	−2.50	−1.50	−2.50	−1.50	−2.50	−1.50	−2.50	−1.50	−2.50	−1.50	−2.50	−1.50	−2.50	−1.50	−5/2	−3/2
7	−1.50	−0.50	−1.50	−0.50	−1.50	−0.50	−1.50	−0.50	−1.50	−0.50	−1.50	−0.50	−1.50	−0.50	−3/2	−1/2
8	−1.00	0	−1.00	0	−1.00	0	−1.00	0	−1.00	0	−1.00	0	−1.00	0	−1	0
9	4.65	4.65	4.95	4.95	5.27	5.27	5.60	5.60	5.95	5.95	6.31	6.31	6.68	6.68	7/16 N	7/16 N
10	3.32	3.32	3.54	3.54	3.76	3.76	4.00	4.00	4.25	4.25	4.51	4.51	4.77	4.77	5/16 N	5/16 N
11	1.99	1.99	2.12	2.12	2.26	2.26	2.40	2.40	2.55	2.55	2.70	2.70	2.86	2.86	3/16 N	3/16 N
12	0.66	0.66	0.71	0.71	0.75	0.75	0.80	0.80	0.85	0.85	0.90	0.90	0.95	0.95	1/16 N	1/16 N

	13	3.06	3.50	3.94	4.38	4.81	5.25	5.69	7/16 n
	14	5.25	6.00	6.75	7.50	8.25	9.00	9.75	12/16 n
	15	6.56	7.00	8.44	9.38	10.31	11.25	12.19	15/16 n
	16	0	0	0	0	0	0	0	0
	17	−4.00	−4.00	−4.00	−4.00	−4.00	−4.00	−4.00	−4

LENGTH COEFFICIENTS

	K	K	K	K	K	K	K	K
Horizontal	0.875000	1.000000	1.125000	1.250000	1.375000	1.500000	1.625000	1/8 n
Diagonal	1.328768	1.414214	1.505199	1.600781	1.700184	1.802776	1.908042	1/8 N
Vertical	1.000000	1.000000	1.000000	1.000000	1.000000	1.000000	1.000000	1

ANGLES (IN DEGREES) BETWEEN MEMBERS

Between horizontal and diagonal members	49	45	42	39	36	34	32	$8/n = \tan a$
Between diagonal and vertical members	41	45	48	51	54	56	58	$90 - a$
Between vertical and horizontal members	90	90	90	90	90	90	90	90

Table 3-34. Flat Pratt 10 Panels at Top & Bottom

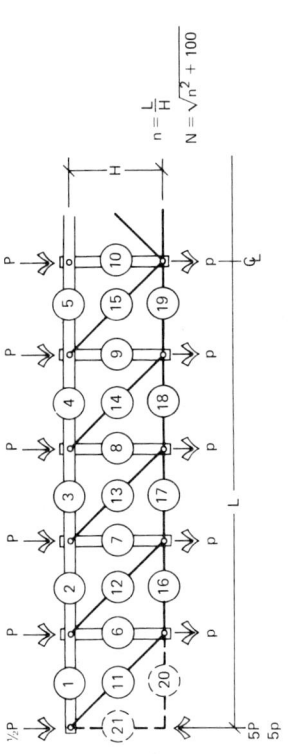

$n = \frac{L}{H}$
$N = \sqrt{n^2 + 100}$

Member	VALUES OF n														General Formulas	
	7		8		9		10		11		12		13			
	K_1	K_2	K_1	K_2	K_1	K_2	K_1	K_2	K_1	K_2	K_1	K_2	K_1	K_2	K_1	K_2
	AXIAL FORCE COEFFICIENTS															
1	−3.15	−3.15	−3.60	−3.60	−4.05	−4.05	−4.50	−4.50	−4.95	−4.95	−5.40	−5.40	−5.85	−5.85	−9/20 n	−9/20 n
2	−5.60	−5.60	−6.40	−6.40	−7.20	−7.20	−8.00	−8.00	−8.90	−8.90	−9.60	−9.60	−10.40	−10.40	−16/20 n	−16/20 n
3	−7.35	−7.35	−8.40	−8.40	−9.45	−9.45	−10.50	−10.50	−11.55	−11.55	−12.60	−12.60	−13.65	−13.65	−21/20 n	−21/20 n
4	−8.40	−8.40	−9.60	−9.60	−10.80	−10.80	−12.00	−12.00	−13.20	−13.20	−14.40	−14.40	−15.60	−15.60	−24/20 n	−24/20 n
5	−8.75	−8.75	−10.00	−10.00	−11.25	−11.25	−12.50	−12.50	−13.75	−13.75	−15.00	−15.00	−16.25	−16.25	−25/20 n	−25/20 n
6	−4.50	−3.50	−4.50	−3.50	−4.50	−3.50	−4.50	−3.50	−4.50	−3.50	−4.50	−3.50	−4.50	−3.50	−9/2	−7/2
7	−3.50	−2.50	−3.50	−2.50	−3.50	−2.50	−3.50	−2.50	−3.50	−2.50	−3.50	−2.50	−3.50	−2.50	−7/2	−5/2
8	−2.50	−1.50	−2.50	−1.50	−2.50	−1.50	−2.50	−1.50	−2.50	−1.50	−2.50	−1.50	−2.50	−1.50	−5/2	−3/2
9	−1.50	−0.50	−1.50	−0.50	−1.50	−0.50	−1.50	−0.50	−1.50	−0.50	−1.50	−0.50	−1.50	−0.50	−3/2	−1/2
10	−1.00	0	−1.00	0	−1.00	0	−1.00	0	−1.00	0	−1.00	0	−1.00	0	−1	0
11	5.49	5.49	5.76	5.76	6.05	6.05	6.36	6.36	6.69	6.69	7.03	7.03	7.38	7.38	9/20 N	9/20 N
12	4.27	4.27	4.48	4.48	4.71	4.71	4.95	4.95	5.20	5.20	5.47	5.47	5.74	5.74	7/20 N	7/20 N

13	3.05	3.05	3.20	3.20	3.36	3.36	3.54	3.54	3.72	3.72	3.91	3.91	4.10	4.10	5/20 N	5/20 N
14	1.83	1.83	1.92	1.92	2.02	2.02	2.12	2.12	2.23	2.23	2.34	2.34	2.46	2.46	3/20 N	3/20 N
15	0.61	0.61	0.64	0.64	0.67	0.67	0.71	0.71	0.74	0.74	0.78	0.78	0.82	0.82	1/20 N	1/20 N
16	3.15	3.15	3.60	3.60	4.05	4.05	4.50	4.50	4.95	4.95	5.40	5.40	5.85	5.85	9/20 n	9/20 n
17	5.60	5.60	6.40	6.40	7.20	7.20	8.00	8.00	8.80	8.80	9.60	9.60	10.40	10.40	16/20 n	16/20 n
18	7.35	7.35	8.40	8.40	9.45	9.45	10.50	10.50	11.55	11.55	12.60	12.60	13.65	13.65	21/20 n	21/20 n
19	8.40	8.40	9.60	9.60	10.80	10.80	12.00	12.00	13.20	13.20	14.40	14.40	15.60	15.60	24/20 n	24/20 n
20	0	0	0	0	0	0	0	0	0	0	0	0	0	0	0	0
21	-5.00	-5.00	-5.00	-5.00	-5.00	-5.00	-5.00	-5.00	-5.00	-5.00	-5.00	-5.00	-5.00	-5.00	-5	0

LENGTH COEFFICIENTS

	K	K	K	K	K	K	K	K	K
Horizontal	0.700000	0.800000	0.900000	1.000000	1.000000	1.100000	1.200000	1.300000	1/10 n
Diagonal	1.220656	1.280625	1.345362	1.414214	1.414214	1.486607	1.562050	1.640122	1/10 N
Vertical	1.000000	1.000000	1.000000	1.000000	1.000000	1.000000	1.000000	1.000000	1

ANGLES (IN DEGREES) BETWEEN MEMBERS

Between horizontal and diagonal members	55	51	48	45	45	42	40	38	10/n = tan a
Between diagonal and vertical members	35	39	42	45	45	48	50	52	90 – a
Between vertical and horizontal members	90	90	90	90	90	90	90	90	90

Table 3-35. Flat Pratt 12 Panels at Top & Bottom

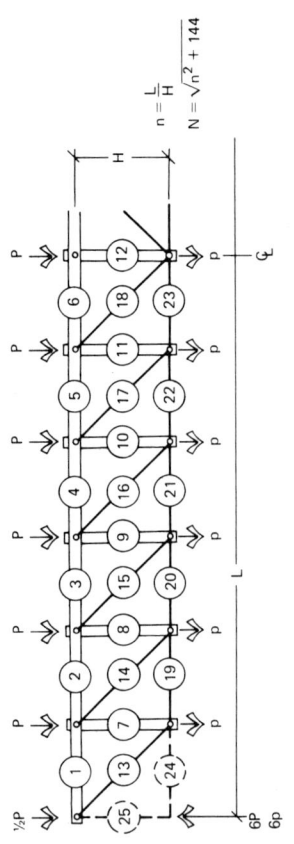

$n = \dfrac{L}{H}$

$N = \sqrt{n^2 + 144}$

Member	7		8		9		10		11		12		13		General Formulas	
	K_1	K_2	K_1	K_2	K_1	K_2	K_1	K_2	K_1	K_2	K_1	K_2	K_1	K_2	K_1	K_2
							VALUES OF n									
							AXIAL FORCE COEFFICIENTS									
1	−3.21	−3.21	−3.67	−3.67	−4.13	−4.13	−4.58	−4.58	−5.04	−5.04	−5.50	−5.50	−5.96	−5.96	−11/24 n	−11/24 n
2	−5.83	−5.83	−6.67	−6.67	−7.50	−7.50	−8.33	−8.33	−9.17	−9.17	−10.00	−10.00	−10.83	−10.83	−20/24 n	−20/24 n
3	−7.88	−7.88	−9.00	−9.00	−10.13	−10.13	−11.25	−11.25	−12.38	−12.38	−13.50	−13.50	−14.63	−14.63	−27/24 n	−27/24 n
4	−9.33	−9.33	−10.67	−10.67	−12.00	−12.00	−13.33	−13.33	−14.67	−14.67	−16.00	−16.00	−17.33	−17.33	−32/24 n	−32/24 n
5	−10.21	−10.21	−11.67	−11.67	−13.13	−13.13	−14.58	−14.58	−16.04	−16.04	−17.50	−17.50	−18.96	−18.96	−35/24 n	−35/24 n
6	−10.50	−10.50	−12.00	−12.00	−13.50	−13.50	−15.00	−15.00	−16.50	−16.50	−18.00	−18.00	−19.50	−19.50	−36/24 n	−36/24 n
7	−5.50	−4.50	−5.50	−4.50	−5.50	−4.50	−5.50	−4.50	−5.50	−4.50	−5.50	−4.50	−5.50	−4.50	−11/2	−9/2
8	−4.50	−3.50	−4.50	−3.50	−4.50	−3.50	−4.50	−3.50	−4.50	−3.50	−4.50	−3.50	−4.50	−3.50	−9/2	−7/2
9	−3.50	−2.50	−3.50	−2.50	−3.50	−2.50	−3.50	−2.50	−3.50	−2.50	−3.50	−2.50	−3.50	−2.50	−7/2	−5/2
10	−2.50	−1.50	−2.50	−1.50	−2.50	−1.50	−2.50	−1.50	−2.50	−1.50	−2.50	−1.50	−2.50	−1.50	−5/2	−3/2
11	−1.50	−0.50	−1.50	−0.50	−1.50	−0.50	−1.50	−0.50	−1.50	−0.50	−1.50	−0.50	−1.50	−0.50	−3/2	−1/2
12	−1.00	0	−1.00	0	−1.00	0	−1.00	0	−1.00	0	−1.00	0	−1.00	0	−1	0
13	6.37	6.37	6.61	6.61	6.88	6.88	7.16	7.16	7.46	7.46	7.78	7.78	8.11	8.11	11/24 N	11/24 N

14	5.21	5.21	5.41	5.41	5.63	5.63	5.86	5.86	6.10	6.10	6.36	6.36	6.63	6.63	9/24 N	9/24 N
15	4.05	4.05	4.21	4.21	4.38	4.38	4.56	4.56	4.75	4.75	4.95	4.95	5.16	5.16	7/24 N	7/24 N
16	2.89	2.89	3.00	3.00	3.13	3.13	3.25	3.25	3.39	3.39	3.54	3.54	3.69	3.69	5/24 N	5/24 N
17	1.74	1.74	1.80	1.80	1.88	1.88	1.95	1.95	2.03	2.03	2.12	2.12	2.21	2.21	3/24 N	3/24 N
18	0.58	0.58	0.60	0.60	0.63	0.63	0.65	0.65	0.68	0.68	0.71	0.71	0.74	0.74	1/24 N	1/24 N
19	3.21	3.21	3.67	3.67	4.13	4.13	4.58	4.58	5.04	5.04	5.50	5.50	5.96	5.96	11/24 n	11/24 n
20	5.83	5.83	6.67	6.67	7.50	7.50	8.33	8.33	9.17	9.17	10.00	10.00	10.83	10.83	20/24 n	20/24 n
21	7.88	7.88	9.00	9.00	10.13	10.13	11.25	11.25	12.38	12.38	13.50	13.50	14.63	14.63	27/24 n	27/24 n
22	9.33	9.33	10.67	10.67	12.00	12.00	13.33	13.33	14.67	14.67	16.00	16.00	17.33	17.33	32/24 n	32/24 n
23	10.21	10.21	11.67	11.67	13.13	13.13	14.58	14.58	16.04	16.04	17.50	17.50	18.96	18.96	35/24 n	35/24 n
24	0	0	0	0	0	0	0	0	0	0	0	0	0	0	0	0
25	−6.00	−6.00	−6.00	−6.00	−6.00	−6.00	−6.00	−6.00	−6.00	−6.00	−6.00	−6.00	−6.00	−6.00	0	−6

LENGTH COEFFICIENTS

	K	K	K	K	K	K	K	K
Horizontal	0.583333	0.666667	0.750000	0.833333	0.016667	1.000000	1.083333	$1/12\, n$
Diagonal	1.157704	1.201850	1.250000	1.301708	1.356562	1.414214	1.474317	$1/12\, N$
Vertical	1.000000	1.000000	1.000000	1.000000	1.000000	1.000000	1.000000	1

ANGLES (IN DEGREES) BETWEEN MEMBERS

Between horizontal and diagonal members	60	56	53	50	47	45	43	$12/n = \tan a$
Between diagonal and vertical members	30	34	37	40	43	45	47	$90 - a$
Between vertical and horizontal members	90	90	90	90	90	90	90	90

Table 3-36. Flat Pratt 4 Panels at Top & Bottom

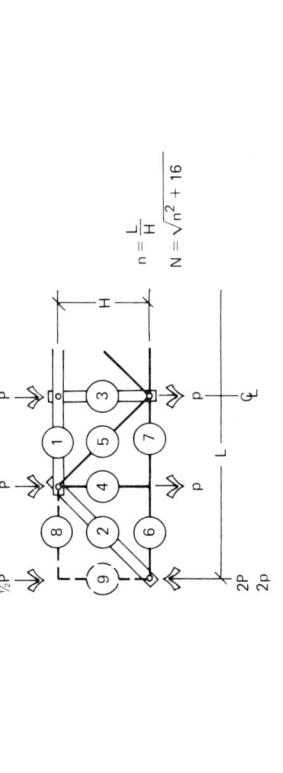

$n = \frac{L}{H}$
$N = \sqrt{n^2 + 16}$

Member	7		8		9		10		11		12		13		General Formulas	
	K_1	K_2	K_1	K_2	K_1	K_2	K_1	K_2	K_1	K_2	K_1	K_2	K_1	K_2	K_1	K_2
	\multicolumn{14}{c}{VALUES OF n}		
	\multicolumn{14}{c}{AXIAL FORCE COEFFICIENTS}		
1	−3.50	−3.50	−4.00	−4.00	−4.50	−4.50	−5.00	−5.00	−5.50	−5.50	−6.00	−6.00	−6.50	−6.50	−1/2 n	−1/2 n
2	−3.02	−3.02	−3.35	−3.35	−3.69	−3.69	−4.04	−4.04	−4.39	−4.39	−4.74	−4.74	−5.10	−5.10	−3/8 N	−3/8 N
3	−1.00	0	−1.00	0	−1.00	0	−1.00	0	−1.00	0	−1.00	0	−1.00	0	−1	0
4	0	1.00	0	1.00	0	1.00	0	1.00	0	1.00	0	1.00	0	1.00	1	1
5	1.01	1.01	1.12	1.12	1.23	1.23	1.35	1.35	1.46	1.46	1.58	1.58	1.70	1.70	1/8 N	1/8 N
6	2.63	2.63	3.00	3.00	3.38	3.38	3.75	3.75	4.13	4.13	4.50	4.50	4.88	4.88	3/8 n	3/8 n
7	2.63	2.63	3.00	3.00	3.38	3.38	3.75	3.75	4.13	4.13	4.50	4.50	4.88	4.88	3/8 n	3/8 n
8	0	0	0	0	0	0	0	0	0	0	0	0	0	0	0	0
9	−0.50	0	−0.50	0	−0.50	0	−0.50	0	−0.50	0	−0.50	0	−0.50	0	−1/2	0

LENGTH COEFFICIENTS

	K	K	K	K	K	K	K	K	K
Horizontal	1.750000	2.000000	2.250000	2.500000	2.750000	3.000000	3.250000		1/4 n
Diagonal	2.015564	2.236068	2.462214	2.692582	2.926175	3.162278	3.400368		1/4 N
Vertical	1.000000	1.000000	1.000000	1.000000	1.000000	1.000000	1.000000		1

ANGLES (IN DEGREES) BETWEEN MEMBERS

Between horizontal and diagonal members	30	27	24	22	20	18	17		$4/n = \tan a$
Between diagonal and vertical members	60	63	66	68	70	72	73		$90 - a$
Between vertical and horizontal members	90	90	90	90	90	90	90		90

Table 3-37. Modified Flat Pratt 6 Panels at Top & Bottom

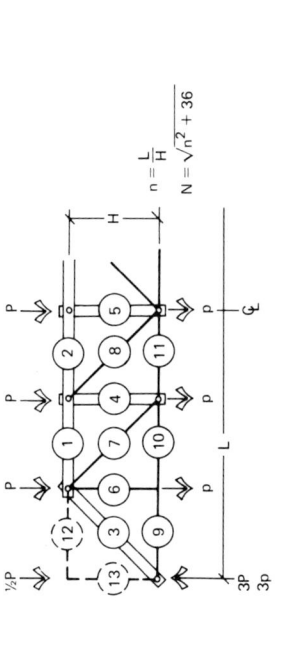

$n = \dfrac{L}{H}$

$N = \sqrt{n^2 + 36}$

VALUES OF n

AXIAL FORCE COEFFICIENTS

Member	7		8		9		10		11		12		13		General Formulas	
	K_1	K_2	K_1	K_2	K_1	K_2	K_1	K_2	K_1	K_2	K_1	K_2	K_1	K_2	K_1	K_2
1	-4.67	-4.67	-5.33	-5.33	-6.00	-6.00	-6.67	-6.67	-7.33	-7.33	-8.00	-8.00	-8.67	-8.67	$-2/3\,n$	$-2/3\,n$
2	-5.25	-5.25	-6.00	-6.00	-6.75	-6.75	-7.50	-7.50	-8.25	-8.25	-9.00	-9.00	-9.75	-9.75	$-3/4\,n$	$-3/4\,n$
3	-3.84	-3.84	-4.17	-4.17	-4.51	-4.51	-4.86	-4.86	-5.22	-5.22	-5.59	-5.59	-5.97	-5.97	$-5/12\,N$	$-5/12\,N$
4	-1.50	-0.50	-1.50	-0.50	-1.50	-0.50	-1.50	-0.50	-1.50	-0.50	-1.50	-0.50	-1.50	-0.50	$-3/2$	$-1/2$
5	-1.00	0	-1.00	0	-1.00	0	-1.00	0	-1.00	0	-1.00	0	-1.00	0	-1	0
6	0	1.00	0	1.00	0	1.00	0	1.00	0	1.00	0	1.00	0	1.00	0	1
7	2.30	2.30	2.50	2.50	2.70	2.70	2.92	2.92	3.13	3.13	3.35	3.35	3.58	3.58	$1/4\,N$	$1/4\,N$
8	0.77	0.77	0.83	0.83	0.90	0.90	0.97	0.97	1.04	1.04	1.12	1.12	1.19	1.19	$1/12\,N$	$1/12\,N$
9	2.92	2.92	3.33	3.33	3.75	3.75	4.17	4.17	4.58	4.58	5.00	5.00	5.42	5.42	$5/12\,n$	$5/12\,n$
10	2.92	2.92	3.33	3.33	3.75	3.75	4.17	4.17	4.58	4.58	5.00	5.00	5.42	5.42	$5/12\,n$	$5/12\,n$
11	4.67	4.67	5.33	5.33	6.00	6.00	6.67	6.67	7.33	7.33	8.00	8.00	8.67	8.67	$2/3\,n$	$2/3\,n$
12	0	0	0	0	0	0	0	0	0	0	0	0	0	0	0	0
13	-0.50	0	-0.50	0	-0.50	0	-0.50	0	-0.50	0	-0.50	0	-0.50	0	$-1/2$	0

LENGTH COEFFICIENTS

	K	K	K	K	K	K	K	K
Horizontal	1.166667	1.333333	1.500000	1.666667	1.833333	2.000000	2.166667	1/6 n
Diagonal	1.536591	1.666667	1.802776	1.943651	2.083327	2.236068	2.386304	1/6 N
Vertical	1.000000	1.000000	1.000000	1.000000	1.000000	1.000000	1.000000	1

ANGLES (IN DEGREES) BETWEEN MEMBERS

	K	K	K	K	K	K	K	K
Between horizontal and diagonal members	41	37	34	31	29	27	25	$6/n = \tan a$
Between diagonal and vertical members	49	53	56	59	61	63	65	$90 - a$
Between vertical and horizontal members	90	90	90	90	90	90	90	90

Table 3-38. Modified Flat Pratt 8 Panels at Top & Bottom

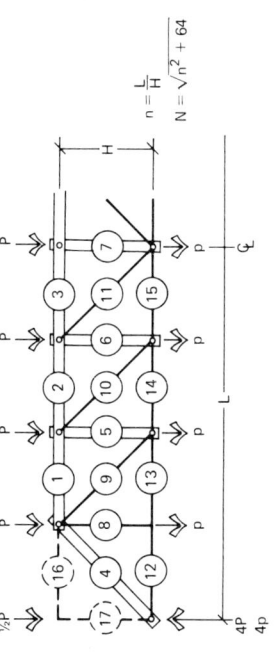

$n = \dfrac{L}{H}$
$N = \sqrt{n^2 + 64}$

VALUES OF n
AXIAL FORCE COEFFICIENTS

Member	7		8		9		10		11		12		13		General Formulas	
	K_1	K_2	K_1	K_2	K_1	K_2	K_1	K_2	K_1	K_2	K_1	K_2	K_1	K_2	K_1	K_2
1	−5.25	−5.25	−6.00	−6.00	−6.75	−6.75	−7.50	−7.50	−8.25	−8.25	−9.00	−9.00	−9.75	−9.75	−3/4 n	−3/4 n
2	−6.56	−6.56	−7.50	−7.50	−8.44	−8.44	−9.38	−9.38	−10.31	−10.31	−11.25	−11.25	−12.19	−12.19	−15/16 n	−15/16 n
3	−7.00	−7.00	−8.00	−8.00	−9.00	−9.00	−10.00	−10.00	−11.00	−11.00	−12.00	−12.00	−13.00	−13.00	−n	−n
4	−4.65	−4.65	−4.95	−4.95	−5.27	−5.27	−5.60	−5.60	−5.95	−5.95	−6.31	−6.31	−6.68	−6.68	−7/16 N	−7/16 N
5	−2.50	−1.50	−2.50	−1.50	−2.50	−1.50	−2.50	−1.50	−2.50	−1.50	−2.50	−1.50	−2.50	−1.50	−5/2	−3/2
6	−1.50	−0.50	−1.50	−0.50	−1.50	−0.50	−1.50	−0.50	−1.50	−0.50	−1.50	−0.50	−1.50	−0.50	−3/2	−1/2
7	−1.00	0	−1.00	0	−1.00	0	−1.00	0	−1.00	0	−1.00	0	−1.00	0	−1	0
8	0	1.00	0	1.00	0	1.00	0	1.00	0	1.00	0	1.00	0	1.00	0	1
9	3.32	3.32	3.54	3.54	3.76	3.76	4.00	4.00	4.25	4.25	4.51	4.51	4.77	4.77	5/16 N	5/16 N
10	1.99	1.99	2.12	2.12	2.26	2.26	2.40	2.40	2.55	2.55	2.70	2.70	2.86	2.86	3/16 N	3/16 N
11	0.66	0.66	0.71	0.71	0.75	0.75	0.80	0.80	0.85	0.85	0.90	0.90	0.95	0.95	1/16 N	1/16 N
12	3.06	3.06	3.50	3.50	3.94	3.94	4.38	4.38	4.81	4.81	5.25	5.25	5.69	5.69	7/16 n	7/16 n

13	3.06	3.06	3.50	3.50	3.94	3.94	4.38	4.38	4.81	4.81	5.25	5.25	5.69	5.69	7/16 n	7/16 n
14	5.25	5.25	6.00	6.00	6.75	6.75	7.50	7.50	8.25	8.25	9.00	9.00	9.75	9.75	3/4 n	3/4 n
15	6.56	6.56	7.50	7.50	8.44	8.44	9.38	9.38	10.31	10.31	11.25	11.25	12.19	12.19	15/16 n	15/16 n
16	0	0	0	0	0	0	0	0	0	0	0	0	0	0	0	0
17	−0.50	0	−0.50	0	−0.50	0	−0.50	0	−0.50	0	−0.50	0	−0.50	0	−1/2	0

LENGTH COEFFICIENTS

	K	K	K	K	K	K	K	K
Horizontal	0.875000	1.000000	1.125000	1.250000	1.375000	1.500000	1.625000	1/8 n
Diagonal	1.328768	1.414214	1.505199	1.600781	1.700184	1.802776	1.908042	1/8 N
Vertical	1.000000	1.000000	1.000000	1.000000	1.000000	1.000000	1.000000	1

ANGLES (IN DEGREES) BETWEEN MEMBERS

Between horizontal and diagonal members	49	45	42	39	36	34	32	$8/n = \tan a$
Between diagonal and vertical members	41	45	48	51	54	56	58	$90 - a$
Between vertical and horizontal members	90	90	90	90	90	90	90	90

Table 3-39. Modified Flat Pratt 10 Panels at Top & Bottom

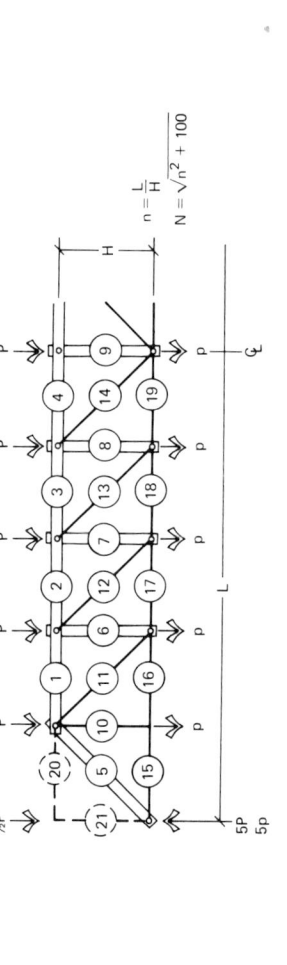

$n = \dfrac{L}{H}$

$N = \sqrt{n^2 + 100}$

Member	\multicolumn{14}{c	}{VALUES OF n — AXIAL FORCE COEFFICIENTS}	General Formulas	
	7		8		9		10		11		12		13		K_1	K_2
	K_1	K_2	K_1	K_2	K_1	K_2	K_1	K_2	K_1	K_2	K_1	K_2	K_1	K_2		
1	−5.60	−5.60	−6.40	−6.40	−7.20	−7.20	−8.00	−8.00	−8.80	−8.80	−9.60	−9.60	−10.40	−10.40	−4/5 n	−4/5 n
2	−7.35	−7.35	−8.40	−8.40	−9.45	−9.45	−10.50	−10.50	−11.55	−11.55	−12.60	−12.60	−13.65	−13.65	−21/20 n	−21/20 n
3	−8.40	−8.40	−9.60	−9.60	−10.80	−10.80	−12.00	−12.00	−13.20	−13.20	−14.40	−14.40	−15.60	−15.60	−6/5 n	−6/5 n
4	−8.75	−8.75	−10.00	−10.00	−11.25	−11.25	−12.50	−12.50	−13.75	−13.75	−15.00	−15.00	−16.25	−16.25	−5/4 n	−5/4 n
5	−5.49	−5.49	−5.76	−5.76	−6.05	−6.05	−6.36	−6.36	−6.69	−6.69	−7.03	−7.03	−7.38	−7.38	−9/20 N	−9/20 N
6	−3.50	−2.50	−3.50	−2.50	−3.50	−2.50	−3.50	−2.50	−3.50	−2.50	−3.50	−2.50	−3.50	−2.50	−7/2	−5/2
7	−2.50	−1.50	−2.50	−1.50	−2.50	−1.50	−2.50	−1.50	−2.50	−1.50	−2.50	−1.50	−2.50	−1.50	−5/2	−3/2
8	−1.50	−0.50	−1.50	−0.50	−1.50	−0.50	−1.50	−0.50	−1.50	−0.50	−1.50	−0.50	−1.50	−0.50	−3/2	−1/2
9	−1.00	0	−1.00	0	−1.00	0	−1.00	0	−1.00	0	−1.00	0	−1.00	0	−1	0
10	0	1.00	0	1.00	0	1.00	0	1.00	0	1.00	0	1.00	0	1.00	0	1
11	4.27	4.27	4.48	4.48	4.71	4.71	4.95	4.95	5.20	5.20	5.47	5.47	5.74	5.74	7/20 N	7/20 N
12	3.05	3.05	3.20	3.20	3.36	3.36	3.54	3.54	3.72	3.72	3.91	3.91	4.10	4.10	1/4 N	1/4 N

	13	14	15	16	17	18	19	20	21
	1.83	0.61	3.15	3.15	5.60	7.35	8.40	0	−0.50
	1.83	0.61	3.15	3.15	5.60	7.35	8.40	0	−0.50
	1.92	0.64	3.60	3.60	6.40	8.40	9.60	0	−0.50
	1.92	0.64	3.60	3.60	6.40	8.40	9.60	0	−0.50
	2.02	0.67	4.05	4.05	7.20	9.45	10.80	0	−0.50
	2.02	0.67	4.05	4.05	7.20	9.45	10.80	0	−0.50
	2.12	0.71	4.50	4.50	8.00	10.50	12.00	0	−0.50
	2.12	0.71	4.50	4.50	8.00	10.50	12.00	0	−0.50
	2.23	0.74	4.95	4.95	8.80	11.55	13.20	0	−0.50
	2.23	0.74	4.95	4.95	8.80	11.55	13.20	0	−0.50
	2.34	0.78	5.40	5.40	9.60	12.60	14.40	0	−0.50
	2.34	0.78	5.40	5.40	9.60	12.60	14.40	0	−0.50
	2.46	0.82	5.85	5.85	10.40	13.65	15.60	0	−0.50
	2.46	0.82	5.85	5.85	10.40	13.65	15.60	0	0
	$3/20\,N$	$1/20\,N$	$9/20\,n$	$9/20\,n$	$4/5\,n$	$21/20\,n$	$6/5\,n$	0	$-1/2$
	$3/20\,N$	$1/20\,N$	$9/20\,n$	$9/20\,n$	$4/5\,n$	$21/20\,n$	$6/5\,n$	0	0

LENGTH COEFFICIENTS

	K	K	K	K	K	K	K	K
Horizontal	0.700000	0.800000	0.900000	1.000000	1.100000	1.200000	1.300000	$1/10\,n$
Diagonal	1.220656	1.280625	1.345362	1.414214	1.486607	1.562050	1.640122	$1/10\,N$
Vertical	1.000000	1.000000	1.000000	1.000000	1.000000	1.000000	1.000000	1

ANGLES (IN DEGREES) BETWEEN MEMBERS

Between horizontal and diagonal members	55	51	48	45	42	40	38	$10/n = \tan a$
Between diagonal and vertical members	35	39	42	45	48	50	52	$90 - a$
Between vertical and horizontal members	90	90	90	90	90	90	90	90

Table 3-40. Modified Flat Pratt 12 Panels at Top & Bottom

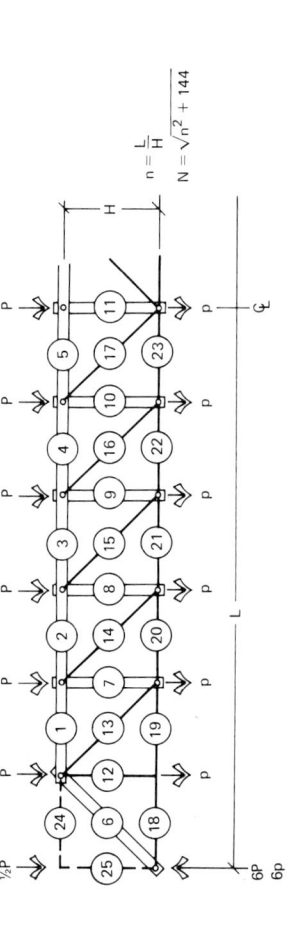

$n = \frac{L}{H}$
$N = \sqrt{n^2 + 144}$

Member	\multicolumn{10}{c}{VALUES OF n — AXIAL FORCE COEFFICIENTS}										General Formulas	
	7		8		9		10		11		12		13		K_1	K_2
	K_1	K_2	K_1	K_2	K_1	K_2	K_1	K_2	K_1	K_2	K_1	K_2	K_1	K_2		
1	−5.83	−5.83	−6.67	−6.67	−7.50	−7.50	−8.33	−8.33	−9.17	−9.17	−10.00	−10.00	−10.83	−10.83	−20/24 n	−20/24 n
2	−7.88	−7.88	−9.00	−9.00	−10.13	−10.13	−11.25	−11.25	−12.38	−12.38	−13.50	−13.50	−14.63	−14.63	−27/24 n	−27/24 n
3	−9.33	−9.33	−10.67	−10.67	−12.00	−12.00	−13.33	−13.33	−14.67	−14.67	−16.00	−16.00	−17.33	−17.33	−32/24 n	−32/24 n
4	−10.21	−10.21	−11.67	−11.67	−13.13	−13.13	−14.58	−14.58	−16.04	−16.04	−17.50	−17.50	−18.96	−18.96	−35/24 n	−35/24 n
5	−10.50	−10.50	−12.00	−12.00	−13.50	−13.50	−15.00	−15.00	−16.50	−16.50	−18.00	−18.00	−19.50	−19.50	−36/24 n	−36/24 n
6	−6.37	−6.37	−6.61	−6.61	−6.88	−6.88	−7.16	−7.16	−7.46	−7.46	−7.78	−7.78	−8.11	−8.11	−11/24 N	−11/24 N
7	−4.50	−3.50	−4.50	−3.50	−4.50	−3.50	−4.50	−3.50	−4.50	−3.50	−4.50	−3.50	−4.50	−3.50	−9/2	−7/2
8	−3.50	−2.50	−3.50	−2.50	−3.50	−2.50	−3.50	−2.50	−3.50	−2.50	−3.50	−2.50	−3.50	−2.50	−7/2	−5/2
9	−2.50	−1.50	−2.50	−1.50	−2.50	−1.50	−2.50	−1.50	−2.50	−1.50	−2.50	−1.50	−2.50	−1.50	−5/2	−3/2
10	−1.50	−0.50	−1.50	−0.50	−1.50	−0.50	−1.50	−0.50	−1.50	−0.50	−1.50	−0.50	−1.50	−0.50	−3/2	−1/2
11	−1.00	0	−1.00	0	−1.00	0	−1.00	0	−1.00	0	−1.00	0	−1.00	0	−1	0
12	0	1.00	0	1.00	0	1.00	0	1.00	0	1.00	0	1.00	0	1.00	0	1

13	5.21	5.21	5.41	5.41	5.63	5.63	5.86	5.86	6.10	6.10	6.36	6.36	6.63	6.63	9/24 N	9/24 N
14	4.05	4.05	4.21	4.21	4.38	4.38	4.56	4.56	4.75	4.75	4.95	4.95	5.16	5.16	7/24 N	7/24 N
15	2.89	2.89	3.00	3.00	3.13	3.13	3.25	3.25	3.39	3.39	3.54	3.54	3.69	3.69	5/24 N	5/24 N
16	1.74	1.74	1.80	1.80	1.88	1.88	1.95	1.95	2.03	2.03	2.12	2.12	2.21	2.21	3/24 N	3/24 N
17	0.58	0.58	0.60	0.60	0.63	0.63	0.65	0.65	0.68	0.68	0.71	0.71	0.74	0.74	1/24 N	1/24 N
18	3.21	3.21	3.67	3.67	4.13	4.13	4.58	4.58	5.04	5.04	5.50	5.50	5.96	5.96	11/24 n	11/24 n
19	3.21	3.21	3.67	3.67	4.13	4.13	4.58	4.58	5.04	5.04	5.50	5.50	5.96	5.96	11/24 n	11/24 n
20	5.83	5.83	6.67	6.67	7.50	7.50	8.33	8.33	9.17	9.17	10.00	10.00	10.83	10.83	20/24 n	20/24 n
21	7.88	7.88	9.00	9.00	10.13	10.13	11.25	11.25	12.38	12.38	13.50	13.50	14.63	14.63	27/24 n	27/24 n
22	9.33	9.33	10.67	10.67	12.00	12.00	13.33	13.33	14.67	14.67	16.00	16.00	17.33	17.33	32/24 n	32/24 n
23	10.21	10.21	11.67	11.67	13.13	13.13	14.58	14.58	16.04	16.04	17.50	17.50	18.96	18.96	35/24 n	35/24 n
24	0	0	0	0	0	0	0	0	0	0	0	0	0	0	0	0
25	−0.50	0	−0.50	0	−0.50	0	−0.50	0	−0.50	0	−0.50	0	−0.50	0	−1/2	0

LENGTH COEFFICIENTS

	K	K	K	K	K	K	K	K
Horizontal	0.583333	0.666667	0.750000	0.833333	0.916667	1.000000	1.083333	1/12 n
Diagonal	1.157704	1.201850	1.250000	1.301708	1.356568	1.414214	1.474317	1/12 N
Vertical	1.000000	1.000000	1.000000	1.000000	1.000000	1.000000	1.000000	1

ANGLES (IN DEGREES) BETWEEN MEMBERS

Between horizontal and diagonal members	60	56	53	50	47	45	43	12/n = tan a
Between diagonal and vertical members	30	34	37	40	43	45	47	90 − a
Between vertical and horizontal members	90	90	90	90	90	90	90	90

Table 3.41. Flat Warren 2 Panels at Bottom

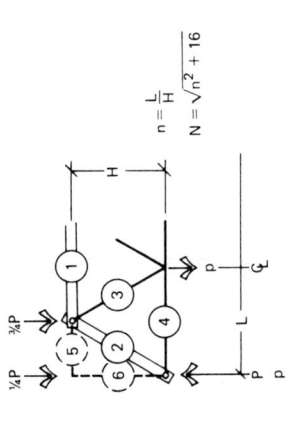

$n = \dfrac{L}{H}$
$N = \sqrt{n^2 + 16}$

Member	VALUES OF n													General Formulas	
	7		8		9		10		11		12		13		
	K_1	K_2	K_1	K_2	K_1	K_2	K_1	K_2	K_1	K_2	K_1	K_2	K_1	K_2	K_1	K_2
	AXIAL FORCE COEFFICIENTS															
1	-1.31	-1.75	-1.50	-2.00	-1.69	-2.25	-1.88	-2.50	-2.05	-2.75	-2.25	-3.00	-2.44	-3.25	-3/16 n	-1/4 n
2	-1.51	-1.01	-1.68	-1.12	-1.85	-1.23	-2.02	-1.35	-2.19	-1.46	-2.37	-1.58	-2.55	-1.70	-3/16 N	-1/8 N
3	0	1.01	0	1.12	0	1.23	0	1.35	0	1.46	0	1.58	0	1.70	0	1/8 N
4	1.31	0.88	1.50	1.00	1.69	1.13	1.88	1.25	2.06	1.38	2.25	1.50	2.44	1.63	3/16 n	1/8 n
5	0	0	0	0	0	0	0	0	0	0	0	0	0	0	0	0
6	-0.25	0	-0.25	0	-0.25	0	-0.25	0	-0.25	0	-0.25	0	-0.25	0	-1/4	0

LENGTH COEFFICIENTS

	K	K	K	K	K	K	K	K
1, 4	3.500000	4.000000	4.500000	5.000000	5.500000	6.000000	6.500000	$1/2\,n$
2, 3	2.015564	2.236068	2.462214	2.692582	2.926175	3.162278	3.400368	$1/4\,N$
5	1.750000	2.000000	2.250000	2.500000	2.750000	3.000000	3.250000	$1/4\,n$
6	1.000000	1.000000	1.000000	1.000000	1.000000	1.000000	1.000000	1

ANGLES (IN DEGREES) BETWEEN MEMBERS

Between horizontal and diagonal members	30	27	24	22	20	18	17	$4/n = \tan a$
Between diagonal members	120	126	132	136	140	144	146	$180 - 2a$
Between vertical and diagonal members	60	63	66	68	70	72	73	$90 - a$
Between vertical and horizontal members	90	90	90	90	90	90	90	90

Table 3.42. Flat Warren 3 Panels at Bottom

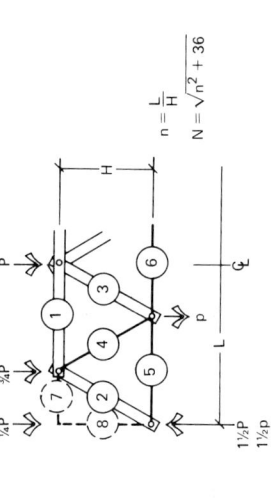

$n = \dfrac{L}{H}$

$N = \sqrt{n^2 + 36}$

Member	\multicolumn{14}{c}{VALUES OF n}	\multicolumn{2}{c}{General Formulas}
	7		8		9		10		11		12		13			
	K_1	K_2	K_1	K_2	K_1	K_2	K_1	K_2	K_1	K_2	K_1	K_2	K_1	K_2	K_1	K_2
							AXIAL FORCE COEFFICIENTS									
1	−2.04	−2.33	−2.33	−2.67	−2.63	−3.00	−2.92	−3.33	−3.21	−3.67	−3.50	−4.00	−3.79	−4.33	−7/24 n	−1/3 n
2	−1.92	−1.54	−2.08	−1.67	−2.25	−1.80	−2.43	−1.94	−2.61	−2.09	−2.80	−2.24	−2.98	−2.39	−5/24 N	−1/6 N
3	−0.77	0	−0.83	0	−0.90	0	−0.97	0	−1.04	0	−1.12	0	−1.19	0	−1/12 N	0
4	0.77	1.54	0.83	1.67	0.90	1.80	0.97	1.94	1.04	2.09	1.12	2.24	1.19	2.39	1/12 N	1/6 N
5	1.46	1.17	1.67	1.33	1.88	1.50	2.08	1.67	2.29	1.83	2.50	2.00	2.71	2.17	5/24 n	1/6 n
6	2.63	2.33	3.00	2.67	3.38	3.00	3.75	3.33	4.13	3.67	4.50	4.00	4.88	4.33	3/8 n	1/3 n
7	0	0	0	0	0	0	0	0	0	0	0	0	0	0	0	0
8	−0.25	0	−0.25	0	−0.25	0	−0.25	0	−0.25	0	−0.25	0	−0.25	0	−1/4	0

LENGTH COEFFICIENTS

	K	K	K	K	K	K	K	K
1, 5, 6	2.333333	2.666667	3.000000	3.333333	3.666667	4.000000	4.333333	1/3 n
2, 3, 4	1.536591	1.666667	1.802776	1.943651	2.083327	2.236068	2.386304	1/6 N
7	1.166667	1.333333	1.500000	1.666667	1.833333	2.000000	2.166667	1/6 n
8	1.000000	1.000000	1.000000	1.000000	1.000000	1.000000	1.000000	1

ANGLES (IN DEGREES) BETWEEN MEMBERS

	K	K	K	K	K	K	K	K
Between horizontal and diagonal members	41	37	34	31	29	27	25	$6/n = \tan a$
Between diagonal members	98	106	112	118	122	126	130	$180 - 2a$
Between vertical and diagonal members	49	53	56	59	61	63	65	$90 - a$
Between vertical and horizontal members	90	90	90	90	90	90	90	90

Table 3.43. Flat Warren 4 Panels at Bottom

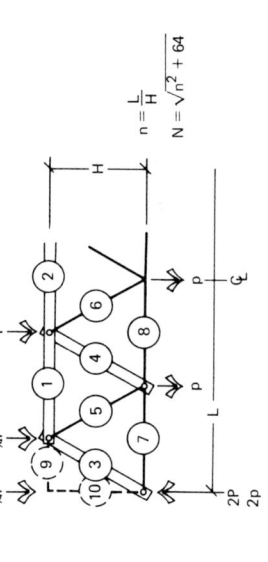

$n = \dfrac{L}{H}$

$N = \sqrt{n^2 + 64}$

VALUES OF n
AXIAL FORCE COEFFICIENTS

Member	7		8		9		10		11		12		13		General Formulas	
	K_1	K_2	K_1	K_2	K_1	K_2	K_1	K_2	K_1	K_2	K_1	K_2	K_1	K_2	K_1	K_2
1	-2.41	-2.63	-2.75	-3.00	-3.09	-3.38	-3.44	-3.75	-3.78	-4.13	-4.13	-4.50	-4.47	-4.88	-11/32 n	-3/8 n
2	-3.28	-3.50	-3.75	-4.00	-4.22	-4.50	-4.69	-5.00	-5.16	-5.50	-5.68	-6.00	-6.09	-6.50	-15/32 n	-1/2 n
3	-2.33	-1.99	-2.47	-2.12	-2.63	-2.26	-2.80	-2.40	-2.98	-2.55	-3.15	-2.70	-3.34	-2.86	-7/32 N	-3/16 N
4	-1.33	-0.66	-1.41	-0.71	-1.51	-0.75	-1.60	-0.80	-1.70	-0.85	-1.80	-0.90	-1.91	-0.95	-1/8 N	-1/16 N
5	1.33	1.99	1.41	2.12	1.51	2.26	1.60	2.40	1.70	2.55	1.80	2.70	1.91	2.86	1/8 N	3/16 N
6	0	0.66	0	0.71	0	0.75	0	0.80	0	0.85	0	0.90	0	0.95	0	1/16 N
7	1.53	1.31	1.75	1.50	1.97	1.69	2.19	1.88	2.41	2.06	2.63	2.25	2.84	2.44	7/32 n	3/16 n
8	3.28	3.06	3.75	3.50	4.22	3.94	4.69	4.38	5.16	4.81	5.63	5.25	6.09	5.69	15/32 n	7/16 n
9	0	0	0	0	0	0	0	0	0	0	0	0	0	0	0	0
10	-0.25	0	-0.25	0	-0.25	0	-0.25	0	-0.25	0	-0.25	0	-0.25	0	-1/4	0

LENGTH COEFFICIENTS

	K	K	K	K	K	K	K	K
1, 2, 7, 8	1.750000	2.000000	2.250000	2.500000	2.275000	3.000000	3.250000	1/4 n
3, 4, 5, 6	1.328768	1.414214	1.505199	1.600781	1.700184	1.802776	1.908042	1/8 N
9	0.875000	1.000000	1.125000	1.250000	1.375000	1.500000	1.625000	1/8 n
10	1.000000	1.000000	1.000000	1.000000	1.000000	1.000000	1.000000	1

ANGLES (IN DEGREES) BETWEEN MEMBERS

Between horizontal and diagonal members	49	45	42	39	36	34	32	8/n = tan a
Between diagonal members	82	90	96	102	106	112	116	180 − 2a
Between vertical and diagonal members	41	45	48	51	54	56	58	90 − a
Between vertical and horizontal members	90	90	90	90	90	90	90	90

Table 3.44. Flat Warren 5 Panels at Bottom

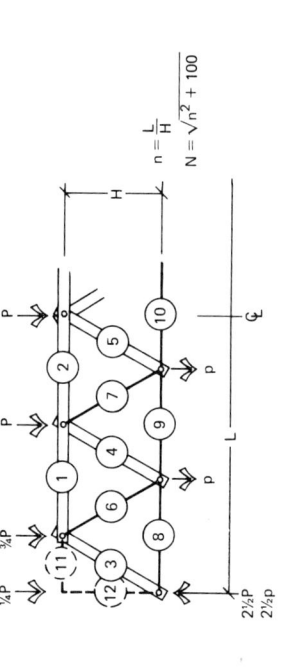

Member	VALUES OF n													General Formulas	
	7		8		9		10		11		12		13		
	K_1	K_2	K_1	K_2	K_1	K_2	K_1	K_2	K_1	K_2	K_1	K_2	K_1	K_2	K_1	K_2
1	-2.63	-2.80	-3.00	-3.20	-3.38	-3.60	-3.75	-4.00	-4.13	-4.40	-4.50	-4.80	-4.88	-5.20	$-15/40\,n$	$-16/40\,n$
2	-4.03	-4.20	-4.60	-4.80	-5.18	-5.40	-5.75	-6.00	-6.33	-6.60	-6.90	-7.20	-7.48	-7.80	$-23/40\,n$	$-24/40\,n$
3	-2.75	-2.44	-2.88	-2.56	-3.03	-2.69	-3.18	-2.83	-3.35	-2.97	-3.52	-3.12	-3.69	-3.28	$-9/40\,N$	$-8/40\,N$
4	-1.83	-1.22	-1.92	-1.28	-2.02	-1.35	-2.12	-1.41	-2.23	-1.49	-2.34	-1.56	-2.46	-1.64	$-6/40\,N$	$-4/40\,N$
5	-0.61	0	-0.64	0	-0.67	0	-0.71	0	-0.74	0	-0.78	0	-0.82	0	$-2/40\,N$	0
6	1.83	2.44	1.92	2.56	2.02	2.69	2.12	2.83	2.23	2.97	2.34	3.12	2.46	3.28	$6/40\,N$	$8/40\,N$
7	0.61	1.22	0.64	1.28	0.67	1.35	0.71	1.41	0.74	1.49	0.78	1.56	0.82	1.64	$2/40\,N$	$4/40\,N$
8	1.58	1.40	1.80	1.60	2.03	1.80	2.25	2.00	2.48	2.20	2.70	2.40	2.93	2.60	$9/40\,n$	$8/40\,n$
9	3.68	3.50	4.20	4.00	4.73	4.50	5.25	5.00	5.78	5.50	6.30	6.00	6.83	6.50	$21/40\,n$	$20/40\,n$
10	4.38	4.20	5.00	4.80	5.63	5.40	6.25	6.00	6.88	6.60	7.50	7.20	8.13	7.80	$25/40\,n$	$24/40\,n$
11	0	0	0	0	0	0	0	0	0	0	0	0	0	0	0	0
12	-0.25	0	-0.25	0	-0.25	0	-0.25	0	-0.25	0	-0.25	0	-0.25	0	$-1/4$	0

AXIAL FORCE COEFFICIENTS

$n = \dfrac{L}{H}$

$N = \sqrt{n^2 + 100}$

LENGTH COEFFICIENTS

	K	K	K	K	K	K	K	K
1, 2, 8, 9, 10	1.400000	1.600000	1.800000	2.000000	2.200000	2.400000	2.600000	1/5 n
3, 4, 5, 6, 7	1.220656	1.280625	1.345362	1.414214	1.486607	1.562050	1.640122	1/10 N
11	0.700000	0.800000	0.900000	1.000000	1.100000	1.200000	1.300000	1/10 n
12	1.000000	1.000000	1.000000	1.000000	1.000000	1.000000	1.000000	1

ANGLES (IN DEGREES) BETWEEN MEMBERS

Between horizontal and diagonal members	55	51	48	45	42	40	38	$10/n = \tan a$
Between diagonal members	70	78	84	90	96	100	104	$180 - 2a$
Between vertical and diagonal members	35	39	42	45	48	50	52	$90 - a$
Between vertical and horizontal members	90	90	90	90	90	90	90	90

Table 3.45. Flat Warren 6 Panels at Bottom

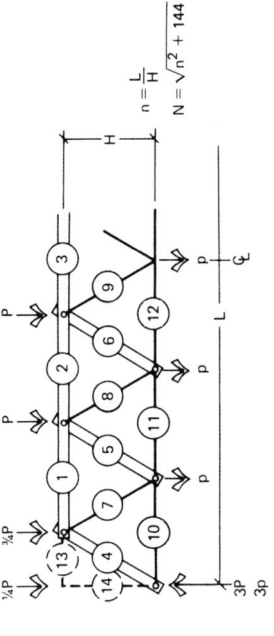

$n = \dfrac{L}{H}$

$N = \sqrt{n^2 + 144}$

| Member | \multicolumn{12}{c|}{VALUES OF n} | \multicolumn{2}{c|}{General Formulas} |
| | 7 | | 8 | | 9 | | 10 | | 11 | | 12 | | 13 | | | |
	K_1	K_2	K_1	K_2	K_1	K_2	K_1	K_2	K_1	K_2	K_1	K_2	K_1	K_2	K_1	K_2
1	−2.77	−2.92	−3.17	−3.33	−3.56	−3.75	−3.96	−4.17	−4.35	−4.58	−4.75	−5.00	−5.15	−5.42	−19/48 n	−20/48 n
2	−4.52	−4.67	−5.17	−5.33	−5.81	−6.00	−6.46	−6.67	−7.10	−7.33	−7.75	−8.00	−8.40	−8.67	−31/48 n	−32/48 n
3	−5.10	−5.25	−5.83	−6.00	−6.56	−6.75	−7.29	−7.50	−8.02	−8.25	−8.75	−9.00	−9.48	−9.75	−35/48 n	−36/48 n
4	−3.18	−2.89	−3.31	−3.00	−3.44	−3.13	−3.58	−3.25	−3.73	−3.39	−3.89	−3.54	−4.05	−3.69	−11/48 N	−10/48 N
5	−2.32	−1.74	−2.40	−1.80	−2.50	−1.88	−2.60	−1.95	−2.71	−2.03	−2.83	−2.12	−2.95	−2.21	−8/48 N	−6/48 N
6	−1.16	−0.58	−1.20	−0.60	−1.25	−0.63	−1.30	−0.65	−1.36	−0.68	−1.41	−0.71	−1.47	−0.74	−4/48 N	−2/48 N
7	2.32	2.89	2.40	3.00	2.50	3.13	2.60	3.25	2.71	3.39	2.83	3.54	2.95	3.69	8/48 N	10/48 N
8	1.16	1.74	1.20	1.80	1.25	1.88	1.30	1.95	1.36	2.03	1.41	2.12	1.47	2.21	4/48 N	6/48 N
9	0	0.58	0	0.60	0	0.63	0	0.65	0	0.68	0	0.71	0	0.74	0	2/48 N
10	1.60	1.46	1.83	1.67	2.06	1.87	2.29	2.08	2.52	2.29	2.75	2.50	2.98	2.71	11/48 n	10/48 n
11	3.94	3.79	4.50	4.33	5.06	4.87	5.62	5.42	6.19	5.96	6.75	6.50	7.31	7.04	27/48 n	26/48 n

AXIAL FORCE COEFFICIENTS

	5.10	4.96	5.83	5.67	6.56	6.37	7.29	7.08	8.02	7.79	8.75	8.50	9.48	9.21	35/48 n	34/48 n
12	0	0	0	0	0	0	0	0	0	0	0	0	0	0	0	0
13	0	0	0	0	0	0	0	0	0	0	0	0	0	0	0	0
14	−0.25	0	−0.25	0	−0.25	0	−0.25	0	−0.25	0	−0.25	0	−0.25	0	−1/4	0

LENGTH COEFFICIENTS

	5.10	5.83	6.56	7.29	8.02	8.75	9.48	35/48 n
	K	K	K	K	K	K	K	K
1, 2, 3, 10, 11, 12	1.166667	1.333333	1.500000	1.666667	1.833333	2.000000	2.166667	1/6 n
4, 5, 6, 7, 8, 9	1.157704	1.201850	1.250000	1.301708	1.356568	1.414214	1.474317	1/12 N
13	0.583333	0.666667	0.750000	0.833333	0.916667	1.000000	1.083333	1/12 n
14	1.000000	1.000000	1.000000	1.000000	1.000000	1.000000	1.000000	1

ANGLES (IN DEGREES) BETWEEN MEMBERS

	5.10	5.83	6.56	7.29	8.02	8.75	9.48	35/48 n
Between horizontal and diagonal members	60	56	53	50	47	45	43	12/n = tan a
Between diagonal members	60	68	74	80	86	90	94	180 − 2a
Between vertical and diagonal members	30	34	37	40	43	45	47	90 − a
Between vertical and horizontal members	90	90	90	90	90	90	90	90

Table 3.46. Flat Warren 7 Panels at Bottom

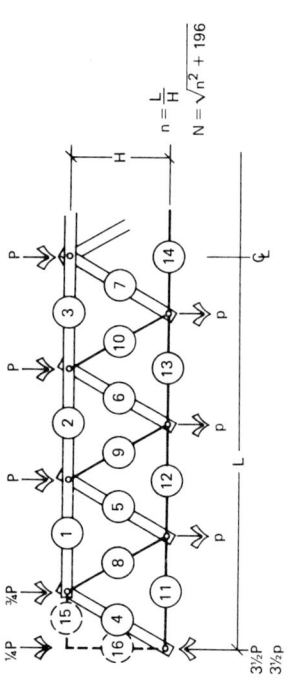

$n = \dfrac{L}{H}$

$N = \sqrt{n^2 + 196}$

Member	VALUES OF n														General Formulas	
	7		8		9		10		11		12		13			
	K_1	K_2	K_1	K_2	K_1	K_2	K_1	K_2	K_1	K_2	K_1	K_2	K_1	K_2	K_1	K_2
1	−2.88	−3.00	−3.29	−3.43	−3.70	−3.86	−4.11	−4.29	−4.52	−4.71	−4.93	−5.14	−5.34	−5.57	−23/56 n	−24/56 n
2	−4.88	−5.00	−5.57	−5.71	−6.27	−6.43	−6.96	−7.14	−7.66	−7.86	−8.36	−8.57	−9.05	−9.29	−39/56 n	−40/56 n
3	−5.88	−6.00	−6.71	−6.86	−7.55	−7.71	−8.39	−8.57	−9.23	−9.43	−10.07	−10.29	−10.91	−11.14	−47/56 n	−48/56 n
4	−3.63	−3.35	−3.74	−3.46	−3.86	−3.57	−3.99	−3.69	−4.13	−3.82	−4.28	−3.95	−4.44	−4.09	−13/56 N	−12/56 N
5	−2.80	−2.24	−2.88	−2.30	−2.97	−2.38	−3.07	−2.46	−3.18	−2.54	−3.29	−2.63	−3.41	−2.73	−10/56 N	−8/56 N
6	−1.68	−1.12	−1.73	−1.15	−1.78	−1.19	−1.84	−1.23	−1.91	−1.27	−1.98	−1.32	−2.05	−1.36	−6/56 N	−4/56 N
7	−0.56	0	−0.58	0	−0.59	0	−0.61	0	−0.64	0	−0.66	0	−0.68	0	−2/56 N	0
8	2.80	3.35	2.88	3.46	2.97	3.57	3.07	3.69	3.18	3.82	3.29	3.95	3.41	4.09	10/56 N	12/56 N
9	1.68	2.24	1.73	2.30	1.78	2.38	1.84	2.46	1.91	2.54	1.98	2.63	2.05	2.73	6/56 N	8/56 N
10	0.56	1.12	0.58	1.15	0.59	1.19	0.61	1.23	0.64	1.27	0.66	1.32	0.68	1.36	2/56 N	4/56 N
11	1.63	1.50	1.86	1.71	2.09	1.93	2.32	2.14	2.55	2.36	2.79	2.57	3.02	2.79	13/56 n	12/56 n
12	4.13	4.00	4.71	4.57	5.30	5.14	5.89	5.71	6.48	6.29	7.07	6.86	7.66	7.43	33/56 n	32/56 n

AXIAL FORCE COEFFICIENTS

13	5.63	5.50	6.43	6.29	7.23	7.07	8.04	7.86	8.84	8.64	9.64	9.43	10.45	10.21	45/56 n	44/56 n
14	6.13	6.00	7.00	6.86	7.88	7.71	8.75	8.57	9.63	9.43	10.50	10.29	11.38	11.14	49/56 n	48/56 n
15	0	0	0	0	0	0	0	0	0	0	0	0	0	0	0	0
16	−0.25	0	0	0	−0.25	0	−0.25	0	−0.25	0	−0.25	0	−0.25	0	−1/4	0

LENGTH COEFFICIENTS

	K	K	K	K	K	K	K	K
1, 2, 3, 11, 12, 13, 14	1.000000	1.142857	1.285714	1.428571	1.571429	1.714286	1.857143	1/7 n
4, 5, 6, 7, 8, 9, 10	1.118034	1.151751	1.188808	1.228904	1.271750	1.317078	1.364641	1/14 N
15	0.500000	0.571429	0.642857	0.714286	0.785714	0.857143	0.928571	1/14 n
16	1.000000	1.000000	1.000000	1.000000	1.000000	1.000000	1.000000	1

ANGLES (IN DEGREES) BETWEEN MEMBERS

Between horizontal and diagonal members	63	60	57	54	52	49	47	14/n = tan a
Between diagonal members	54	60	66	72	76	82	86	180 − 2a
Between diagonal and vertical members	27	30	33	36	38	41	43	90 − a
Between vertical and horizontal members	90	90	90	90	90	90	90	90

Table 3.47. Flat Warren 8 Panels at Bottom

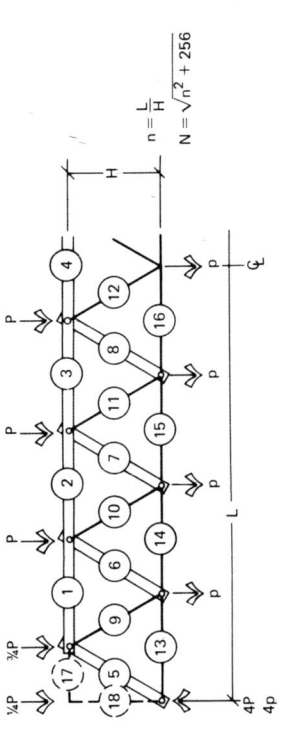

$n = \dfrac{L}{H}$

$N = \sqrt{n^2 + 256}$

Member	VALUES OF n													General Formulas	
	7		8		9		10		11		12		13		
	K_1	K_2	K_1	K_2	K_1	K_2	K_1	K_2	K_1	K_2	K_1	K_2	K_1	K_2	K_1	K_2
	AXIAL FORCE COEFFICIENTS															
1	-2.95	-3.06	-3.38	-3.50	-3.80	-3.94	-4.22	-4.38	-4.64	-4.81	-5.06	-5.25	-5.48	-5.69	-27/64 n	-28/64 n
2	-5.14	-5.25	-5.88	-6.00	-6.61	-6.75	-7.34	-7.50	-8.08	-8.25	-8.81	-9.00	-9.55	-9.75	-47/64 n	-48/64 n
3	-6.45	-6.56	-7.38	-7.50	-8.30	-8.44	-9.22	-9.38	-10.14	-10.31	-11.06	-11.25	-11.98	-12.19	-59/64 n	-60/64 n
4	-6.89	-7.00	-7.88	-8.00	-8.86	-9.00	-9.84	-10.00	-10.83	-11.00	-11.81	-12.00	-12.80	-13.00	-63/64 n	-64/64 n
5	-4.09	-3.82	-4.19	-3.91	-4.30	-4.02	-4.42	-4.13	-4.55	-4.25	-4.69	-4.38	-4.83	-4.51	-15/64 N	-14/64 N
6	-3.27	-2.73	-3.35	-2.80	-3.44	-2.87	-3.54	-2.95	-3.64	-3.03	-3.75	-3.13	-3.87	-3.22	-12/64 N	-10/64 N
7	-2.18	-1.64	-2.24	-1.68	-2.29	-1.72	-2.36	-1.77	-2.43	-1.82	-2.50	-1.88	-2.58	-1.93	-8/64 N	-6/64 N
8	-1.09	-0.55	-1.12	-0.56	-1.15	-0.57	-1.18	-0.59	-1.21	-0.61	-1.25	-0.63	-1.29	-0.64	-4/64 N	-2/64 N
9	3.27	3.82	3.35	3.91	3.44	4.02	3.54	4.13	3.64	4.25	3.75	4.38	3.87	4.51	12/64 N	14/64 N
10	2.18	2.73	2.24	2.80	2.29	2.87	2.36	2.95	2.43	3.03	2.50	3.13	2.58	3.22	8/64 N	10/64 N
11	1.09	1.64	1.12	1.68	1.15	1.72	1.18	1.77	1.21	1.82	1.25	1.88	1.29	1.93	4/64 N	6/64 N
12	0	0.55	0	0.56	0	0.57	0	0.59	0	0.61	0	0.63	0	0.64	0	2/64 N
13	1.64	1.53	1.88	1.75	2.11	1.97	2.34	2.19	2.58	2.41	2.81	2.63	3.05	2.84	15/64 n	14/64 n

14	4.27	4.16	4.88	4.75	5.48	5.34	6.09	5.94	6.70	6.53	7.31	7.13	7.92	7.72	39/64 n	38/64 n
15	6.02	5.91	6.88	6.75	7.73	7.59	8.59	8.44	9.45	9.28	10.31	10.13	11.17	10.97	55/64 n	54/64 n
16	6.89	6.78	7.88	7.75	8.86	8.72	9.84	9.69	10.83	10.66	11.81	11.63	12.80	12.59	63/64 n	62/64 n
17	0	0	0	0	0	0	0	0	0	0	0	0	0	0	0	0
18	−0.25	0	−0.25	0	−0.25	0	−0.25	0	−0.25	0	−0.25	0	−0.25	0	−1/4	0

LENGTH COEFFICIENTS

	K	K	K	K	K	K	K	K
1, 2, 3, 4, 13, 14, 15, 16	0.875000	1.000000	1.125000	1.250000	1.375000	1.500000	1.625000	$1/8\,n$
5, 6, 7, 8, 9, 10, 11, 12	1.091516	1.118034	1.147347	1.179248	1.213530	1.250000	1.288471	$1/16\,N$
17	0.437500	0.500000	0.562500	0.625000	0.687500	0.750000	0.812500	$1/16\,n$
18	1.000000	1.000000	1.000000	1.000000	1.000000	1.000000	1.000000	1

ANGLES (IN DEGREES) BETWEEN MEMBERS

Between horizontal and diagonal members	66	63	61	58	55	53	51	$16/n = \tan a$
Between diagonal members	48	54	58	64	70	74	78	$180 - 2a$
Between diagonal and vertical members	24	27	29	32	35	37	39	$90 - a$
Between vertical and horizontal members	90	90	90	90	90	90	90	90

Table 3.48. Flat Warren 9 Panels at Bottom

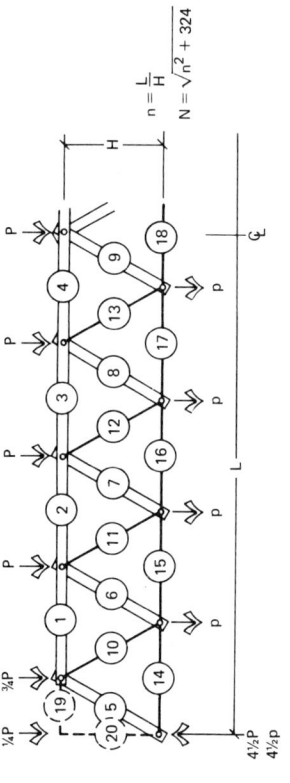

$n = \dfrac{L}{H}$
$N = \sqrt{n^2 + 324}$

VALUES OF n
AXIAL FORCE COEFFICIENTS

Member	7		8		9		10		11		12		13		General Formulas	
	K_1	K_2	K_1	K_2	K_1	K_2	K_1	K_2	K_1	K_2	K_1	K_2	K_1	K_2	K_1	K_2
1	-3.01	-3.11	-3.44	-3.56	-3.88	-4.00	-4.31	-4.44	-4.74	-4.89	-5.17	-5.33	-5.60	-5.78	$-31/72\,n$	$-32/72\,n$
2	-5.35	-5.44	-6.11	-6.22	-6.88	-7.00	-7.64	-7.78	-8.40	-8.56	-9.17	-9.33	-9.93	-10.11	$-55/72\,n$	$-56/72\,n$
3	-6.90	-7.00	-7.89	-8.00	-8.88	-9.00	-9.86	-10.00	-10.85	-11.00	-11.83	-12.00	-12.82	-13.00	$-71/72\,n$	$-72/72\,n$
4	-7.68	-7.78	-8.78	-8.89	-9.88	-10.00	-10.97	-11.11	-12.07	-12.22	-13.17	-13.33	-14.26	-14.44	$-79/72\,n$	$-80/72\,n$
5	-4.56	-4.29	-4.65	-4.38	-4.75	-4.47	-4.86	-4.58	-4.98	-4.69	-5.11	-4.81	-5.24	-4.93	$-17/72\,N$	$-16/72\,N$
6	-3.76	-3.22	-3.83	-3.28	-3.91	-3.35	-4.00	-3.43	-4.10	-3.52	-4.21	-3.61	-4.32	-3.70	$-14/72\,N$	$-12/72\,N$
7	-2.68	-2.15	-2.74	-2.19	-2.80	-2.24	-2.86	-2.29	-2.93	-2.34	-3.00	-2.40	-3.08	-2.47	$-10/72\,N$	$-8/72\,N$
8	-1.61	-1.07	-1.64	-1.09	-1.68	-1.12	-1.72	-1.14	-1.76	-1.17	-1.80	-1.20	-1.85	-1.23	$-6/72\,N$	$-4/72\,N$
9	-0.54	0	-0.55	0	-0.56	0	-0.57	0	-0.59	0	-0.60	0	-0.62	0	$-2/72\,N$	0
10	3.76	4.29	3.83	4.38	3.91	4.47	4.00	4.58	4.10	4.69	4.21	4.81	4.32	4.93	$14/72\,N$	$16/72\,N$
11	2.68	3.22	2.74	3.28	2.80	3.35	2.86	3.43	2.93	3.52	3.00	3.61	3.08	3.70	$10/72\,N$	$12/72\,N$
12	1.61	2.15	1.64	2.19	1.68	2.24	1.72	2.29	1.76	2.34	1.80	2.40	1.85	2.47	$6/72\,N$	$8/72\,N$

LENGTH COEFFICIENTS

	0.54	1.07	0.55	1.09	0.56	1.12	0.57	1.14	0.59	1.17	0.60	1.20	0.62	1.23	2/72 N	4/72 N
13	1.65	1.56	1.89	1.78	2.13	2.00	2.36	2.22	2.60	2.44	2.83	2.67	3.07	2.89	17/72 n	16/72 n
14	4.38	4.28	5.00	4.89	5.63	5.50	6.25	6.11	6.88	6.72	7.50	7.33	8.13	7.94	45/72 n	44/72 n
16	6.32	6.22	7.22	7.11	8.13	8.00	9.03	8.89	9.93	9.78	10.83	10.67	11.74	11.56	65/72 n	64/72 n
17	7.49	7.39	8.56	8.44	9.63	9.50	10.69	10.56	11.76	11.61	12.83	12.67	13.90	13.72	77/72 n	76/72 n
18	7.88	7.78	9.00	8.89	10.13	10.00	11.25	11.11	12.38	12.22	13.50	13.33	14.63	14.44	81/72 n	80/72 n
19	0	0	0	0	0	0	0	0	0	0	0	0	0	0	0	0
20	−0.25	0	−0.25	0	−0.25	0	−0.25	0	−0.25	0	−0.25	0	−0.25	0	−1/4	0

	K	K	K	K	K	K	K	K	K
1, 2, 3, 4, 14, 15, 16, 17, 18	0.777778	0.888889	1.000000	1.111111	1.222222	1.333333	1.444444		1/9 n
5, 6, 7, 8, 9	1.072956	1.094318	1.118034	1.143959	1.171946	1.201850	1.233534		1/18 N
10, 11, 12, 13	0.388889	0.444444	0.500000	0.555556	0.611111	0.666667	0.722222		1/18 n
19	1.000000	1.000000	1.000000	1.000000	1.000000	1.000000	1.000000		1
20									

ANGLES (IN DEGREES) BETWEEN MEMBERS

	K	K	K	K	K	K	K	K
Between horizontal and diagonal members	69	66	63	61	59	56	54	18/n = tan a
Between diagonal members	42	48	54	58	62	68	72	180 − 2a
Between diagonal and vertical members	21	24	27	29	31	34	36	90 − a
Between vertical and horizontal members	90	90	90	90	90	90	90	90

Table 3-49. Flat Warren 10 Panels at Bottom

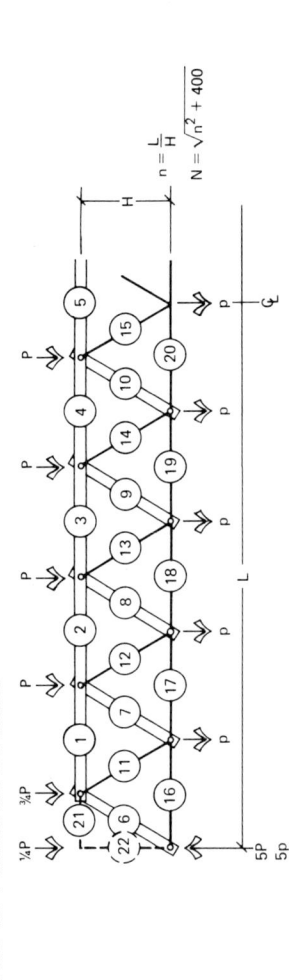

$n = \dfrac{L}{H}$
$N = \sqrt{n^2 + 400}$

AXIAL FORCE COEFFICIENTS

| Member | \multicolumn{14}{c|}{VALUES OF n} | \multicolumn{2}{c|}{General Formulas} |
| | 7 | | 8 | | 9 | | 10 | | 11 | | 12 | | 13 | | | |
	K_1	K_2	K_1	K_2	K_1	K_2	K_1	K_2	K_1	K_2	K_1	K_2	K_1	K_2	K_1	K_2
1	−3.06	−3.15	−3.50	−3.60	−3.94	−4.05	−4.38	−4.50	−4.81	−4.95	−5.25	−5.40	−5.69	−5.85	−35/80 n	−36/80 n
2	−5.51	−5.60	−6.30	−6.40	−7.09	−7.20	−7.88	−8.00	−8.66	−8.80	−9.45	−9.60	−10.24	−10.40	−63/80 n	−64/80 n
3	−7.26	−7.35	−8.30	−8.40	−9.34	−9.45	−10.38	−10.50	−11.41	−11.55	−12.45	−12.60	−13.49	−13.65	−83/80 n	−84/80 n
4	−8.31	−8.40	−9.50	−9.60	−10.69	−10.80	−11.88	−12.00	−13.06	−13.20	−14.25	−14.40	−15.44	−15.60	−95/80 n	−96/80 n
5	−8.66	−8.75	−9.90	−10.00	−11.14	−11.25	−12.38	−12.50	−13.61	−13.75	−14.85	−15.00	−16.09	−16.25	−99/80 N	−100/80 N
6	−5.03	−4.77	−5.12	−4.85	−5.21	−4.93	−5.31	−5.03	−5.42	−5.14	−5.54	−5.25	−5.67	−5.37	−19/80 N	−18/80 N
7	−4.24	−3.71	−4.31	−3.77	−4.39	−3.84	−4.47	−3.91	−4.57	−3.99	−4.66	−4.08	−4.77	−4.17	−16/80 N	−14/80 N
8	−3.18	−2.65	−3.23	−2.69	−3.29	−2.74	−3.35	−2.80	−3.42	−2.85	−3.50	−2.92	−3.58	−2.98	−12/80 N	−10/80 N
9	−2.12	−1.59	−2.15	−1.62	−2.19	−1.64	−2.24	−1.68	−2.28	−1.71	−2.33	−1.75	−2.39	−1.79	−8/80 N	−6/80 N
10	−1.06	−0.53	−1.08	−0.54	−1.10	−0.55	−1.12	−0.56	−1.14	−0.57	−1.17	−0.58	−1.19	−0.60	−4/80 N	−2/80 N
11	4.24	4.77	4.31	4.85	4.39	4.93	4.47	5.03	4.57	5.14	4.66	5.25	4.77	5.37	16/80 N	18/80 N
12	3.18	3.71	3.23	3.77	3.29	3.84	3.35	3.91	3.42	3.99	3.50	4.08	3.58	4.17	12/80 N	14/80 N
13	2.12	2.65	2.15	2.69	2.19	2.74	2.24	2.80	2.28	2.85	2.33	2.92	2.39	2.98	8/80 N	10/80 N
14	1.06	1.59	1.08	1.62	1.10	1.64	1.12	1.68	1.14	1.71	1.17	1.75	1.19	1.79	4/80 N	6/80 N

	0	0.53	0	0.54	0	0.55	0	0.56	0	0.57	0	0.58	0	0.60	0	
15	0	.	0	0	0	0	0	0	0	0	0	0	0	0	0	2/80 N
16	1.66	1.58	1.90	1.80	2.14	2.03	2.38	2.25	2.61	2.48	2.85	2.70	3.09	2.93	19/80 n	18/80 n
17	4.46	4.38	5.10	5.00	5.74	5.63	6.38	6.25	7.01	6.88	7.65	7.50	8.29	8.13	51/80 n	50/80 n
18	6.56	6.48	7.50	7.40	8.44	8.33	9.38	9.25	10.31	10.18	11.25	11.10	12.19	12.03	75/80 n	74/80 n
19	7.96	7.88	9.10	9.00	10.24	10.13	11.38	11.25	12.51	12.38	13.65	13.50	14.79	14.63	91/80 n	90/80 n
20	8.66	8.58	9.90	9.80	11.14	11.03	12.38	12.25	13.61	13.48	14.85	14.70	16.09	15.93	99/80 n	98/80 n
21	0	0	0	0	0	0	0	0	0	0	0	0	0	0	0	0
22	-0.25	0	-0.25	0	-0.25	0	-0.25	0	-0.25	0	-0.25	0	-0.25	0	-1/4	0

LENGTH COEFFICIENTS

	K	K	K	K	K	K	K	K
1, 2, 3, 4, 5, 16, 17, 18, 19, 20 6, 7, 8, 9, 10, 11, 12, 13	0.700000	0.800000	0.900000	1.000000	1.100000	1.200000	1.300000	1/10 n
14, 15	1.059481	1.077033	1.096586	1.118034	1.141271	1.166190	1.192686	1/20 N
21	0.350000	0.400000	0.450000	0.500000	0.550000	0.600000	0.650000	1/20 n
22	1.000000	1.000000	1.000000	1.000000	1.000000	1.000000	1.000000	1

ANGLES (IN DEGREES) BETWEEN MEMBERS

Between horizontal and diagonal members	71	68	66	63	61	59	57	20/n = tan a
Between diagonal members	38	44	48	54	58	62	66	180 − 2a
Between diagonal and vertical members	19	22	24	27	29	31	33	90 − a
Between vertical and horizontal members	90	90	90	90	90	90	90	90

4
Analysis by Computer

For the designer working with trusses, a computer is a medium of analysis which he should not ignore or fear, since the beast has been tamed for the non-initiated. A major advantage in using prepackaged computer programs in truss analysis is that computer programs pose practically no limits concerning the complexity of the truss including statically determinate or statically indeterminate types with any degree of indeterminacy. Any type of loading conditions, including temperature changes, settlements of the supports, etc., can easily be considered. However, because computer programs often require special expertise and training, the designer may shy away from using them. This chapter examines how STRUDL II may be applied to truss analysis without any special computer training, and thus without the complications that conventional programming may cause.

STRUDL II is a wide-range computer program applicable to a large spectrum of structures in civil engineering in addition to plane trusses. Such a program, easily available through IBM representatives around the world, is particularly helpful because once the user has familiarized himself with it, he can solve almost any structural problem without having to switch to another program. Even more significant is the fact that the program does not require the user to have any familiarity with computers, computer programming, or methods of analysis. STRUDL II was produced at M.I.T. It was based on a unique language (ICETRAN) that allows the designer to interact with the computer on the same basis he would with conventional calculations.

The basic operations for the solution of problems using STRUDL are as follows.

1. Drawing the truss with its geometrical characteristics, dimensions, and loads, without restrictions in the geometry of the truss and size and type of load.

244 SIMPLIFIED TRUSS DESIGN

2. Preparing the "input" in terms of all data and commands as required by the program, then punching the cards the computer will analyze.

3. Reading and interpreting the computer output.

PROBLEM 1

Let us consider a six-panel flat Howe steel truss, with a 60 ft span and a height of 7.5 ft. Consider, also, two loading conditions acting independently of each other. (See Figure 4-1.)

1. Vertical load $P = 20K$ applied to the top chord at joints 2 through 6, and vertical load $\frac{1}{2}P = 10K$ applied at joints 1 and 7.

2. Vertical load $p = 1K$, applied on the bottom chord at joints 9 through 13.

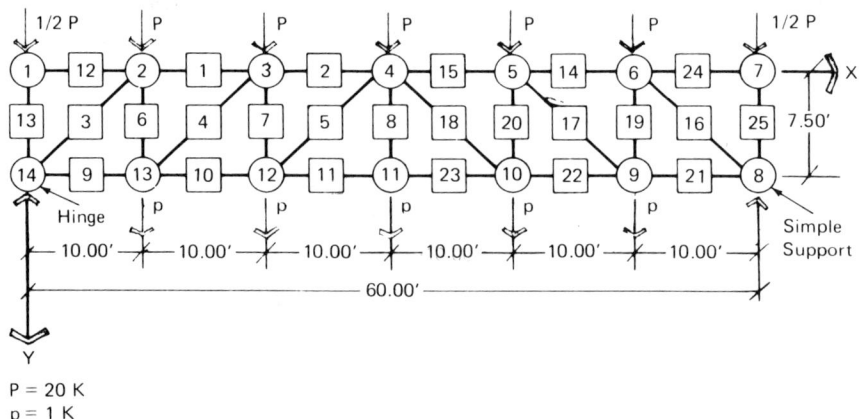

Figure 4-1. Problem 1: Schematic layout of the truss, showing loading conditions.

We want to find the axial forces in each member of the truss, the reactions, the vertical and horizontal displacements of each joint as the truss deflects under loads, and the distortion—contraction (−) and elongation (+)—of each member due to the axial forces.

Let us draw the truss and the loads applied to it (Figure 4-1). We shall number each joint from 1 to 14; the members, 1 to 25. The "input" we must prepare consists of a series of commands. Each command gives the computer a specific datum of the problem or the type and order in which to perform a certain calculation. Each command, written in a language that uses letters and numbers, is transferred to individual cards (see Plate 4-1) by inserting the card in a card-puncher machine and typing the command on it (see Plates 4-2 and 4-3). The machine itself punches the card in accordance to a logic that the computer can read.

Input. The following commands constitute the "input" for the solution of our problem.

Line 1 specifies the program to be used (STRUDL II). It indicates the problem-identifier in quotation marks (no more than eight arbitrary numbers and/or letters), followed by the title of the problem in quotation marks (no more than 64 arbitrary numbers and/or letters).

Line 2 designates that the type of structure is a truss (therefore, all joints are assumed to be hinged) and that it is planar and contained in the x–y plane.

Line 3 specifies that length is given as ft. (If we do not specify feet, the computer will automatically assume length is given as inches.) We assume that the other units are those already assumed by the program: Kip, radians, °F, and sec.

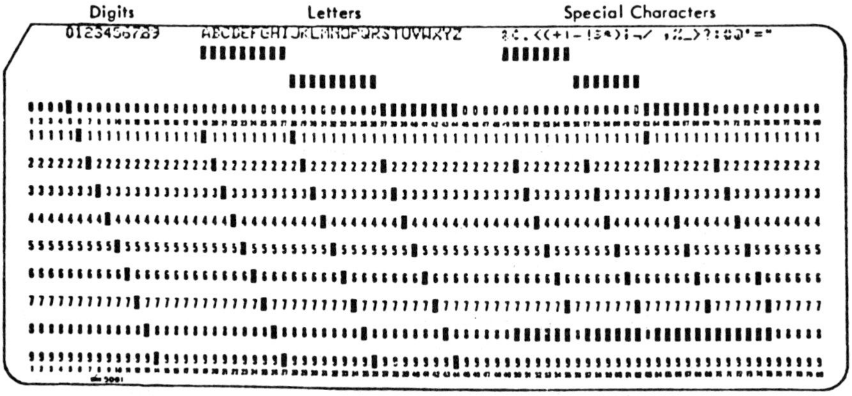

246 SIMPLIFIED TRUSS DESIGN

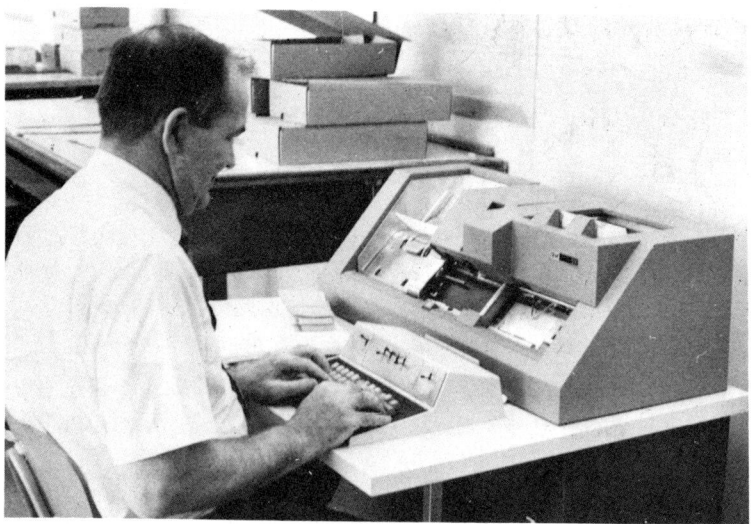

Plate 4-1. A typical IBM card on which the operator will record one of the commands constituting the input. The typing on the upper edge corresponds to the perforations on the rest of the card. While the typing is for the benefit of the reader, the perforation is the language that the computer will need.

Plate 4-2. Operator punching cards. The operation is almost like that of conventional typing. Cards to be punched are stored in the machine and are automatically fed and expelled. Duplication of cards can be done automatically.

ANALYSIS BY COMPUTER 247

Plate 4-3. Keyboard of a typical IBM keypunch machine. Its appearance is very similar to that of a conventional typewriter.

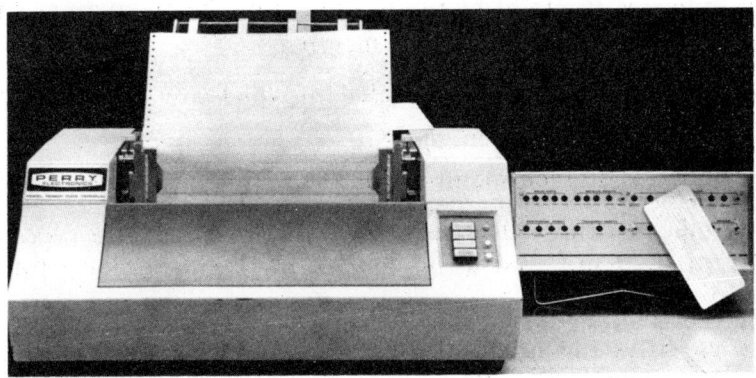

Plate 4-4. Final stage of the process: The readout coming out from the printer.

248 SIMPLIFIED TRUSS DESIGN

Line 4 indicates that the following commands will furnish the coordinates of each joint (ft).

Lines 5 through 18 give the joint coordinates based on the assumption that the point of origin coincides with joint #1. The x-x axis extends horizontally from left to right, and the y-y axis extends vertically downward. Notice that for joints 8 and 4, the letter S after the coordinates identifies these joints as exterior supports for the truss.

Line 19 is a command that changes joint #8 from a fixed hinge to a simple support.

Line 21 indicates that the following commands describe the members of the truss by numbering each member and identifying the number of the joints the member connects (members' incidence).

In lines 22 through 46, the first number indicates the member. The second number indicates the joint at the beginning of the member, and the third indicates the joint at the end of the member.

Line 47 indicates that the units of length which are used from this point on are changed from ft to in.

Line 48 shows that each member has a constant cross-section (prismatic).

Line 49 indicates that all members from 1 to 25 have a cross-sectional area AX of 6.45 in.^{2} and a moment of inertia about the strong axis of 21.95 in.^{4}

Line 50 specifies the modulus of elasticity of steel that is used for this truss, in Ksi. Notice that this information is necessary for calculating the deflection and the distortion of the truss as requested in the problem. It would also be required for the determination of the axial forces if the truss is statically indeterminate.

Line 51 indicates that the following commands are part of loading condition #1. The statement in quotation marks is an arbitrary statement for the benefit of the reader; it is ignored by the computer.

Line 52 shows that a vertical force of $10K$ directed downward is applied at joints 2 through 6.

Line 54 indicates that the following commands refer to another loading condition #2. The arbitrary statement in quotation marks is for the benefit of the reader only, and is not read by the computer.

Line 55 indicates that joints 9 through 13 are all loaded with a vertical downward force $p = 1K$.

Line 56 is a command for the computer to perform the analysis of the structure on the basis of the geometrical and elastic characteristics that have been described.

Line 57 is a command that specifies the results which the analysis must compute, including the axial forces in each member, the elongation or contraction of each member, the external reactions, and the displacement of each joint.

"INPUT" (Problem 1)

```
LINE  1.    STRUDL 'MGM A' 'SIMPLE TRUSS ANALYSIS'
LINE  2.    TYPE PLANE TRUSS XY
LINE  3.    UNITS FEET
LINE  4.    JOINT COORDINATES
LINE  5.    1     X  0   Y 0
LINE  6.    2     X 10   Y 0
LINE  7.    3     X 20   Y 0
```

Continued

LINE 8.	4	X 30	Y 0		
LINE 9.	5	X 40	Y 0		
LINE 10.	6	X 50	Y 0		
LINE 11.	7	X 60	Y 0		
LINE 12.	8	X 60	Y 7.5	S	
LINE 13.	9	X 50	Y 7.5		
LINE 14.	10	X 40	Y 7.5		
LINE 15.	11	X 30	Y 7.5		
LINE 16.	12	X 20	Y 7.5		
LINE 17.	13	X 10	Y 7.5		
LINE 18.	14	X 0	Y 7.5	S	
LINE 19.	JOINT RELEASE FORCE X				
LINE 20.	8				
LINE 21.	MEMBER INCIDENCE				
LINE 22.	1	2	3		
LINE 23.	2	3	4		
LINE 24.	3	2	14		
LINE 25.	4	3	13		
LINE 26.	5	4	12		
LINE 27.	6	2	13		
LINE 28.	7	3	12		
LINE 29.	8	4	11		
LINE 30.	9	14	13		
LINE 31.	10	13	12		
LINE 32.	11	12	11		
LINE 33.	12	1	2		
LINE 34.	13	1	14		
LINE 35.	14	5	6		
LINE 36.	15	4	5		
LINE 37.	16	6	8		

LINE 38. 17   5   9
LINE 39. 18   4   10
LINE 40. 19   6   9
LINE 41. 20   5   10
LINE 42. 21   8   9
LINE 43. 22   9   10
LINE 44. 23   10 11
LINE 45. 24   6   7
LINE 46. 25   7   8
LINE 47. UNITS KIPS INCHES
LINE 48. MEMBER PROPERTIES PRISMATIC
LINE 49. 1 TO 25 AX 6.45   IZ 21.95
LINE 50. CONSTANT E 29000 ALL (STEEL)
LINE 51. LOADING 1 'VERTICAL ON TOP CHORD'
LINE 52. JOINT 1.7 LOAD FORCE Y +10
LINE 53. JOINT 2 TO 6 LOAD FORCE Y +20
LINE 54. LOADING 2 'VERTICAL ON BOTTOM CHORD'
LINE 55. JOINT 9 TO 13 LOAD FORCE Y +1
LINE 56. STIFFNESS ANALYSIS
LINE 57. LIST FORCES DISTORTIONS LOADS REACTIONS DISPLACEMENT

Output. The printout below is the result of the analysis. Its significance is straightforward and self-evident. Loading condition #1 is examined first, followed by the results of this loading condition, in the order given below.

1. The axial forces in Kips for each member.

2. The axial distortion (in.) for each member; that is, elongation (+) or contraction (−).

3. The reactions at the supports in Kips.

4. The indication of the loads.

5. The displacements of the joints (in.), given for all the joints, with respect to the system of coordinates that has been used to describe the truss.

The same type of data in the same order is then furnished for loading condition #2.

"OUTPUT" (Problem 1)

```
**********************************
* RESULTS OF LATEST ANALYSES *
**********************************
```

PROBLEM – MGM A   TITLE – SIMPLE TRUSS ANALYSIS

ACTIVE UNITS   INCH KIP   RAD   DEGF SEC

ACTIVE STRUCTURE TYPE   PLANE TRUSS

ACTIVE COORDINATE AXES   X Y

LOADING – 1   VERTICAL ON TOP CHORD

MEMBER FORCES

MEMBER	JOINT	FORCE
		AXIAL
1	3	-66.6666107
2	4	-106.6666107
3	14	-83.3332977
4	13	-49.9999847
5	12	-16.6666565
6	13	29.9999847
7	12	9.9999990
8	11	0.0000000
9	13	66.6666107
10	12	106.6666107
11	11	119.9999237
12	2	0.0
13	14	-9.9999952
14	6	-66.6666107
15	5	-106.6666107
16	8	-83.3332977
17	9	-49.9999847
18	10	-16.6666565
19	9	29.9999847
20	10	9.9999990
21	9	66.6666107
22	10	106.6666107
23	11	119.9999237
24	7	0.0000000
25	8	-9.9999952

MEMBER DISTORTIONS

MEMBER	DISTORTION
	AXIAL
1	-0.0427693
2	-0.0684309
3	0.0668271
4	0.0400963
5	0.0133654
6	-0.0144346
7	-0.0048115
8	-0.0000000
9	0.0199590
10	0.0684309
11	0.0769848
12	0.0
13	-0.0048115
14	-0.0427693
15	-0.0684309
16	-0.0668271
17	-0.0400963
18	-0.0133654
19	0.0144346
20	0.0048115
21	0.0427693
22	0.0684309
23	0.0769848
24	0.0000000
25	-0.0048115

RESULTANT JOINT
                    SUPPORTS
            LOADS —

JOINT		FORCE	
		X FORCE	Y FORCE
8	GLOBAL	0.0	-59.9999847
14	GLOBAL	0.0000000	-59.9999847

RESULTANT JOINT
                    FREE JOINTS
            LOADS —

JOINT		FORCE	
		X FORCE	Y FORCE
1	GLOBAL	0.0	9.9999952
2	GLOBAL	-0.0000000	19.9999847
3	GLOBAL	-0.0000000	19.9999847
4	GLOBAL	-0.0000000	19.9999847
5	GLOBAL	0.0000000	19.9999847
6	GLOBAL	-0.0000000	19.9999847
7	GLOBAL	0.0000000	9.9999952
9	GLOBAL	-0.0000000	0.0000000
10	GLOBAL	0.0000000	0.0000000
11	GLOBAL	-0.0000000	0.0000000
12	GLOBAL	-0.0000000	0.0000000
13	GLOBAL	-0.0000000	-0.0000000

RESULTANT JOINT
DISPLACEMENTS — SUPPORTS

JOINT		X DISP.	Y DISP.
8	GLOBAL	0.3763702	0.0
14	GLOBAL	0.0	0.0

RESULTANT JOINT
DISPLACEMENTS — FREE JOINTS

JOINT		X DISP.	Y DISP.
1	GLOBAL	0.2993854	0.0048115
2	GLOBAL	0.2993854	0.5105590
3	GLOBAL	0.2566160	0.8769497
4	GLOBAL	0.1881851	1.0066833
5	GLOBAL	0.1197541	0.8769497
6	GLOBAL	0.0769848	0.5105590
7	GLOBAL	0.0769848	0.0048115
9	GLOBAL	0.3336008	0.5249937
10	GLOBAL	0.2651699	0.8817613
11	GLOBAL	0.1881851	1.0066833
12	GLOBAL	0.1112003	0.8817613
13	GLOBAL	0.0427693	0.5249937

LOADING — 2   VERTICAL ON BOTTOM CHORD

MEMBER FORCES

MEMBER	JOINT	FORCE
		AXIAL
1	3	-3.3333330
2	4	-5.3333311
3	14	-4.1666632
4	13	-2.4999990
5	12	-0.8333332
6	13	2.4999990
7	12	1.4999990
8	11	0.9999999
9	13	0.3333330
10	12	5.3333311
11	11	5.9999952
12	2	0.0
13	14	0.0
14	6	-3.3333330
15	5	-5.3333311
16	8	-4.1666632
17	9	-2.4999990
18	10	-0.8333332
19	9	2.4999990
20	10	1.4999990
21	9	3.3333330
22	10	5.3333311
23	11	5.9999952
24	7	0.0000000
25	8	0.0

258 SIMPLIFIED TRUSS DESIGN

MEMBER DISTORTION*   MEMBER DISTORTIONS

MEMBER	DISTORTION
	AXIAL
1	−0.0021385
2	−0.0034215
3	−0.0033414
4	−0.0020048
5	−0.0006683
6	0.0012029
7	0.0007217
8	0.0004812
9	0.0021385
10	0.0034215
11	0.0038492
12	0.0
13	0.0
14	−0.0021385
15	−0.0034215
16	−0.0033414
17	−0.0020048
18	−0.0006683
19	0.0012029
20	0.0007217
21	0.0021385
22	0.0034215
23	0.0038492
24	0.0000000
25	0.0

RESULTANT JOINT
          SUPPORTS
     LOADS –

JOINT		X FORCE	Y FORCE
8	GLOBAL	0.0000000	-2.4999990
14	GLOBAL	0.0000000	-2.4999990

RESULTANT JOINT
          FREE JOINTS
     LOADS –

JOINT		X FORCE	Y FORCE
1	GLOBAL	0.0	0.0
2	GLOBAL	-0.0000000	0.0000000
3	GLOBAL	-0.0000000	-0.0000000
4	GLOBAL	0.0000000	-0.0000000
5	GLOBAL	0.0000000	-0.0000000
6	GLOBAL	-0.0000000	-0.0000000
7	GLOBAL	0.0000000	0.0
9	GLOBAL	-0.0000000	0.9999996
10	GLOBAL	0.0000000	0.9999999
11	GLOBAL	0.0000000	0.9999999
12	GLOBAL	-0.0000000	0.9999999
13	GLOBAL	-0.0000000	0.9999999

RESULTANT JOINT DISPLACEMENTS — SUPPORTS

JOINT		DISPLACEMENT	
		X DISP.	Y DISP.
8	GLOBAL	0.0	0.0
14	GLOBAL	0.0	0.0

RESULTANT JOINT DISPLACEMENTS — FREE JOINTS

JOINT		DISPLACEMENT	
		X DISP.	Y DISP.
1	GLOBAL	0.0149693	0.0
2	GLOBAL	0.0149693	0.0255280
3	GLOBAL	0.0128308	0.0443286
4	GLOBAL	0.0094093	0.0512965
5	GLOBAL	0.0059877	0.0443286
6	GLOBAL	0.0038492	0.0255280
7	GLOBAL	0.0038492	0.0
9	GLOBAL	0.0166800	0.0267308
10	GLOBAL	0.0132585	0.0450504
11	GLOBAL	0.0094093	0.0517776
12	GLOBAL	0.0055600	0.0450504
13	GLOBAL	0.0021385	0.0267308

PROBLEM 2

Let us consider the truss in Figure 4-2. This is a typical truss for stiffening steel frames against the effects of lateral forces. The structure is statically determinate, internally and externally (three-hinged arch). The material used is steel. Assumptions have been made for selecting the cross-sectional area of the various members and also the moments of inertia. Three loading conditions have been established. (See Figure 4-2.)

1. Concentrated vertical forces of $20K$ applied at each joint.

2. Horizontal forces of $20K$ applied at joints 1, 4, 5, 8, 9, and 12.

3. Combination of loadings #1 and #2.

Data to be determined are the following: Axial forces in each member; elongation and contraction of each member; deflection of each joint; and reactions at supports. The system of coordinates x and y, used for describing the truss, has its origin at joint #1.

After having drawn the scheme of the truss and numbered joints and members, the "input" is prepared as described in the previous example.

262 SIMPLIFIED TRUSS DESIGN

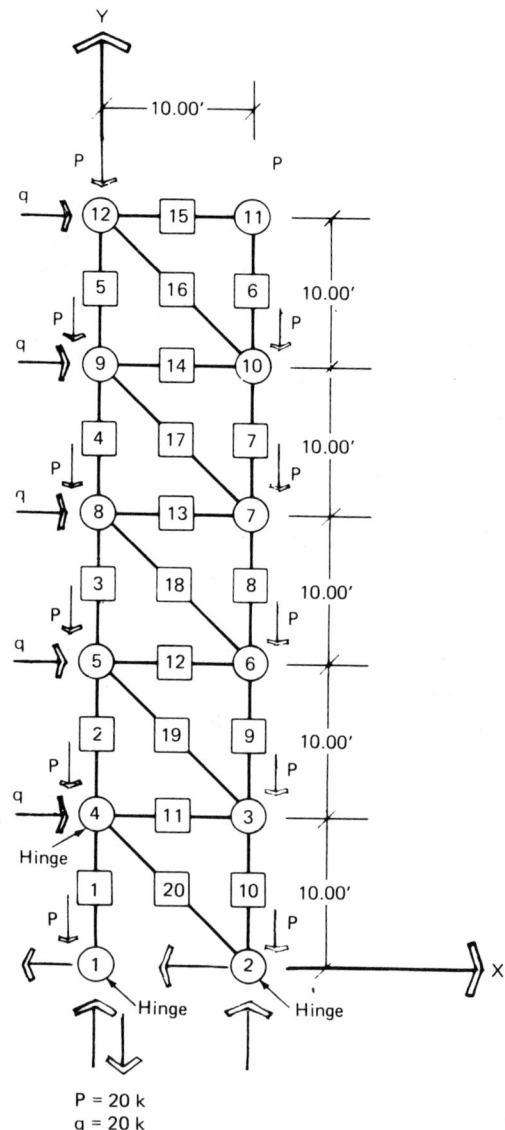

Figure 4-2. Problem 2: Schematic layout of the truss, showing both loading conditions. Note the "three-hinged arch" scheme that includes joints 1, 2, and 4.

"INPUT" (Problem 2)

STRUDL 'MGM 2' 'VERTICAL TRUSS ANALYSIS'

```
************************************************
*        THE STRUCTURAL DESIGN LANGUAGE         *
*                                               *
*     CIVIL ENGINEERING SYSTEMS LABORATORY      *
*     MASSACHUSETTS INSTITUTE OF TECHNOLOGY     *
*           CAMBRIDGE, MASSACHUSETTS            *
*           V2 M2         JUNE, 1972            *
*           11:18:26      10/07/77              *
************************************************
```

TYPE PLANE TRUSS XY
UNITS FEET KIPS
JOINT COORDINATES
1   X   0   Y   0 SUPPORT
2   X 10   Y   0 SUPPORT
3   X 10   Y 10
4   X   0   Y 10
5   X   0   Y 20
6   X 10   Y 20
7   X 10   Y 30
8   X   0   Y 30
9   X   0   Y 40
10 X 10   Y 40
11 X 10   Y 50
12 X   0   Y 50

264 SIMPLIFIED TRUSS DESIGN

MEMBER INCIDENCES
1   1   4
2   4   5
3   5   8
4   8   9
5   9   12
6   10 11
7   7   10
8   6   7
9   3   6
10 2   3
14 9   10
15 12 11
16 12 10
17 9   7
18 8   6
19 5   3
20 4   2

UNITS INCHES KIPS
MEMBER PROPERTIES PRISMATIC
1 TO 10   AX 7.37   IZ 18.7
11 TO 20 AX 4.14   IZ 88.2
CONSTANT E 29000 ALL (STEEL)
LOADING 1 'VERTICAL'
JOINT 1, 4, 5, 8, 9, 12, 11, 10, 7, 6, 3, 2 LOAD FORCE Y -20
LOADING 2 'HORIZONTAL'
JOINT 1, 4, 5, 8, 9, 12 LOAD FORCE X 20
STIFFNESS ANALYSIS
LOADING COMBINATION 'A'
COMBINE 'A' 1 1.0   2 1.0
LIST FORCES, REACTIONS, DISTORSIONS, DISPLACEMENTS

Let us read some of the answers from the "output" below so the reader can familiarize himself with the format of the computer readout.

1. The axial forces in member #1:
   a. Due to loading condition #1: $-99.99K$ compression;
   b. Due to loading condition #2: $299.99K$ tension; and
   c. Due to loading condition #3: $199.99K$ tension.

2. The axial distortion of member #1:
   a. Due to loading condition #1: -0.0561 in. contraction;
   b. Due to loading condition #2: 0.1684 in. elongation; and
   c. Due to loading condition #3: 0.1122 in. elongation.

3. The deflection of joint #12:
   a. Due to loading condition #1: $x = -0.1684$ in., $y = -0.1684$ in.;
   b. Due to loading condition #2: $x = 3.3499$ in., $y = 0.3930$ in.; and
   c. Due to loading condition #3: $x = 3.1815$ in., $y = 0.2245$ in.

4. The reaction at joint #2:
   a. Due to loading condition #1: $x = 0$, $y = 0$;
   b. Due to loading condition #2: $x = -99.99K$, $y = 299.99K$; and
   c. Due to loading condition #3: $x = -99.99K$, $y = 399.99K$.

"OUTPUT" (Problem 2)

@M|

```
*******************************
* RESULTS OF LATEST ANALYSES *
*******************************
```

PROBLEM – MGM 2   TITLE – VERTICAL TRUSS ANALYSIS

266 SIMPLIFIED TRUSS DESIGN

```
ACTIVE UNITS   INCH  KIP   RAD   DEGF SEC
ACTIVE STRUCTURE TYPE   PLANE   TRUSS
ACTIVE COORDINATE AXES   X Y
```

LOADING – 1   VERTICAL

MEMBER FORCES

MEMBER	JOINT	FORCE AXIAL
1	4	-99.9999847
2	5	-79.9999847
3	8	-59.9999847
4	9	-39.9999847
5	12	-19.9999847
6	11	-19.9999847
7	10	-39.9999847
8	7	-59.9999847
9	6	-79.9999847
10	3	-99.9999237
11	3	-0.0000000
12	6	-0.0000000
13	7	0.0000000
14	10	0.0000000
15	11	-0.0000000
16	10	0.0000000
17	7	0.0000000
18	6	0.0000000
19	3	0.0000000
20	2	0.0000000

MEMBER DISTORTIONS

MEMBER	DISTORTION
	AXIAL
1	-0.0561456
2	-0.0449165
3	-0.0336874
4	-0.0224582
5	-0.0112291
6	-0.0112291
7	-0.0224582
8	-0.0336874
9	-0.0449165
10	-0.0561456
11	-0.0000000
12	-0.0000000
13	0.0000000
14	0.0000000
15	-0.0000000
16	0.0000000
17	0.0000000
18	0.0000000
19	0.0000000
20	0.0000000

RESULTANT JOINT SUPPORTS LOADS –

JOINT	FORCE

268 SIMPLIFIED TRUSS DESIGN

		X FORCE	Y FORCE
1	GLOBAL	0.0	99.9999847
2	GLOBAL	0.0000000	99.9999237

RESULTANT JOINT DISPLACEMENTS — SUPPORTS

JOINT		DISPLACEMENT	
		X DISP.	Y DISP.
1	GLOBAL	0.0	0.0
2	GLOBAL	0.0	0.0

RESULTANT JOINT DISPLACEMENTS — FREE JOINTS

JOINT		DISPLACEMENT	
		X DISP.	Y DISP.
3	GLOBAL	-0.0561456	-0.0561456
4	GLOBAL	-0.0561456	-0.0561456
5	GLOBAL	-0.1010621	-0.1010621
6	GLOBAL		-0.1010621
7	GLOBAL		-0.1347495
8	GLOBAL	-0.1347495	-0.1347495
9	GLOBAL	-0.1572077	-0.1572077
10	GLOBAL	-0.1572077	-0.1572077
11	GLOBAL	-0.1684368	-0.1684368
12	GLOBAL	-0.1684368	-0.1684368

LOADING – 2   HORIZONTAL

MEMBER FORCES

MEMBER	JOINT	FORCE
		AXIAL
1	4	299.9997559
2	5	199.9999847
3	8	119.9999847
4	9	59.9999847
5	12	19.9999847
6	11	0.0000000
7	10	-19.9999847
8	7	-59.9999847
9	6	-119.9999847
10	3	-199.9999847
11	3	79.9999847
12	6	59.9999847
13	7	39.9999847
14	10	19.9999847
15	11	0.0
16	10	-28.2842560
17	7	-56.5685272
18	6	-84.8527985
19	3	-113.1370544
20	2	-141.4212952

MEMBER DISTORTIONS

MEMBER	DISTORTION
	AXIAL
1	0.1684368
2	0.1122912
3	0.0673747
4	0.0336874
5	0.0112291
6	0.0000000
7	-0.0112291
8	-0.0336874
9	-0.0673747
10	-0.1122912
11	0.0799600
12	0.0599700
13	0.0399800
14	0.0199900
15	0.0
16	-0.0399800
17	-0.0799600
18	-0.1199400
19	-0.1599200
20	-0.1999001

RESULTANT JOINT SUPPORTS LOADS –

JOINT		FORCE	
		X FORCE	Y FORCE
1	GLOBAL	0.0	-299.9997559
2	GLOBAL	-99.9999847	299.9997559

RESULTANT JOINT DISPLACEMENTS – SUPPORTS

JOINT		DISPLACEMENT		
		X DISP.	Y DISP.	Z DISP.
1	GLOBAL	0.0	0.0	
2	GLOBAL	0.0	0.0	

RESULTANT JOINT DISPLACEMENTS – FREE JOINTS

JOINT		DISPLACEMENT	
		X DISP.	Y DISP.
3	GLOBAL	0.5310983	-0.1122912
4	GLOBAL	0.4511383	0.1684368
5	GLOBAL	1.1502781	0.2807280
6	GLOBAL	1.2102480	-0.1796659
7	GLOBAL	1.9476185	-0.2133533
8	GLOBAL	1.9076376	0.3481028
9	GLOBAL	2.6558418	0.3817902
10	GLOBAL	2.6758318	-0.2245824
11	GLOBAL	3.3499746	-0.2245824
12	GLOBAL	3.3499746	0.3930193

LOADING – 3

MEMBER FORCES

MEMBER	JOINT	FORCE		
		AXIAL	SHEAR Y	SHEAR Z
1	4	199.9999847		
2	5	119.9999847		
3	8	59.9999847		
4	9	20.0000000		
5	12	0.0		
6	11	-19.9999847		
7	10	-59.9999847		
8	7	-119.9999237		
9	6	-199.9999847		
10	3	-299.9997559		
11	3	79.9999847		
12	6	59.9999847		
13	7	39.9999847		
14	10	19.9999847		
15	11	-0.0000000		
16	10	-28.2842560		
17	7	-56.5685272		
18	6	-84.8527985		
19	3	-113.1370544		
20	2	-141.4212952		

MEMBER DISTORTIONS

MEMBER	DISTORTION
	AXIAL
1	0.1122912
2	0.0673747
3	0.0336873
4	0.0112291
5	0.0
6	-0.0112291
7	-0.0336874
8	-0.0673747
9	-0.1122912
10	-0.1684368
11	0.0799600
12	0.0599700
13	0.0399800
14	0.0199900
15	-0.0000000
16	-0.0399800
17	-0.0799600
18	-0.1199400
19	-0.1599200
20	-0.1999001

RESULTANT JOINT SUPPORTS LOADS –

JOINT		FORCE	
		X FORCE	Y FORCE
2	GLOBAL	-99.9999847	399.9997559

RESULTANT JOINT SUPPORTS DISPLACEMENTS –

JOINT		DISPLACEMENT	
		X DISP.	Y DISP.
1	GLOBAL	0.0	0.0
2	GLOBAL	0.0	0.0

RESULTANT JOINT
DISPLACEMENTS – FREE JOINTS

JOINT		DISPLACEMENT	
		X DISP.	Y DISP.
3	GLOBAL	0.4749526	-0.1684368
4	GLOBAL	0.3949926	0.1122912
5	GLOBAL	1.0492153	0.1796660
6	GLOBAL	1.1091852	-0.2807280
7	GLOBAL	1.8128681	-0.3481028
8	GLOBAL	1.7728872	0.2133533
9	GLOBAL	2.4986334	0.2245825
10	GLOBAL	2.5186234	-0.3817901
11	GLOBAL	3.1815376	-0.3930193
12	GLOBAL	3.1815376	0.2245824

Comparison of Methods

Let us compare the results obtained using the computer methods and the tables previously seen.

Example: Let us consider Problem 1, already analyzed by STRUDL II. If we observe member #2 under loading condition #1, we read in the "output" that the axial compression force is $106.66K$. If we use the tables, we find that for $L/H = 60$ ft/7.5 ft $= 8$, the table furnishes a coefficient; $K_1 = -5.33$. Thus, for $P = 20K$, we compute the axial stresses: $K_1 \times P = -5.33 \times 20K = -106.60K$ compression.

PROBLEM 3

The following problem illustrates a statically indeterminate truss. Such a truss is the same as that in Problem 2, with the addition of five diagonal members (see Figure 4-3). The resulting truss is statically indeterminate to the fifth degree; that is, four degrees for internal conditions, and one degree for external supports. The readouts for the "input" and "output" are the same as in the previous examples. It may be interesting, however, to compare the axial forces (under the same loading conditions) in Problems 2 and 3, in order to understand the effects of the additional five diagonals on the trusses' behavior.

For an explanation of the readouts for the "input" and "output" of this problem, the reader is referred to Problem 1 because of the similarity between problems. (See Figure 4-3.)

ANALYSIS BY COMPUTER 277

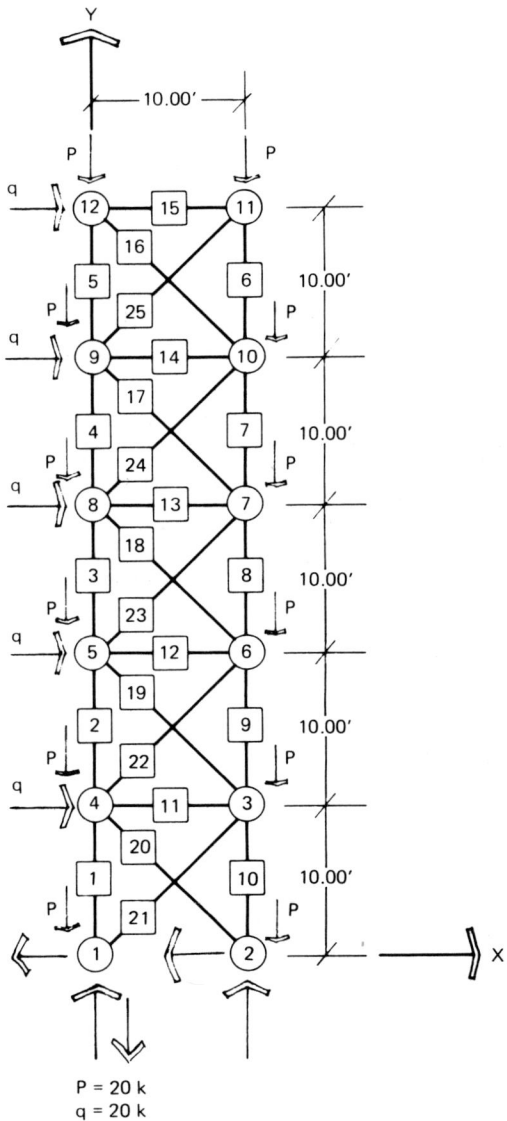

Figure 4-3. Problem 3: statically indeterminate truss.

"INPUT" (Problem 3)

STRUDL 'MGM 3' 'VER TRUSS STAT INDET'

```
****************************************************
*                                                  *
*              ICES STRUDL – II                    *
*        THE STRUCTURAL DESIGN LANGUAGE            *
*                                                  *
*        CIVIL ENGINEERING SYSTEMS LABORATORY      *
*        MASSACHUSETTS INSTITUTE OF TECHNOLOGY     *
*             CAMBRIDGE, MASSACHUSETTS             *
*             V2 M2        JUNE, 1972              *
*             11:38:19      10/17/77               *
*                                                  *
****************************************************
```

TYPE PLANE TRUSS XY
UNITS FEET KIPS
JOINT COORDINATES
1   X   0   Y   0   SUPPORT
2   X 10   Y   0   SUPPORT
3   X 10   Y 10
4   X   0   Y 10
5   X   0   Y 20
6   X 10   Y 20
7   X 10   Y 30
8   X   0   Y 30
9   X   0   Y 40
10 X 10   Y 40
11 X 10   Y 50
12 X   0   Y 50

MEMBER	INCIDENCES	
1	1	4
2	4	5
3	5	8
4	8	9
5	9	12
6	10	11
7	7	10
8	6	7
9	3	6
10	2	3
11	4	3
12	5	6
13	8	7
14	9	10
15	12	11
16	12	10
17	9	7
18	8	6
19	5	3
20	4	2
21	1	3
22	4	6
23	5	7
24	8	10
25	9	11

UNITS INCHES KIPS
MEMBER PROPERTIES PRISMATIC
1 TO 10   AX 7.37   IZ 18.7
11 TO 25   AX 4.14   IZ 88.2
CONSTANT E 29000 ALL (STEEL)
LOADING 1 'VERTICAL'
JOINT 1, 4, 5, 8, 9, 12, 11, 10, 7, 6, 3, 2 LOAD FORCE Y -20
LOADING 2 'HORIZONTAL'
JOINT 1, 4, 5, 8, 9, 12 LOAD FORCE X 20
STIFFNESS ANALYSIS
LOADING COMBINATION 'A'
COMBINE 'A' 1 1.0   2 1.0
LIST FORCES, REACTIONS, DISTORSIONS, DISPLACEMENTS

"OUTPUT" (Problem 3)

```
**********************************
* RESULTS OF LATEST ANALYSES *
**********************************
```

PROBLEM – MGM 3   TITLE – VER TRUSS STAT INDET

ACTIVE UNITS   INCH KIP RAD DEGF SEC

ACTIVE STRUCTURE TYPE   PLANE TRUSS

ACTIVE COORDINATE AXES   X Y

LOADING – 1   VERTICAL

MEMBER FORCES

MEMBER   JOINT   ——————— FORCE ———————————

		AXIAL
1	4	-86.5872345
2	5	-72.0079193
3	8	-53.7072449
4	9	-35.8359833
5	12	-17.9151611
6	11	-17.9151611
7	10	-35.8359833
8	7	-53.7072449
9	6	-72.0079193
10	3	-86.5872345
11	3	21.4047089
12	6	14.2847481
13	7	10.4567528
14	10	6.2488308
15	11	2.0848255
16	10	-2.9483891
17	7	-5.8887959
18	6	-8.8992882
19	3	-11.3023968
20	2	-18.9684296
21	3	-18.9684296
22	6	-11.3023968
23	7	-8.8992882
24	10	-5.8887959
25	11	-2.9483891

282 SIMPLIFIED TRUSS DESIGN

MEMBER DISTORTIONS

MEMBER	DISTORTION AXIAL
1	-0.0486150
2	-0.0404293
3	-0.0301543
4	-0.0201203
5	-0.0100586
6	-0.0100586
7	-0.0201203
8	-0.0301543
9	-0.0404293
10	-0.0486150
11	0.0213940
12	0.0142776
13	0.0104515
14	0.0062457
15	0.0020838
16	-0.0041676
17	-0.0083239
18	-0.0125792
19	-0.0159760
20	-0.0268120
21	-0.0268120
22	-0.159760
23	-0.0125792
24	-0.0083239
25	-0.0041676

RESULTANT JOINT LOADS — SUPPORTS

JOINT	FORCE

		X FORCE	Y FORCE
1	GLOBAL	13.4127131	99.9999847
2	GLOBAL	-13.4127131	99.9999237

RESULTANT JOINT DISPLACEMENTS — SUPPORTS

JOINT		X DISP.	Y DISP.	Z DISP.
1	GLOBAL	0.0	0.0	
2	GLOBAL	0.0	0.0	

RESULTANT JOINT DISPLACEMENTS — FREE JOINTS

JOINT		X DISP.	Y DISP.
3	GLOBAL	0.0106970	-0.0486150
4	GLOBAL	-0.0106970	-0.0486150
5	GLOBAL	-0.0071388	
6	GLOBAL	0.0071288	-0.0890440
7	GLOBAL	0.0052258	-0.1191986
8	GLOBAL	-0.0052258	-0.1191986
9	GLOBAL	-0.0031229	-0.1393189
10	GLOBAL	0.0031229	-0.1393189
11	GLOBAL	0.0010419	-0.1493775
12	GLOBAL	-0.0010419	-0.1493775

LOADING – 2   HORIZONTAL

MEMBER FORCES

MEMBER	JOINT	FORCE
		AXIAL
1	4	251.0346680
2	5	161.9495392
3	8	91.8471069
4	9	41.8317719
5	12	12.0691938
6	11	-7.9307995
7	10	-38.1681976
8	7	-88.1528015
9	6	-158.0503540
10	3	-248.9652252
11	3	-7.0156822
12	6	-6.2032413
13	7	-6.3210611
14	10	-6.0990191
15	11	-7.9307995
16	10	-17.0684052
17	7	-30.8747864
18	6	-45.0386658
19	3	-59.3256836
20	2	-72.1739197
21	3	69.2473602
22	6	53.8113861
23	7	39.8141327
24	10	25.6937256
25	11	11.2158461

MEMBER DISTORTIONS

MEMBER	DISTORTION
	AXIAL
1	0.1409450
2	0.0909275
3	0.0515681
4	0.0234867
5	0.0067763
6	-0.0044528
7	-0.0214298
8	-0.0494940
9	-0.0887383
10	-0.1387831
11	-0.0070122
12	-0.0062001
13	-0.0063179
14	-0.0060960
15	-0.0079268
16	-0.0241263
17	-0.0436418
18	-0.0636625
19	-0.0838573
20	-0.1020184
21	0.0978816
22	0.0760627
23	0.0562776
24	0.0363183
25	0.0158537

RESULTANT JOINT SUPPORTS LOADS —

JOINT		FORCE	
		X FORCE	Y FORCE
1	GLOBAL	-48.9652710	-299.9997559
2	GLOBAL	-51.0346985	299.9997559

RESULTANT JOINT SUPPORTS DISPLACEMENTS —

JOINT		DISPLACEMENT	
		X DISP.	Y DISP.
1	GLOBAL	0.0	0.0
2	GLOBAL	0.0	0.0

RESULTANT JOINT FREE JOINTS DISPLACEMENTS —

JOINT		DISPLACEMENT	
		X DISP.	Y DISP.
3	GLOBAL	0.2782086	-0.1397831
4	GLOBAL	0.2852208	0.1409450
5	GLOBAL	0.7684565	0.2318726
6	GLOBAL	0.7622563	-0.2285214
7	GLOBAL	1.3579321	-0.2780154
8	GLOBAL	1.3642502	0.2834407
9	GLOBAL	2.0045938	0.3069274
10	GLOBAL	1.9984980	-0.2994452
11	GLOBAL	2.6378403	-0.3038980
12	GLOBAL	2.6457672	0.3137037

MEMBER FORCES

MEMBER	JOINT	FORCE
		AXIAL
1	4	164.4474182
2	5	89.9416046
3	8	38.1398621
4	9	5.9957924
5	12	-5.8459759
6	11	-25.8459625
7	10	-74.0041809
8	7	-141.8600464
9	6	-230.0582886
10	3	-335.5522461
11	3	14.3890333
12	6	8.0815067
13	7	4.1356907
14	10	0.1498125
15	11	-5.8459721
16	10	-20.0167999
17	7	-36.7635803
18	6	-53.9379578
19	3	-70.6280518
20	2	-91.1423645
21	3	50.2789154
22	6	42.5089874
23	7	30.9148407
24	10	19.8049316
25	11	8.2674561

MEMBER DISTORTIONS

MEMBER	DISTORTION
	AXIAL
1	0.0923300
2	0.0504982
3	0.0214139
4	0.0033664
5	-0.0032823
6	-0.0145114
7	-0.0415501
8	-0.0796482
9	-0.1291676
10	-0.1883980
11	0.0143818
12	0.0080775
13	0.0041336
14	0.0001497
15	-0.0058431
16	-0.0282939
17	-0.0519656
18	-0.0762417
19	-0.0998333
20	-0.1288304
21	-0.0710695
22	0.0600867
23	0.0436984
24	0.0279944
25	0.0116861

RESULTANT JOINT SUPPORTS LOADS –

JOINT		FORCE	
		X FORCE	Y FORCE
1	GLOBAL	-35.5525665	-199.9999847
2	GLOBAL	-64.4474182	399.9997559

RESULTANT JOINT SUPPORTS DISPLACEMENTS –

JOINT		DISPLACEMENT	
		X DISP.	Y DISP.
1	GLOBAL	0.0	0.0
2	GLOBAL	0.0	0.0

RESULTANT JOINT FREE JOINTS DISPLACEMENTS –

JOINT		DISPLACEMENT	
		X DISP.	Y DISP.
3	GLOBAL	0.2889056	-0.1883980
4	GLOBAL	0.2745237	0.0923300
5	GLOBAL	0.7613176	0.1428283
6	GLOBAL	0.7693951	-0.3175657
7	GLOBAL	1.3631573	-0.3972139
8	GLOBAL	1.3590240	0.1642421
9	GLOBAL	2.0014706	0.1676085
10	GLOBAL	2.0016203	-0.4387640
11	GLOBAL	2.6388817	-0.4532754
12	GLOBAL	2.6447248	0.1643263

PROBLEM 4

This problem illustrates the effects of cantilevers on a simply supported truss. The truss is equal to that analyzed in Problem 1 with the exception that the supports have been moved from joint 14 to 13 and from 8 to 9, which produces a 10 ft cantilevered part on each end of the truss. The reader would find it interesting to compare the results of this problem with those of Problem 1. (See Figure 4-4.)

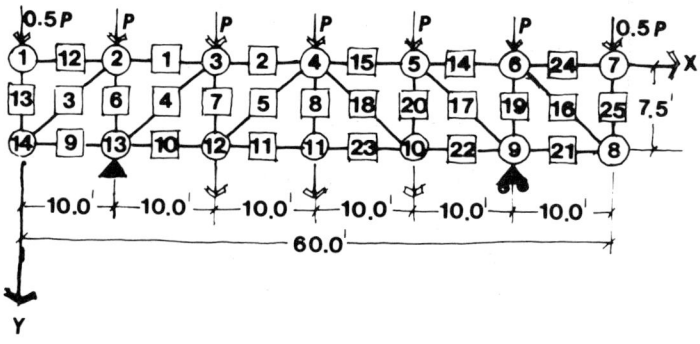

Figure 4-4.

"INPUT" (Problem 4)

```
STRUDL 'MGM' 'SIMPLE TRUSS WITH CANTILEVERS'
TYPE PLANE TRUSS XY
UNITS FEET
JOINT COORDINATES
1      X 0    Y 0
2      X 10   Y 0
3      X 20   Y 0
4      X 30   Y 0
5      X 40   Y 0
6      X 50   Y 0
7      X 60   Y 0
8      X 60   Y 7.5
9      X 50   Y 7.5    S
10     X 40   Y 7.5
11     X 30   Y 7.5
12     X 20   Y 7.5
13     X 10   Y 7.5    S
14     X 0    Y 7.5
JOINT RELEASE FORCE X
9
MEMBER INCIDENCE
1    2    3
2    3    4
3    2    14
4    3    13
5    4    12
6    2    13
7    3    12
```

Continued

8	4	11
9	14	13
10	13	12
11	12	11
12	1	2
13	1	14
14	5	6
15	4	5
16	6	8
17	5	9
18	4	10
19	6	9
20	5	10
21	8	9
22	9	10
23	10	11
24	6	7
25	7	8

UNITS KIPS INCHES
MEMBER PROPERTIES PRISMATIC
1 TO 25   AX 6.45   IZ 21.95
CONSTANT E 29000 ALL (STEEL)
LOADING 1 'VERTICAL ON TOP CHORD'
JOINT 1.7 LOAD FORCE Y 10
JOINT 2 TO 6 LOAD FORCE Y 20
LOADING 2 'VERTICAL ON BOTTOM CHORD'
JOINT 9 TO 13 LOAD FORCE Y 1
STIFFNESS ANALYSIS
LIST FORCES DISTORTIONS LOADS REACTIONS DISPLACEMENT

"OUTPUT" (Problem 4)

```
**********************************
* RESULTS OF LATEST ANALYSES *
**********************************
```

LOADING – 1

MEMBER FORCES

MEMBER	JOINT	AXIAL	
1	3	13.3333311	(TENSION)
2	4	-26.6666565	(COMPRESSION)
3	14	16.6666565	(TENSION)
4	13	-49.9999847	(COMPRESSION)
5	12	-16.6666565	(COMPRESSION)
6	13	-29.9999847	(COMPRESSION)
7	12	9.9999990	(TENSION)
8	11	-0.0000000	(–)
9	13	-13.3333311	(COMPRESSION)
10	12	26.6666565	(TENSION)
11	11	39.9999847	(TENSION)
12	2	0.0	(–)
13	14	-9.9999990	(COMPRESSION)
14	6	13.3333311	(TENSION)
15	5	-26.6666565	(COMPRESSION)
16	8	16.6666565	(TENSION)
17	9	-49.9999847	(COMPRESSION)
18	10	-16.6666565	(COMPRESSION)
19	9	-29.9999847	(COMPRESSION)
20	10	9.9999990	(TENSION)
21	9	-13.3333311	(COMPRESSION)
22	10	26.6666565	(TENSION)
23	11	39.9999847	(TENSION)
24	7	0.0000000	(–)
25	8	-9.9999990	(COMPRESSION)

MEMBER DISTORTIONS

MEMBER	AXIAL	
1	0.0085539	(ELONGATION)
2	-0.0171077	(SHORTENING)
3	0.0133654	(ELONGATION)
4	-0.0400963	(SHORTENING)
5	-0.0133654	(SHORTENING)
6	-0.0144346	(SHORTENING)
7	0.0048115	(ELONGATION)
8	-0.0000000	(SHORTENING)
9	-0.0085539	(SHORTENING)
10	0.0171077	(ELONGATION)
11	0.0256616	(ELONGATION)
12	0.0	(—)
13	-0.0048115	(SHORTENING)
14	0.0085539	(ELONGATION)
15	-0.0171077	(SHORTENING)
16	0.0133654	(ELONGATION)
17	-0.0400963	(SHORTENING)
18	-0.0133654	(SHORTENING)
19	-0.0144346	(SHORTENING)
20	0.0048115	(ELONGATION)
21	-0.0085539	(SHORTENING)
22	0.0171077	(ELONGATION)
23	0.0256616	(ELONGATION)
24	0.0000000	(—)
25	-0.0048115	(SHORTENING)

SUPPORTS

		X FORCE	Y FORCE
9		-0.0000000	-59.9999847
13		0.0000000	-59.9999847

JOINT		———————— DISPLACEMENT ————————	
		X DISP.	Y DISP.
1	GLOBAL	0.0513232	-0.0155039
2	GLOBAL	0.0513232	0.0144346
3	GLOBAL	0.0598771	0.1466632
4	GLOBAL	0.0427693	0.2079659
5	GLOBAL	0.0256616	0.1466632
6	GLOBAL	0.0342155	0.0144346
7	GLOBAL	0.0342155	-0.0155039
8	GLOBAL	0.0769848	-0.0203154
10	GLOBAL	0.0684309	0.1514747
11	GLOBAL	0.0427693	0.2079659
12	GLOBAL	0.0171077	0.1514747
14	GLOBAL	0.0085539	-0.0203154

LOADING – 2   VERTICAL ON BOTTOM CHORD

MEMBER FORCES

MEMBER	JOINT	AXIAL	
1	3	0.0000000	(—)
2	4	-1.9999990	(COMPRESSION)
3	14	0.0000000	(—)
4	13	-2.4999990	(COMPRESSION)
5	12	-0.8333332	(COMPRESSION)
6	13	-0.0000000	(—)
7	12	1.4999990	(TENSION)
8	11	0.9999996	(TENSION)
9	13	-0.0000000	(—)
10	12	1.9999990	(TENSION)
11	11	2.6666660	(TENSION)
12	2	0.0	(—)
13	14	-0.0000000	(—)
14	6	-0.0000000	(—)
15	5	-1.9999990	(COMPRESSION)
16	8	-0.0000000	(—)
17	9	-2.4999990	(COMPRESSION)
18	10	-0.8333332	(COMPRESSION)
19	9	0.0000000	(—)
20	10	1.4999990	(TENSION)
21	9	-0.0000000	(—)
22	10	1.9999990	(TENSION)
23	11	2.6666660	(TENSION)
24	7	0.0000000	(—)
25	8	-0.0000000	(—)

MEMBER DISTORTIONS

MEMBER	AXIAL	
1	0.0000000	(—)
2	-0.0012831	(SHORTENING)
3	0.0000000	(—)
4	-0.0020048	(SHORTENING)
5	-0.0006683	(SHORTENING)
6	-0.0000000	(SHORTENING)
7	0.0007217	(ELONGATION)
8	0.0004812	(ELONGATION)
9	-0.0000000	(—)
10	0.0012831	(ELONGATION)
11	0.0017108	(ELONGATION)
12	0.0	(—)
13	-0.0000000	(—)
14	-0.0000000	(—)
15	-0.0012831	(SHORTENING)
16	-0.0000000	(SHORTENING)
17	-0.0020048	(SHORTENING)
18	-0.0006683	(SHORTENING)
19	0.0000000	(—)
20	0.0007217	(ELONGATION)
21	-0.0000000	(SHORTENING)
22	0.0012831	(ELONGATION)
23	0.0017108	(ELONGATION)
24	0.0000000	(—)
25	-0.0000000	(SHORTENING)

SUPPORTS

JOINT	X FORCE	Y FORCE
9	-0.0000000	-2.4999990
13	0.0000000	-1.4999990

DISPLACEMENTS –

JOINT		X DISP.	Y DISP.
1	GLOBAL	0.0042769	-0.0057026
2	GLOBAL	0.0042769	0.0000000
3	GLOBAL	0.0042769	0.0090439
4	GLOBAL	0.0029939	0.0131605
5	GLOBAL	0.0017108	0.0090439
6	GLOBAL	0.0017108	-0.0000000
7	GLOBAL	0.0017108	-0.0057026
8	GLOBAL	0.0059877	-0.0057026
10	GLOBAL	0.0047046	0.0097657
11	GLOBAL	0.0029939	0.0136416
12	GLOBAL	0.0012831	0.0097657
14	GLOBAL	0.0000000	-0.0057026

PROBLEM 5

The vertical truss in this problem can be visualized as a structural lattice incorporated in the exterior wall of a building. Its 20 joints and 37 members indicate the truss to be statically determinate according to the equation: $m = k$, where $k = 2j - r$, but the configuration of the interior panels (squares instead of triangles) creates some suspicion of geometrical instability. The following analysis will remove the doubts. If instability does exist, it can be immediately detected with a minimum amount of computer time and cost. Made of steel (E = 29,000 Ksi), the truss is analyzed first for vertical loads (loading 1) and then for horizontal loads (loading 2). Reactions, axial forces, elongations and shortening of members (distortions), and displacements of the joints are shown in the readout of the "output" for both loading conditions. (See Figure 4-5.)

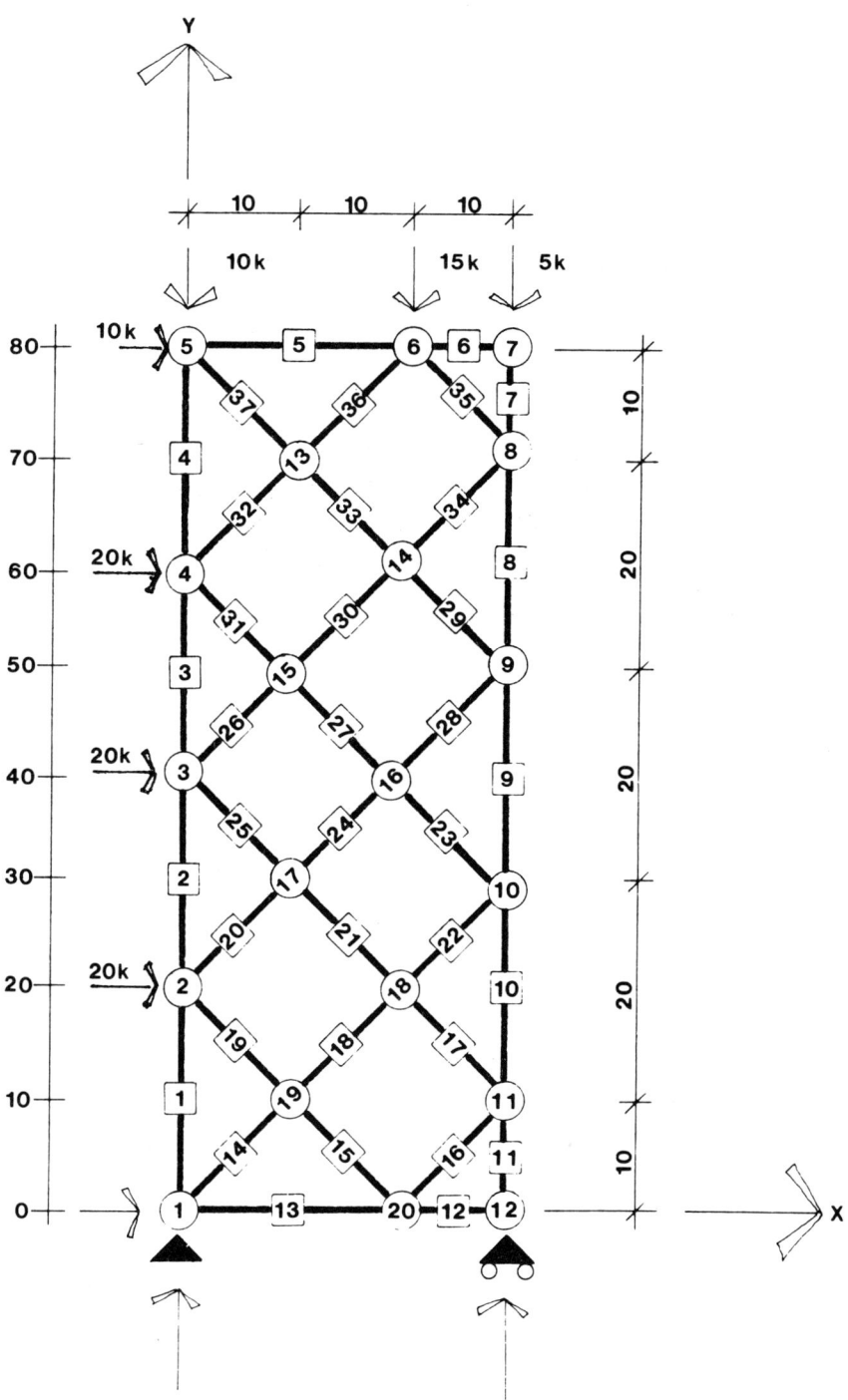

Figure 4-5.

"INPUT" (Problem 5)

STRUDL 'MGM' 'VERTICAL LATTICE'

```
**************************************************
*                  ICES STRUDL – II               *
*           THE STRUCTURAL DESIGN LANGUAGE        *
*                                                 *
*           CIVIL ENGINEERING SYSTEMS LABORATORY  *
*          MASSACHUSETTS INSTITUTE OF TECHNOLOGY  *
*                 CAMBRIDGE, MASSACHUSETTS        *
*               V2 M2         JUNE, 1972          *
*                13:09:41         12/08/72        *
**************************************************
```

TYPE PLANE TRUSS
UNIT FEET
JOINT COORDINATES
1   X   0   Y   0   S
2   X   0   Y   20
3   X   0   Y   40
4   X   0   Y   60
5   X   0   Y   80
6   X   20 Y   80
7   X   30 Y   80
8   X   30 Y   70
9   X   30 Y   50
10 X   30 Y   30
11 X   30 Y   10
12 X   30 Y   0   S
13 X   10 Y   70
14 X   20 Y   60
15 X   10 Y   50
16 X   20 Y   40
17 X   10 Y   30
18 X   20 Y   20
19 X   10 Y   10
20 X   20 Y   0

JOINT RELEASE FORCE X
12

MEMBER INCIDENCE
1   1   2
2   2   3
3   3   4
4   4   5
5   5   6
6   6   7
7   7   8
8   8   9
9   9   10
10 10 11
11 11 12
12 12 20
13 20 1
14 1   19
15 19 20
16 20 11
17 11 18
18 18 19
19 19 2
20 2   17
21 17 18
22 18 10
23 10 16
24 16 17
25 17 3
26 3   15
27 15 16
28 16 9
29 9   14
30 14 15
31 15 4

Continued

32   4   13
33 13   14
34 14   8
35   8   6
36   6   13
37 13   5

UNITS KIPS INCHES
MEMBER PROPERTIES PRISMATIC
1 TO 37   AX 1   IZ 0.0833
CONSTANT E 29000 ALL
LOADING 1 'VERTICAL'
JOINT 5 LOAD FORCE Y -10
JOINT 6 LOAD FORCE Y -15
JOINT 7 LOAD FORCE Y -5
LOADING 2 'HORIZONTAL'
JOINT 2 TO 4 LOAD FORCE X 20
JOINT 5 LOAD FORCE X 10
STIFFNESS ANALYSIS
LIST   FORCES DISTORTIONS REACTIONS DISPLACEMENTS

"OUTPUT" (Problem 5)

****
*   RESULTS OF LATEST ANALYSES   *
****

PROBLEM – MGM   TITLE – VERTICAL LATTICE

ACTIVE UNITS   INCH KIP   RAD   DEGF SEC

ACTIVE STRUCTURE TYPE PLANE   TRUSS

ACTIVE COORDINATE AXES   X Y

LOADING – 1   VERTICAL

MEMBER FORCES

MEMBER	JOINT	FORCE
		AXIAL
1	2	-4.9999990
2	3	-14.9999981
3	4	-24.9999847
4	5	-4.9999990
5	6	4.9999952
6	7	0.0000000
7	8	-4.9999952
8	9	-14.9999981
9	10	-24.9999847
10	11	-4.9999990
11	12	-14.9999943
12	20	-0.0000000
13	1	9.9999952
14	19	-14.1421318
15	20	-7.0710659
16	1	7.0710659
17	18	-7.0710659
18	19	-14.1421318
19	2	-7.0710659
20	17	7.0710659
21	18	-7.0710659
22	10	-14.1421318
23	16	14.1421318
24	17	7.0710659
25	3	-7.0710659
26	15	7.0710659
27	16	14.1421318
28	9	7.0710659
29	14	-7.0710659
30	15	7.0710659
31	4	14.1421318
32	13	-14.1421318
33	14	-7.0710659
34	8	7.0710659
35	6	-7.0710659
36	13	-14.1421318
37	5	-7.0710659

MEMBER DISTORTIONS

MEMBER	DISTORTION	
	AXIAL	Y
1	-0.0413793	
2	-0.1241379	
3	-0.2068965	
4	-0.0413793	
5	0.0413793	
6	0.0000000	
7	-0.0206897	
8	-0.1241379	
9	-0.2068965	
10	-0.0413793	
11	-0.0620690	
12	-0.0000000	
13	0.0827586	
14	-0.0827586	
15	-0.0413793	
16	0.0413793	
17	-0.0413793	
18	-0.0827586	
19	-0.0413793	
20	0.0413793	
21	-0.0413793	
22	-0.0827586	
23	0.0827586	
24	0.0413793	
25	-0.0413793	
26	0.0413793	
27	0.0827586	
28	0.0413793	
29	-0.0413793	
30	0.0413793	
31	0.0827586	
32	-0.0827586	
33	-0.0413793	
34	0.0413793	
35	-0.0413793	
36	-0.0827586	
37	-0.0413793	

RESULTANT JOINT SUPPORTS LOADS —

JOINT		FORCE	
		X FORCE	Y FORCE
1	GLOBAL	0.0000000	14.9999981
12	GLOBAL	-0.0000000	14.9999943

RESULTANT JOINT SUPPORTS DISPLACEMENTS —

JOINT		DISPLACEMENT	
		X DISP.	Y DISP.
1	GLOBAL	0.0	0.0
12	GLOBAL	0.0827586	0.0

RESULTANT JOINT FREE JOINTS DISPLACEMENTS —

JOINT		DISPLACEMENT	
		X DISP.	Y DISP.
2	GLOBAL	-0.4062876	-0.0413793
3	GLOBAL	0.8401613	-0.1655172
4	GLOBAL	-0.8677475	-0.3724138
5	GLOBAL	0.1103448	-0.4137931
6	GLOBAL	0.1517242	-1.6259623
7	GLOBAL	0.1517242	-0.4551725
8	GLOBAL	1.2846842	-0.4344828
9	GLOBAL	0.0382356	-0.3103449
10	GLOBAL	-0.2476668	-0.1034483
11	GLOBAL	0.7680520	-0.0620690
13	GLOBAL	-0.4457904	-0.9114093
14	GLOBAL	0.5993910	0.1922914
15	GLOBAL	0.1774340	0.5557292
16	GLOBAL	-0.2959427	-0.0346858
17	GLOBAL	0.2790058	-0.6681535
18	GLOBAL	0.3272817	-0.5613585
19	GLOBAL	-0.2702329	0.1531945
20	GLOBAL	0.0827586	0.5647052

LOADING — 2   HORIZONTAL

MEMBER FORCES

MEMBER	JOINT	FORCE AXIAL	Y
1	2	83.3332977	
2	3	56.6666565	
3	4	29.9999847	
4	5	3.3333330	
5	6	-6.6666632	
6	7	0.0	
7	8	0.0000000	
8	9	-6.6666632	
9	10	-13.3333311	
10	11	-59.9999847	
11	12	-106.6666107	
12	20	-0.0000000	
13	1	46.6666565	
14	19	32.9983063	
15	20	-32.9983063	
16	11	32.9983063	
17	18	-32.9983063	
18	19	32.9983063	
19	2	-32.9983063	
20	17	4.7140417	
21	18	-32.9983063	
22	10	32.9983063	
23	16	-32.9983063	
24	17	4.7140417	
25	3	-32.9983063	
26	15	4.7140417	
27	16	-32.9983063	
28	9	4.7140417	
29	14	-4.7140417	
30	15	4.7140417	
31	4	-32.9983063	
32	13	4.7140417	
33	14	-4.7140417	
34	8	4.7140417	
35	6	-4.7140417	
36	13	4.7140417	
37	5	-4.7140417	

MEMBER DISTORTIONS

MEMBER	DISTORTION	
	AXIAL	Y
1	0.6896552	
2	0.4689656	
3	0.2482759	
4	0.0275862	
5	-0.0551724	
6	0.0	
7	0.0000000	
8	-0.0551724	
9	-0.1103448	
10	-0.4965518	
11	-0.4413794	
12	-0.0000000	
13	0.3862069	
14	0.1931034	
15	-0.1931034	
16	0.1931034	
17	-0.1931034	
18	0.1931034	
19	-0.1931034	
20	0.0275862	
21	-0.1931034	
22	0.1931034	
23	-0.1931034	
24	0.0275862	
25	-0.1931034	
26	0.0275862	
27	-0.1931034	
28	0.0275862	
29	-0.0275862	
30	0.0275862	
31	-0.1931034	
32	0.0275862	
33	-0.0275862	
34	0.0275862	
35	-0.0275862	
36	0.0275862	
37	-0.0275862	

RESULTANT JOINT
                    SUPPORTS
          LOADS —

JOINT		——————— FORCE ———————	
		X FORCE	Y FORCE
1	GLOBAL	-69.9999237	-106.6666107
12	GLOBAL	-0.0000000	106.6666107

RESULTANT JOINT
                    SUPPORTS
          DISPLACEMENTS —

JOINT		——————— DISPLACEMENT ———————	
		X DISP.	Y DISP.
1	GLOBAL	0.0	0.0
12	GLOBAL	0.3862069	0.0

RESULTANT JOINT
                    FREE JOINTS
          DISPLACEMENTS —

JOINT		DISPLACEMENT	
		X DISP.	Y DISP.
2	GLOBAL	2.0901947	0.6896552
3	GLOBAL	3.0517912	1.1586208
4	GLOBAL	4.9212961	1.4068966
5	GLOBAL	6.5449619	1.4344826
6	GLOBAL	6.4897890	-0.0835707
7	GLOBAL	6.4897890	-1.1034479
8	GLOBAL	5.4308987	-1.1034479
9	GLOBAL	3.9451637	-1.0482759
10	GLOBAL	1.7571993	-0.9379312
11	GLOBAL	0.6325223	-0.4413794
13	GLOBAL	5.7193356	0.6478699
14	GLOBAL	4.6604452	-0.3720078
15	GLOBAL	3.7453671	0.5040574
16	GLOBAL	2.9130478	-0.0551724
17	GLOBAL	2.2194710	0.5993910
18	GLOBAL	0.9465852	-0.4004061
19	GLOBAL	0.7002698	-0.4271802
20	GLOBAL	0.3862069	-0.4681535

PROBLEM 6 (INSTABILITY)

The hexagonal truss illustrated in this example is an unstable structure, a fact which could go undetected by the observer because the truss satisfies the equation for stability. Its geometrical instability, however, is immediately detected by computer analysis, as indicated in the following computer printout. After the stiffness analysis command, the messages that follow show static instability in the indicated joints. (See Figure 4-6.)

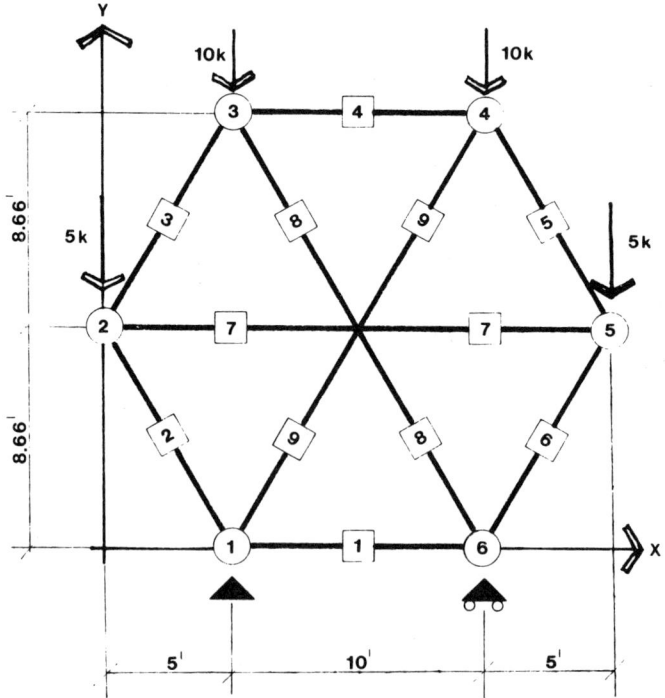

Figure 4-6.

"INPUT" (Problem 6)

```
STRUDL 'MGM' 'HEXAGONAL TRUSS'
TYPE PLANE TRUSS XY
UNITS FEET
JOINT COORDINATES
1  X 5    Y 0         S
2  X 0    Y 8.66
3  X 5    Y 17.32
4  X 15   Y 17.32
5  X 20   Y 8.66
6  X 15   Y 0         S
JOINT RELEASE FORCE X
6
MEMBER INCIDENCE
1  6  1
2  1  2
3  2  3
4  3  4
5  4  5
6  5  6
7  2  5
8  3  6
9  4  1
UNITS INCHES
MEMBER PROPERTIES PRISMATIC
1 TO 9   AX 144   IZ 1728
CONSTANT E 3000 ALL (CONCRETE)
LOADING 1
JOINT 2,5 LOAD FORCE Y -5
JOINT 3,4 LOAD FORCE Y -10
STIFFNESS ANALYSIS
```

NOTE: The following messages after the last command (stiffness analysis) indicate that the statical instability of the truss had already been determined before the last command was executed.

STATICS CHECK FAILED FOR JOINT 2
STATICS CHECK FAILED FOR JOINT 3
STATICS CHECK FAILED FOR JOINT 4
STATICS CHECK FAILED FOR JOINT 5
STATICS CHECK FAILED FOR JOINT 6
LIST FORCES DISTORTIONS LOADS REACTIONS DISPLACEMENTS

PROBLEM 7

This problem is identical to Problem 6 except for the location of joint 5, which has been moved 10 ft to the right. This makes the truss asymmetrical and, as the computer analysis shows, it makes the structure stable and statically determinate. We can conclude, therefore, that the symmetry of the previous hexagonal truss was one of the necessary factors for its geometrical instability. (See Figure 4-7.)

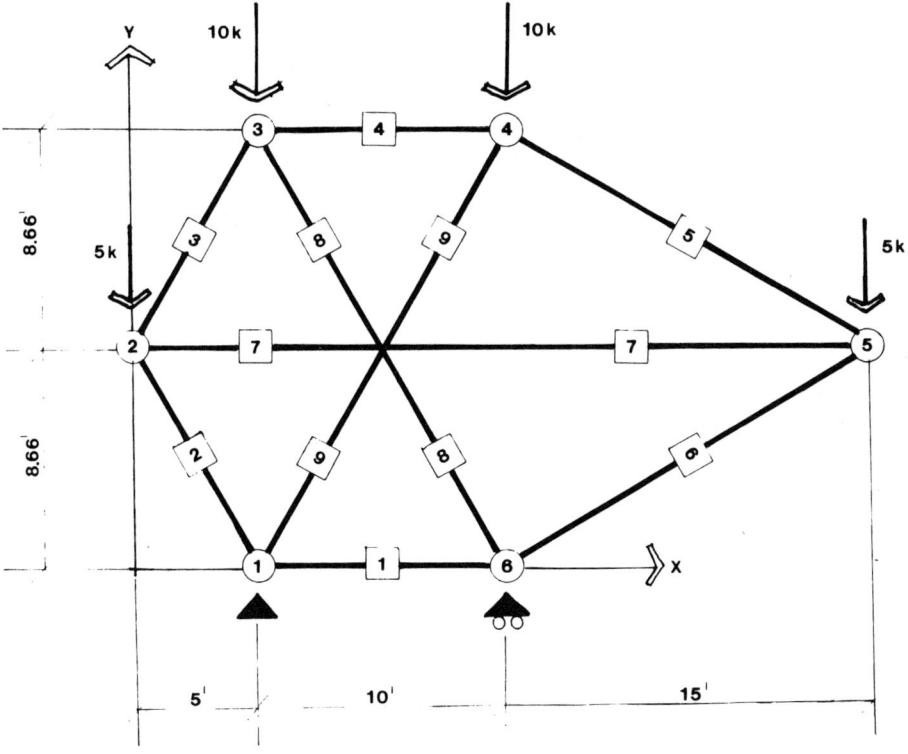

Figure 4-7.

"INPUT" (Problem 7)

STRUDL 'MGM' 'HEXAGONAL TRUSS' (ASIMMETRICAL & STABLE)

```
*****************************************************
*                                                   *
*              ICES STRUDL – II                     *
*         THE STRUCTURAL DESIGN LANGUAGE            *
*                                                   *
*         CIVIL ENGINEERING SYSTEMS LABORATORY      *
*         MASSACHUSETTS INSTITUTE OF TECHNOLOGY     *
*              CAMBRIDGE, MASSACHUSETTS             *
*              V2 M2        JUNE, 1972              *
*                15:10:38     12/12/78              *
*****************************************************
```

TYPE PLANE TRUSS XY
UNITS FEET
JOINT COORDINATES
1   X 5   Y 0   S
2   X 0   Y 8.66
3   X 5   Y 17.32
4   X 15   Y 17.32
5   X 30   Y 8.66
6   X 15   Y 0   S

JOINT RELEASE FORCE X
6
MEMBER INCIDENCE
1 6 1
2 1 2
3 2 3
4 3 4
5 4 5
6 5 6
7 2 5
8 3 6
9 4 1
UNITS INCHES
MEMBER PROPERTIES PRISMATIC
1 TO 9   AX 144   IZ 1728
CONSTANT E 3000 ALL (CONCRETE)
LOADING 1
JOINT 2.5 LOAD FORCE Y -5
JOINT 3.4 LOAD FORCE Y -10
STIFFNESS ANALYSIS
LIST FORCES DISTORTIONS LOADS REACTIONS DISPLACEMENTS

"OUTPUT" (Problem 7)

```
*************************************
*   RESULTS OF LATEST ANALYSES   *
*************************************
```

PROBLEM — MGM   TITLE — HEXAGONAL TRUSS

ACTIVE UNITS INCH LB RAD DEGF SEC

ACTIVE STRUCTURE TYPE PLANE TRUSS

ACTIVE COORDINATE AXES X Y

LOADING — 1

MEMBER FORCES

MEMBER	JOINT	FORCE AXIAL	Y
1	1	11.5473452	
2	2	5.7735424	
3	3	11.5470886	
4	4	17.3210144	
5	5	10.0002241	
6	6	-0.0000022	
7	5	-8.6605110	
8	6	-23.0941772	
9	1	-17.3206329	

MEMBER DISTORTIONS

MEMBER	DISTORTION
	AXIAL
1	0.0032076
2	0.0016037
3	0.0032075
4	0.0048114
5	0.0048113
6	-0.0000000
7	-0.0072171
8	-0.0128298
9	-0.0096224

RESULTANT JOINT SUPPORTS LOADS –

JOINT		FORCE	
		X FORCE	Y FORCE
1	GLOBAL	-0.0000000	10.0000000
6	GLOBAL	-0.0000000	19.9999847

RESULTANT JOINT
LOADS – FREE JOINTS

JOINT		FORCE	
		X FORCE	Y FORCE
2	GLOBAL	-0.0000000	-4.9999990
3	GLOBAL	0.0000000	-10.0000000
4	GLOBAL	0.0000000	-9.9999990
5	GLOBAL	0.0000000	-5.0000000

RESULTANT JOINT
DISPLACEMENTS – SUPPORTS

JOINT		DISPLACEMENT	
		X DISP.	Y DISP.
1	GLOBAL	0.0	0.0
6	GLOBAL	0.0032076	0.0

RESULTANT JOINT
DISPLACEMENTS – FREE JOINTS

JOINT		DISPLACEMENT	
		X DISP.	Y DISP.
2	GLOBAL	0.0925298	0.0552755
3	GLOBAL	0.1117741	0.0478681
4	GLOBAL	0.1165856	-0.0784237
5	GLOBAL	0.0853127	-0.1422144

PROBLEM 8

This problem analyzes a space truss. As the reader will notice, the differences in input between planar and space truss problems are minimal. The command, "TYPE SPACE TRUSS," is the first major difference. (See Figure 4-8.)

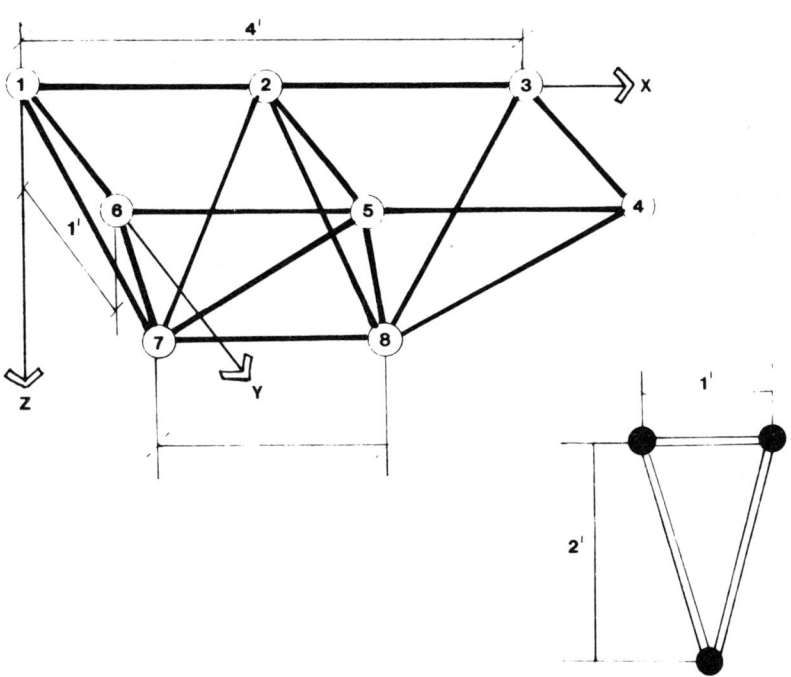

Figure 4-8.

"INPUT" (Problem 8)

STRUDL 'MGM' 'SPACE TRUSS'

```
************************************************
*                                               *
*            ICES STRUDL – II                   *
*      THE STRUCTURAL DESIGN LANGUAGE           *
*                                               *
*      CIVIL ENGINEERING SYSTEMS LABORATORY     *
*      MASSACHUSETTS INSTITUTE OF TECHNOLOGY    *
*           CAMBRIDGE, MASSACHUSETTS            *
*           V2 M2        JUNE, 1972             *
*           14:37:01      12/11/78              *
************************************************
```

DEBUG ALL
TYPE SPACE TRUSS
UNIT KIP FEET
JOINT COORDINATES
1   X 0   Y 0   Z 0   S
2   X 2   Y 0   Z 0
3   X 4   Y 0   Z 0   S
4   X 4   Y 1   Z 0   S
5   X 2   Y 1   Z 0
6   X 0   Y 1   Z 0   S
7   X 1   Y 0   Z 2
8   X 3   Y 0   Z 2
MEMBER INCIDENCES
1   6   1
2   1   2
3   2   3
4   3   4
5   4   5
6   5   6

```
 7   6   7
 8   1   7
 9   3   8
10   4   8
11   7   8
12   7   5
13   8   5
14   7   2
15   8   2
16   2   5
17   6   2
18   4   2
```
UNITS KIP INCHES
MEMBER PROPERTIES PRISMATIC
1 TO 18   AX 1
CONSTANT E 29000 ALL (STEEL)
LOADING 1
JOINT 2, 5, 7, 8 LOAD FORCE Z 1
STIFFNESS ANALYSIS
LIST FORCES DISTORTIONS LOADS REACTIONS DISPLACEMENT

"OUTPUT" (Problem 8)

```
************************************
*   RESULTS OF LATEST ANALYSES   *
************************************
```

PROBLEM – MGM   TITLE – SPACE TRUSS

ACTIVE UNITS INCH KIP RAD DEGF SEC

ACTIVE STRUCTURE TYPE   SPACE TRUSS

ACTIVE COORDINATE AXES   X Y Z

LOADING – 1

MEMBER FORCES

MEMBER	JOINT	FORCE AXIAL
1	1	0.0
2	2	0.0000000
3	3	-0.0000000
4	4	0.0
5	5	-0.0000000
6	6	0.0000000
7	7	0.6123722
8	7	1.6770496
9	8	1.6770496
10	8	0.6123722
11	8	1.4999990
12	5	-0.6123722
13	5	-0.6123722
14	2	-0.5590168
15	2	-0.5590168
16	5	0.4999999
17	2	-0.5590168
18	2	-0.5590168

MEMBER DISTORTIONS

MEMBER	DISTORTION
	AXIAL
1	0.0
2	0.0000000
3	-0.0000000
4	0.0
5	-0.0000000
6	0.0000000
7	0.0006207
8	0.0015517
9	0.0015517
10	0.0006207
11	0.0012414
12	-0.0006207
13	-0.0006207
14	-0.0005172
15	-0.0005172
16	0.0002069
17	-0.0005172
18	-0.0005172

RESULTANT JOINT LOADS – SUPPORTS

JOINT		FORCE		
		X FORCE	Y FORCE	Z FORCE
1	GLOBAL	-0.7499997	0.0	-1.4999990
3	GLOBAL	0.7499997	0.0	-1.4999990
4	GLOBAL	-0.2500000	-0.0000000	-0.4999999
6	GLOBAL	0.2500000	-0.0000000	-0.4999999

RESULTANT JOINT LOADS – FREE JOINTS

JOINT		FORCE		
		X FORCE	Y FORCE	Z FORCE
2	GLOBAL	-0.0000000	0.0000000	0.9999996
5	GLOBAL	-0.0000000	0.0000000	0.9999996
7	GLOBAL	-0.0000000	-0.0000000	0.9999999
8	GLOBAL	0.0000000	-0.0000000	0.9999999

RESULTANT JOINT SUPPORTS
DISPLACEMENTS –

JOINT		DISPLACEMENT		
		X DISP.	Y DISP.	Z DISP.
1	GLOBAL	0.0	0.0	0.0
3	GLOBAL	0.0	0.0	0.0
4	GLOBAL	0.0	0.0	0.0
6	GLOBAL	0.0	0.0	0.0

RESULTANT JOINT FREE JOINTS
DISPLACEMENTS –

JOINT		DISPLACEMENT		
		X DISP.	Y DISP.	Z DISP.
2	GLOBAL	0.0000000	0.0011566	0.0029339
5	GLOBAL	0.0000000	0.0013635	0.0028228
7	GLOBAL	-0.0006207	0.0019494	0.0020452
8	GLOBAL	0.0006207	0.0019494	0.0020452

5
Member and Connection Design

After a panoramic overview of truss development up to present day applications, the reader should familiarize himself with the main factors that characterize the most common truss types. He will then be ready to draw a geometrical configuration for the truss he wants to design. After he has established the member forces, using either the tables in Chapter 3 or the computer analysis in Chapter 4, he has to design members and connections with the materials and techniques he chooses.

Chapter 5, therefore, furnishes some examples and tables that should help the reader to size struts and ties in wood, steel, and reinforced concrete, and design criteria for latticework. Also included are most common examples of connections using the above mentioned materials.

WOOD COMPRESSION MEMBERS

The design of wood compression members is based on the following basic formula:

$$P = Af_c \qquad (5.1)$$

where:

P = maximum axial load on the wood member (lb),
A = cross-sectional area of the member (in.^{2}),
f_c = allowable compression stress (psi) given in Table 5-1.

A major consideration at this point is that f_c must be checked according to two criteria and the smallest of the two values will be used in formula 5.1. Such criteria are as follows:

1. f_c is found in Table 5-1 for any given species and grade of lumber.
2. f_c is computed using the modified Euler's formula (5.4) given below.

Table 5-1. Allowable Stresses and Modulus of Elasticity for the Design of Wood Members

Wood Species and Grades	Grading Agency	Compression Parallel to Fibers (psi)	Tension Parallel to Fibers (psi)	Modulus of Elasticity E (psi)
Ash, White				
2150 f Grade	National Hardwood	1,700	2,050	1,650,000
1900 f Grade	Lumber Association	1,500	1,900	1,650,000
1700 f Grade	1943	1,325	1,700	1,650,000
1450 f Grade		1,150	1,450	1,650,000
1300 f Grade		1,050	1,300	1,650,000
1450 c Grade		1,450		1,650,000
1200 c Grade		1,200		1,650,000
1075 c Grade		1,075		1,650,000
Beech				
Birch				
2150 f Grade	National Hardwood	1,750	2,150	1,760,000
1900 f Grade	Lumber Association	1,525	1,900	1,760,000
1700 f Grade	1943	1,350	1,700	1,760,000
1450 f Grade		1,150	1,450	1,760,000
1550 c Grade		1,550		1,760,000
1450 c Grade		1,450		1,760,000
1200 c Grade		1,200		1,760,000
Chestnut				
1450 f Grade	National Hardwood	1,200	1,450	1,100,000
1200 f Grade	Lumber Association	950	1,200	1,100,000
1075 c Grade	1943	1,075		1,100,000
Cypress, Southern, Coast Type (Tidewater Red)				
1700 f Grade	Southern Cypress	1,425	1,700	1,320,000
1300 f Grade	Manufacturers	1,125	1,300	1,320,000
1450 c Grade	Association	1,450		1,320,000
1200 c Grade	1953	1,200		1,320,000
Cypress, Southern, Inland Type				
1700 f Grade	National Hardwood	1,425	1,700	1,320,000
1300 f Grade	Lumber Association	1,125	1,300	1,320,000
1450 c Grade	1943	1,450		1,320,000
1200 c Grade		1,200		1,320,000

Table 5-1. (Continued)

Wood Species and Grades	Grading Agency	Compression Parallel to Fibers (psi)	Tension Parallel to Fibers (psi)	Modulus of Elasticity E (psi)
Douglas Fir, Coast Region				
Dense Select Structural	West Coast Lumber	1,500	2,050	1,760,000
Select Structural	Inspection Bureau	1,400	1,900	1,760,000
1500 f Industrial	1956	1,200	1,500	1,760,000
1200 f Industrial		1,000	1,200	1,760,000
Dense Select Structural		1,650	1,050	1,760,000
Select Structural		1,500	1,900	1,760,000
Dense Construction		1,400	1,750	1,760,000
Construction		1,200	1,500	1,760,000
Standard		1,000	1,200	1,760,000
Dense Select Structural		1,500	2,050	1,760,000
Select Structural		1,400	1,900	1,760,000
Dense Construction		1,200	1,750	1,760,000
Construction		1,000	1,500	1,760,000
Dense Select Structural		1,650	1,900	1,760,000
Select Structural		1,500	1,750	1,760,000
Dense Construction		1,400	1,500	1,760,000
Construction		1,200	1,200	1,760,000
Douglas Fir				
Dense Select Structural	Western Pine	1,650	2,050	1,760,000
Select Structural	Association	1,500	1,900	1,760,000
Dense Structural	1957	1,400	1,750	1,760,000
Structural		1,200	1,500	1,760,000
Standard Structural		1,000	1,200	1,760,000
Dense Select Structural		1,650	1,900	1,760,000
Select Structural		1,500	1,750	1,760,000
Dense Structural		1,400	1,600	1,760,000
Structural		1,200	1,200	1,760,000
Elm, Rock				
2150 f Grade	National Hardwood	1,750	2,150	1,430,000
1900 f Grade	Lumber Association	1,525	1,900	1,430,000
1700 f Grade	1943	1,350	1,700	1,430,000
1450 f Grade		1,150	1,450	1,430,000
1550 c Grade		1,550		1,430,000
1450 c Grade		1,450		1,430,000
1200 c Grade		1,200		1,430,000

Table 5-1. (Continued)

Wood Species and Grades	Grading Agency	Compression Parallel to Fibers (psi)	Tension Parallel to Fibers (psi)	Modulus of Elasticity E (psi)
Hemlock, West Coast				
Select Structural	West Coast Lumber	1,100	1,600	1,540,000
1500 f Industrial	Inspection Bureau	1,000	1,500	1,540,000
1200 f Industrial	1956	900	1,200	1,540,000
Select Structural		1,200	1,600	1,540,000
Construction		1,100	1,500	1,540,000
Standard		1,000	1,200	1,540,000
Construction		1,000	1,500	1,540,000
Construction		1,100	1,200	1,540,000
Hickory				
Pecan				
2150 f Grade	National Hardwood	1,725	2,150	1,980,000
1900 f Grade	Lumber Association	1,550	1,900	1,980,000
1700 f Grade	1943	1,350	1,700	1,980,000
1550 c Grade		1,550		1,980,000
1450 c Grade		1,450		1,980,000
1325 c Grade		1,325		1,980,000
Larch				
Dense Select Structural	Western Pine	1,650	2,050	1,760,000
Select Structural	Association	1,500	1,900	1,760,000
Dense Structural	1957	1,400	1,750	1,760,000
Structural		1,200	1,500	1,760,000
Standard Structural		1,000	1,200	1,760,000
Dense Select Structural		1,650	1,900	1,760,000
Select Structural		1,500	1,750	1,760,000
Dense Structural		1,400	1,500	1,760,000
Structural		1,200	1,200	1,760,000
Maple Hard				
2150 f Grade	National Hardwood	1,750	2,150	1,760,000
1900 f Grade	Lumber Association	1,525	1,900	1,760,000
1700 f Grade	1943	1,350	1,700	1,760,000
1450 f Grade		1,150	1,450	1,760,000
1550 c Grade		1,550		1,760,000
1450 c Grade		1,450		1,760,000
1200 c Grade		1,200		1,760,000

Wood Species and Grades	Grading Agency	Compression Parallel to Fibers (psi)	Tension Parallel to Fibers (psi)	Modulus of Elasticity E (psi)
Pine Southern				
Dense Structural 86 KD	Southern Pine Inspection Bureau 1956	2,250	3,000	1,760,000
Dense Structural 72 KD		1,950	2,500	1,760,000
Dense Structural 65 KD		1,800	2,250	1,760,000
Dense Structural 58 KD		1,650	2,050	1,760,000
No. 1 Dense KD		1,750	2,050	1,760,000
No. 1 KD		1,500	1,750	1,760,000
No. 2 Dense KD		1,300	1,750	1,760,000
No. 2 KD		1,100	1,500	1,760,000
Dense Structural 86		2,200	2,900	1,760,000
Dense Structural 72		1,800	2,350	1,760,000
Dense Structural 65		1,600	2,050	1,760,000
Dense Structural 58		1,450	1,750	1,760,000
No. 1 Dense		1,550	1,750	1,760,000
No. 1		1,350	1,500	1,760,000
No. 2 Dense		1,050	1,400	1,760,000
No. 2		900	1,200	1,760,000
Dense Structural 86		2,200	2,900	1,760,000
Dense Structural 72		1,800	2,350	1,760,000
Dense Structural 65		1,600	2,050	1,760,000
Dense Structural 58		1,450	1,750	1,760,000
No. 1 Dense SR		1,750	1,750	1,760,000
No. 1 SR		1,500	1,500	1,760,000
No. 2 Dense SR		1,050	1,400	1,760,000
No. 2 SR		900	1,200	1,760,000
Dense Structural 86		1,800	2,400	1,760,000
Dense Structural 72		1,550	2,000	1,760,000
Dense Structural 65		1,400	1,800	1,760,000
Dense Structural 58		1,300	1,600	1,760,000
No. 1 Dense SR		1,500	1,600	1,760,000
No. 1 SR		1,300	1,400	1,760,000
No. 2 Dense SR		1,050	1,400	1,760,000
No. 2 SR		900	1,200	1,760,000
Industrial 86 KD		1,950	2,600	1,760,000
Industrial 72 KD		1,650	2,200	1,760,000
Industrial 65 KD		1,550	2,000	1,760,000
Industrial 58 KD		1,400	1,750	1,760,000
Industrial 50 KD		1,100	1,500	1,760,000
Industrial 86		1,900	2,500	1,760,000
Industrial 72		1,550	2,000	1,760,000
Industrial 65		1,350	1,750	1,760,000
Industrial 58		1,250	1,500	1,760,000
Industrial 50		900	1,200	1,760,000

Table 5-1. (Continued)

Wood Species and Grades	Grading Agency	Compression Parallel to Fibers (psi)	Tension Parallel to Fibers (psi)	Modulus of Elasticity E (psi)
Oak, Red and White				
2150 f Grade	National Hardwood	1,550	2,150	1,650,000
1900 f Grade	Lumber Association	1,375	1,900	1,650,000
1700 f Grade	1943	1,200	1,700	1,650,000
1450 f Grade		1,050	1,450	1,650,000
1300 f Grade		950	1,300	1,650,000
1325 c Grade		1,325		1,650,000
1200 c Grade		1,200		1,650,000
1075 c Grade		1,075		1,650,000
Pine, Norway				
Prime Structural	Northern Hemlock and	900	1,200	1,320,000
Common Structural	Hardwood Manufac-	775	1,100	1,320,000
Utility Structural	turers Association 1947	650	950	1,320,000
Poplar, Yellow				
1500 f Grade	National Hardwood	1,200	1,500	1,210,000
1250 f Grade	Lumber Association	950	1,250	1,210,000
1075 c Grade	1943	1,075		1,210,000
Redwood				
Dense Structural	California Redwood	1,450	1,700	1,320,000
Heart Structural	Association	1,100	1,300	1,320,000
Dense Structural	1955	1,450		1,320,000
Heart Structural		1,100		1,320,000
Spruce, Eastern				
1450 f Structural Grade	Northeastern Lumber	1,050	1,450	1,320,000
1300 f Structural Grade	Manufacturers	975	1,300	1,320,000
1200 f Structural Grade	Association, Inc. 1950	900	1,200	1,320,000

The first Euler's formula, first published in 1757, is given below (5.2).

$$f_c = \mu \frac{\pi^2 E}{\left(\dfrac{l}{r_{min}}\right)^2} \qquad (5.2)$$

where:

f_c = allowable compression stress (psi),
μ = coefficient,
E = modulus of elasticity of the species and grade of wood to be found in Table 5-1 (psi),
l = unsupported length of the member (in.),
r_{min} = minimum radius of gyration of the member's cross-section (in.),
$\dfrac{l}{r_{min}}$ = slenderness ratio of the member.

Table 5-2. Radius of Gyration for Typical Geometrical Shapes

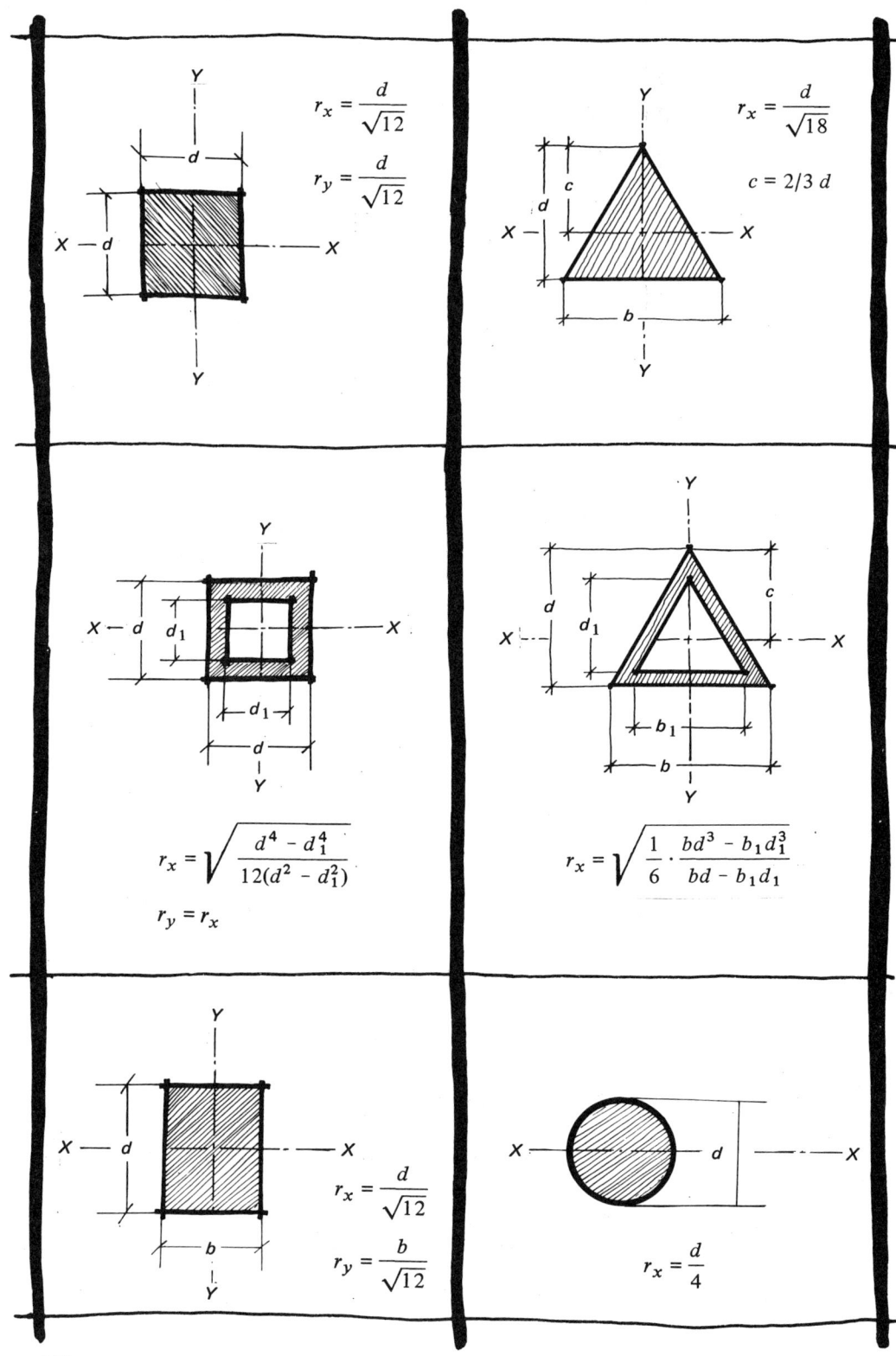

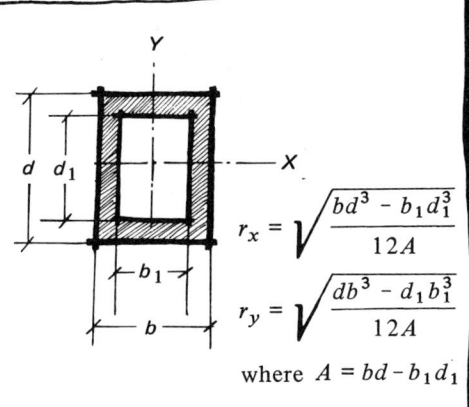

$$r_x = \sqrt{\frac{bd^3 - b_1 d_1^3}{12A}}$$

$$r_y = \sqrt{\frac{db^3 - d_1 b_1^3}{12A}}$$

where $A = bd - b_1 d_1$

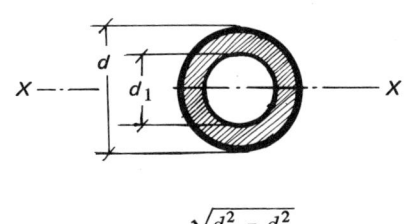

$$r_x = \frac{\sqrt{d^2 - d_1^2}}{4}$$

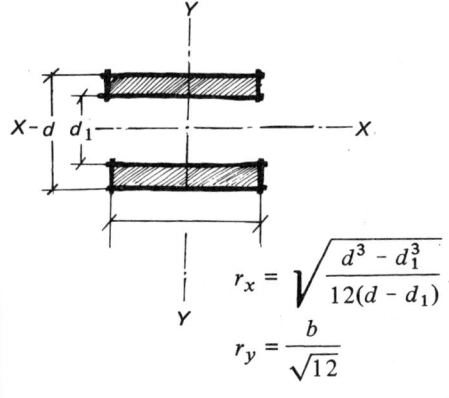

$$r_x = \sqrt{\frac{d^3 - d_1^3}{12(d - d_1)}}$$

$$r_y = \frac{b}{\sqrt{12}}$$

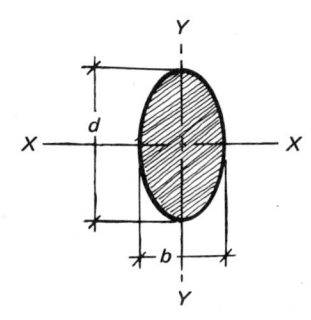

$$r_x = \frac{d}{4}$$

$$r_y = \frac{b}{4}$$

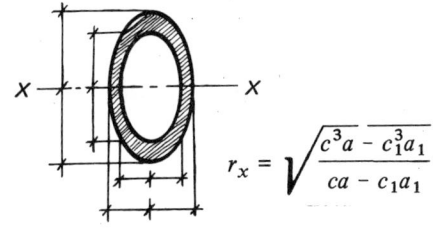

$$r_x = \sqrt{\frac{c^3 a - c_1^3 a_1}{ca - c_1 a_1}}$$

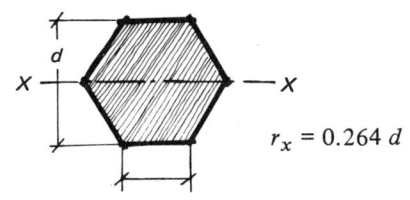

$$r_x = 0.264\, d$$

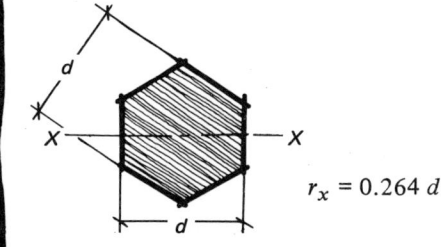

$$r_x = 0.264\, d$$

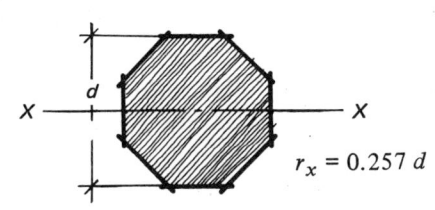

$$r_x = 0.257\, d$$

However, assuming $\mu = 1$, assuming a safety factor of 3.0, and correcting E by a factor of 1.1, formula 5.2 becomes formula 5.3.

$$f_c = \frac{\pi^2 E(1.1)}{3.0 \left(\dfrac{l}{r_{min}}\right)^2} = \frac{\pi^2 E}{2.727 \left(\dfrac{l}{r_{min}}\right)^2} \tag{5.3}$$

Because wood members have mostly rectangular cross-sections, it is more convenient to substitute $d_{min}/\sqrt{12}$ in place of r_{min}. Then, substituting this value in formula 5.3 and simplifying, we obtain formula 5.4.

$$f_c = \frac{0.30\,E}{\left(\dfrac{l}{d_{min}}\right)^2} \tag{5.4}$$

After formula 5.1 has been satisfied, it is necessary to verify that the slenderness ratio is contained within the following upper limit (5.5).

$$\left(\frac{l}{r_{min}}\right) \leq 173 \tag{5.5}$$

Substituting $d_{min}/\sqrt{12}$ in place of r_{min}, formula 5.5 becomes formula 5.6.

$$\left(\frac{l}{d_{min}}\right) \leq 50 \tag{5.6}$$

Example 5.1. Given a wood compression member with the characteristics below, find the maximum load the member can carry.

Material: Douglas Fir, Coastal Region, 1200 f Industrial
Length: 10 ft
Cross-section: 4 in. X 6 in. (effective size)

Solution: From Table 5-1, find $f_c = 1,000$ psi and $E = 1,760,000$ psi. Substituting the data in formula 5.4, we obtain

$$f_c = \frac{(0.30) \times (1,760,000)}{\left(\frac{10 \times 12}{4}\right)^2} = 587 \text{ psi.}$$

Since $587 < 1200$, use $f_c = 1,200$ psi. Substituting this value in formula 5.1, we obtain

$$P = (1,200) \times (6) \times (4) = 14,088 \text{ lb.}$$

Using formula 5.6 for verifying the slenderness ratio, we find

$$\frac{10 \times 12}{4} = 30 < 50 \therefore \text{o.k.}$$

Maximum compression load on the member is 14,088 lb.

STEEL COMPRESSION MEMBERS

In a similar manner to that used for designing wood members, steel compression members are designed with the formula given below.

$$P = Af_a \qquad (5.7)$$

where:

P = maximum axial compression load the member can carry (kips),
A = cross-sectional area of the member (in.^{2}),
f_a = allowable compression stress (ksi) given in Table 5-3.

338 SIMPLIFIED TRUSS DESIGN

The allowable compression stress, f_a, given in Table 5-3 has been tabulated for several types of steel having various yield strength (f_y), as a function of the slenderness ratio Kl/r. Notice that for pin-connected members (as truss members are), $K = 1$.

When

$$\frac{Kl}{r} \leq C_c = \sqrt{\frac{2\pi^2 E}{f_y}} \tag{5.8}$$

f_a is given by formula 5.9.

$$f_a = \frac{\left[1 - \frac{(Kl/r)^2}{2C_c^2}\right] f_y}{\frac{5}{3} + \frac{3(Kl/r)}{8C_c} - \frac{(Kl/r)^3}{8C_c^3}} \tag{5.9}$$

However, when:

$$\frac{Kl}{r} > C_c = \sqrt{\frac{2\pi^2 E}{f_y}} \tag{5.10}$$

f_a is given by formula 5.11.

$$f_a = \frac{12\pi^2 E}{23(Kl/r)^2} \tag{5.11}$$

where:

E = modulus of elasticity of steel equal to 29,000 (ksi),
l = unsupported length of compression member (in.),
r = minimum radius of gyration of the cross-section (in.).

Notice that according to the American Institute of Steel Construction (AISC) specifications, the slenderness ratio is limited to 200 for compression members.

TABLE 5-3. Allowable Stress in Steel

Allowable Stress, f_a (ksi), in Steel for Compression Members of 36 ksi Specified Yield Stress

\multicolumn{6}{c	}{Main Members (Kl/r not over 120)}	\multicolumn{4}{c}{Main Members (Kl/r 121 to 200)}
$\dfrac{Kl}{r}$	f_a (ksi)	$\dfrac{Kl}{r}$	f_a (ksi)	$\dfrac{Kl}{r}$	f_a (ksi)	$\dfrac{Kl}{r}$	f_a (ksi)	$\dfrac{Kl}{r}$	f_a (ksi)
1	21.56	41	19.11	81	15.24	121	10.14	161	5.76
2	21.52	42	19.03	82	15.13	122	9.99	162	5.69
3	21.48	43	18.95	83	15.02	123	9.85	163	5.62
4	21.44	44	18.86	84	14.90	124	9.70	164	5.55
5	21.39	45	18.78	85	14.79	125	9.55	165	5.49
6	21.35	46	18.70	86	14.67	126	9.41	166	5.42
7	21.30	47	18.61	87	14.56	127	9.26	167	5.35
8	21.25	48	18.53	88	14.44	128	9.11	168	5.29
9	21.21	49	18.44	89	14.32	129	8.97	169	5.23
10	21.16	50	18.35	90	14.20	130	8.84	170	5.17
11	21.10	51	18.26	91	14.09	131	8.70	171	5.11
12	21.05	52	18.17	92	13.97	132	8.57	172	5.05
13	21.00	53	18.08	93	13.84	133	8.44	173	4.99
14	20.95	54	17.99	94	13.72	134	8.32	174	4.93
15	20.89	55	17.90	95	13.60	135	8.19	175	4.88
16	20.83	56	17.81	96	13.48	136	8.07	176	4.82
17	20.78	57	17.71	97	13.35	137	7.96	177	4.77
18	20.72	58	17.62	98	13.23	138	7.84	178	4.71
19	20.66	59	17.53	99	13.10	139	7.73	179	4.66
20	20.60	60	17.43	100	12.98	140	7.62	180	4.61
21	20.54	61	17.33	101	12.85	141	7.51	181	4.56
22	20.48	62	17.24	102	12.72	142	7.41	182	4.51
23	20.41	63	17.14	103	12.59	143	7.30	183	4.46
24	20.35	64	17.04	104	12.47	144	7.20	184	4.41
25	20.28	65	16.94	105	12.33	145	7.10	185	4.36
26	20.22	66	16.84	106	12.20	146	7.01	186	4.32
27	20.15	67	16.74	107	12.07	147	6.91	187	4.27
28	20.08	68	16.64	108	11.94	148	6.82	188	4.23
29	20.01	69	16.53	109	11.81	149	6.73	189	4.18
30	19.94	70	16.43	110	11.67	150	6.64	190	4.14
31	19.87	71	16.33	111	11.54	151	6.55	191	4.09
32	19.80	72	16.22	112	11.40	152	6.46	192	4.05
33	19.73	73	16.12	113	11.26	153	6.38	193	4.01
34	19.65	74	16.01	114	11.13	154	6.30	194	3.97
35	19.58	75	15.90	115	10.99	155	6.22	195	3.93
36	19.50	76	15.79	116	10.85	156	6.14	196	3.89
37	19.42	77	15.69	117	10.71	157	6.06	197	3.85
38	19.35	78	15.58	118	10.57	158	5.98	198	3.81
39	19.27	79	15.47	119	10.43	159	5.91	199	3.77
40	19.19	80	15.36	120	10.28	160	5.83	200	3.73

TABLE 5-3. (Continued)

Allowable Stress, f_a (ksi), in Steel for Compression Members of 42 ksi Specified Yield Stress

Main Members (Kl/r not over 120)						Main Members (Kl/r 121 to 200)			
$\dfrac{Kl}{r}$	f_a (ksi)	$\dfrac{Kl}{r}$	f_a (ksi)	$\dfrac{Kl}{r}$	f_a (ksi)	$\dfrac{Kl}{r}$	f_a (ksi)	$\dfrac{Kl}{r}$	f_a (ksi)
1	25.15	41	21.98	81	16.92	121	10.20	161	5.76
2	25.10	42	21.87	82	16.77	122	10.03	162	5.69
3	25.05	43	21.77	83	16.62	123	9.87	163	5.62
4	24.99	44	21.66	84	16.47	124	9.71	164	5.55
5	24.94	45	21.55	85	16.32	125	9.56	165	5.49
6	24.88	46	21.44	86	16.17	126	9.41	166	5.42
7	24.82	47	21.33	87	16.01	127	9.26	167	5.35
8	24.76	48	21.22	88	15.86	128	9.11	168	5.29
9	24.70	49	21.10	89	15.71	129	8.97	169	5.23
10	24.63	50	20.99	90	15.55	130	8.84	170	5.17
11	24.57	51	20.87	91	15.39	131	8.70	171	5.11
12	24.50	52	20.76	92	15.23	132	8.57	172	5.05
13	24.43	53	20.64	93	15.07	133	8.44	173	4.99
14	24.36	54	20.52	94	14.91	134	8.32	174	4.93
15	24.29	55	20.40	95	14.75	135	8.19	175	4.88
16	24.2	56	20.28	96	14.59	136	8.07	176	4.82
17	24.15	57	20.16	97	14.43	137	7.96	177	4.77
18	24.07	58	20.03	98	14.26	138	7.84	178	4.71
19	24.00	59	19.91	99	14.09	139	7.73	179	4.66
20	23.92	60	19.79	100	13.93	140	7.62	180	4.61
21	23.84	61	19.66	101	13.76	141	7.51	181	4.56
22	23.76	62	19.53	102	13.59	142	7.41	182	4.51
23	23.68	63	19.40	103	13.42	143	7.30	183	4.46
24	23.59	64	19.27	104	13.25	144	7.20	184	4.41
25	23.51	65	19.14	105	13.08	145	7.10	185	4.36
26	23.42	66	19.01	106	12.90	146	7.01	186	4.32
27	23.33	67	18.88	107	12.73	147	6.91	187	4.27
28	23.24	68	18.75	108	12.55	148	6.82	188	4.23
29	23.15	69	18.61	109	12.37	149	6.73	189	4.18
30	23.06	70	18.48	110	12.19	150	6.64	190	4.14
31	22.97	71	18.34	111	12.01	151	6.55	191	4.09
32	22.88	72	18.20	112	11.83	152	6.46	192	4.05
33	22.78	73	18.06	113	11.65	153	6.38	193	4.01
34	22.69	74	17.92	114	11.47	154	6.30	194	3.97
35	22.59	75	17.78	115	11.28	155	6.22	195	3.93
36	22.49	76	17.64	116	11.10	156	6.14	196	3.89
37	22.39	77	17.50	117	10.91	157	6.06	197	3.85
38	22.29	78	17.35	118	10.72	158	5.98	198	3.81
39	22.19	79	17.21	119	10.55	159	5.91	199	3.77
40	22.08	80	17.06	120	10.37	160	5.83	200	3.73

TABLE 5-3. (Continued)

Allowable Stress, f_a (ksi), in Steel for Compression Members of 45 ksi Specified Yield Stress

| \multicolumn{6}{c|}{Main Members (Kl/r not over 120)} | \multicolumn{4}{c}{Main Members (Kl/r 121 to 200)} |

$\frac{Kl}{r}$	f_a (ksi)	$\frac{Kl}{r}$	f_a (ksi)	$\frac{Kl}{r}$	f_a (ksi)	$\frac{Kl}{r}$	f_a (ksi)	$\frac{Kl}{r}$	f_a (ksi)
1	26.95	41	23.39	81	17.67	121	10.20	161	5.76
2	26.89	42	23.27	82	17.51	122	10.03	162	5.69
3	26.83	43	23.15	83	17.34	123	9.87	163	5.62
4	26.77	44	23.03	84	17.17	124	9.71	164	5.55
5	26.71	45	22.90	85	17.00	125	9.56	165	5.49
6	26.64	46	22.78	86	16.82	126	9.41	166	5.42
7	26.58	47	22.65	87	16.65	127	9.26	167	5.35
8	26.51	48	22.53	88	16.48	128	9.11	168	5.29
9	26.44	49	22.40	89	16.30	129	8.97	169	5.23
10	26.37	50	22.27	90	16.12	130	8.84	170	5.17
11	26.30	51	22.14	91	15.95	131	8.70	171	5.11
12	26.22	52	22.01	92	15.77	132	8.57	172	5.05
13	26.15	53	21.88	93	15.59	133	8.44	173	4.99
14	26.07	54	21.74	94	15.40	134	8.32	174	4.93
15	25.99	55	21.61	95	15.22	135	8.19	175	4.88
16	25.91	56	21.47	96	15.04	136	8.07	176	4.82
17	25.82	57	21.33	97	14.85	137	7.96	177	4.77
18	25.74	58	21.19	98	14.66	138	7.84	178	4.71
19	25.65	59	21.05	99	14.47	139	7.73	179	4.66
20	25.57	60	20.91	100	14.28	140	7.62	180	4.61
21	25.48	61	20.77	101	14.09	141	7.51	181	4.56
22	25.39	62	20.63	102	13.90	142	7.41	182	4.51
23	25.29	63	20.48	103	13.71	143	7.30	183	4.46
24	25.20	64	20.34	104	13.51	144	7.20	184	4.41
25	25.11	65	20.19	105	13.32	145	7.10	185	4.36
26	25.01	66	20.04	106	13.12	146	7.01	186	4.32
27	24.91	67	19.89	107	12.92	147	6.91	187	4.27
28	24.81	68	19.74	108	12.72	148	6.82	188	4.23
29	24.71	69	19.59	109	12.52	149	6.73	189	4.18
30	24.61	70	19.43	110	12.31	150	6.64	190	4.14
31	24.50	71	19.28	111	12.11	151	6.55	191	4.09
32	24.40	72	19.12	112	11.90	152	6.46	192	4.05
33	24.29	73	18.97	113	11.69	153	6.38	193	4.01
34	24.18	74	18.81	114	11.49	154	6.30	194	3.97
35	24.07	75	18.65	115	11.29	155	6.22	195	3.93
36	23.96	76	18.49	116	11.10	156	6.14	196	3.89
37	23.85	77	18.33	117	10.91	157	6.06	197	3.85
38	23.74	78	18.17	118	10.72	158	5.98	198	3.81
39	23.62	79	18.00	119	10.55	159	5.91	199	3.77
40	23.51	80	17.84	120	10.37	160	5.83	200	3.73

TABLE 5-3. (Continued)

Allowable Stress, f_a (ksi), in Steel for Compression Members of 50 ksi Specified Yield Stress

Main Members (Kl/r not over 120)						Main Members (Kl/r 121 to 200)			
$\dfrac{Kl}{r}$	f_a (ksi)	$\dfrac{Kl}{r}$	f_a (ksi)	$\dfrac{Kl}{r}$	f_a (ksi)	$\dfrac{Kl}{r}$	f_a (ksi)	$\dfrac{Kl}{r}$	f_a (ksi)
1	29.94	41	25.69	81	18.81	121	10.20	161	5.76
2	29.87	42	25.55	82	18.61	122	10.03	162	5.69
3	29.80	43	25.40	83	18.41	123	9.87	163	5.62
4	29.73	44	25.26	84	18.20	124	9.71	164	5.55
5	29.66	45	25.11	85	17.99	125	9.56	165	5.49
6	29.58	46	24.96	86	17.79	126	9.41	166	5.42
7	29.50	47	24.81	87	17.58	127	9.26	167	5.35
8	29.42	48	24.66	88	17.37	128	9.11	168	5.29
9	29.34	49	24.51	89	17.15	129	8.97	169	5.23
10	29.26	50	24.35	90	16.94	130	8.84	170	5.17
11	29.17	51	24.19	91	16.72	131	8.70	171	5.11
12	29.08	52	24.04	92	16.50	132	8.57	172	5.05
13	28.99	53	23.88	93	16.29	133	8.44	173	4.99
14	28.90	54	23.72	94	16.06	134	8.32	174	4.93
15	28.80	55	23.55	95	15.84	135	8.19	175	4.88
16	28.71	56	23.39	96	15.62	136	8.07	176	4.82
17	28.61	57	23.22	97	15.39	137	7.96	177	4.77
18	28.51	58	23.06	98	15.17	138	7.84	178	4.71
19	28.40	59	22.89	99	14.94	139	7.73	179	4.66
20	28.30	60	22.72	100	14.71	140	7.62	180	4.61
21	28.19	61	22.55	101	14.47	141	7.51	181	4.56
22	28.08	62	22.37	102	14.24	142	7.41	182	4.51
23	27.97	63	22.20	103	14.00	143	7.30	183	4.46
24	27.86	64	22.02	104	13.77	144	7.20	184	4.41
25	27.75	65	21.85	105	13.53	145	7.10	185	4.36
26	27.63	66	21.67	106	13.29	146	7.01	186	4.32
27	27.52	67	21.49	107	13.04	147	6.91	187	4.27
28	27.40	68	21.31	108	12.80	148	6.82	188	4.23
29	27.28	69	21.12	109	12.57	149	6.73	189	4.18
30	27.15	70	20.94	110	12.34	150	6.64	190	4.14
31	27.03	71	20.75	111	12.12	151	6.55	191	4.09
32	26.90	72	20.56	112	11.90	152	6.46	192	4.05
33	26.77	73	20.38	113	11.69	153	6.38	193	4.01
34	26.64	74	20.19	114	11.49	154	6.30	194	3.97
35	26.51	75	19.99	115	11.29	155	6.22	195	3.93
36	26.38	76	19.80	116	11.10	156	6.14	196	3.89
37	26.25	77	19.61	117	10.91	157	6.06	197	3.85
38	26.11	78	19.41	118	10.72	158	5.98	198	3.81
39	25.97	79	19.21	119	10.55	159	5.91	199	3.77
40	25.83	80	19.01	120	10.37	160	5.83	200	3.73

TABLE 5-3. (Continued)

Allowable Stress, f_a (ksi), in Steel for Compression Members of 55 ksi Specified Yield Stress

| \multicolumn{6}{c|}{Main Members (Kl/r not over 120)} | \multicolumn{4}{c}{Main Members (Kl/r 121 to 200)} |

$\dfrac{Kl}{r}$	f_a (ksi)	$\dfrac{Kl}{r}$	f_a (ksi)	$\dfrac{Kl}{r}$	f_a (ksi)	$\dfrac{Kl}{r}$	f_a (ksi)	$\dfrac{Kl}{r}$	f_a (ksi)
1	32.93	41	27.94	81	19.80	121	10.20	161	5.76
2	32.85	42	27.78	82	19.56	122	10.03	162	5.69
3	32.77	43	27.61	83	19.32	123	9.87	163	5.62
4	32.69	44	27.43	84	19.08	124	9.71	164	5.55
5	32.60	45	27.26	85	18.83	125	9.56	165	5.49
6	32.51	46	27.08	86	18.58	126	9.41	166	5.42
7	32.42	47	26.91	87	18.34	127	9.26	167	5.35
8	32.33	48	26.73	88	18.08	128	9.11	168	5.29
9	32.23	49	26.55	89	17.83	129	8.97	169	5.23
10	32.14	50	26.36	90	17.58	130	8.84	170	5.17
11	32.03	51	26.18	91	17.32	131	8.70	171	5.11
12	31.93	52	25.99	92	17.06	132	8.57	172	5.05
13	31.82	53	25.80	93	16.80	133	8.44	173	4.99
14	31.72	54	25.61	94	16.53	134	8.32	174	4.93
15	31.61	55	25.42	95	16.27	135	8.19	175	4.88
16	31.49	56	25.23	96	16.00	136	8.07	176	4.82
17	31.38	57	25.03	97	15.73	137	7.96	177	4.77
18	31.26	58	24.83	98	15.46	138	7.84	178	4.71
19	31.14	59	24.63	99	15.19	139	7.73	179	4.66
20	31.02	60	24.43	100	14.91	140	7.62	180	4.61
21	30.89	61	24.23	101	14.35	141	7.51	181	4.56
22	30.76	62	24.03	102	14.35	142	7.41	182	4.51
23	30.63	63	23.82	103	14.08	143	7.30	183	4.46
24	30.50	64	23.61	104	13.81	144	7.20	184	4.41
25	30.37	65	23.40	105	13.54	145	7.10	185	4.36
26	30.23	66	23.19	106	13.29	146	7.01	186	4.32
27	30.09	67	22.98	107	13.04	147	6.91	187	4.27
28	29.95	68	22.76	108	12.80	148	6.82	188	4.23
29	29.81	69	22.54	109	12.57	149	6.73	189	4.18
30	29.67	70	22.33	110	12.34	150	6.64	190	4.14
31	29.52	71	22.11	111	12.12	151	6.55	191	4.09
32	29.37	72	21.88	112	11.90	152	6.46	192	4.05
33	29.22	73	21.66	113	11.69	153	6.38	193	4.01
34	29.07	74	21.43	114	11.49	154	6.30	194	3.97
35	28.91	75	21.21	115	11.29	155	6.22	195	3.93
36	28.76	76	20.98	116	11.10	156	6.14	196	3.89
37	28.60	77	20.75	117	10.91	157	6.06	197	3.85
38	28.44	78	20.51	118	10.72	158	5.98	198	3.81
39	28.28	79	20.28	119	10.55	159	5.91	199	3.77
40	28.11	80	20.04	120	10.37	160	5.83	200	3.73

TABLE 5-3. (Continued)

Allowable Stress, f_a (ksi), in Steel for Compression Members of 60 ksi Specified Yield Stress

Main Members (Kl/r not over 120)						Main Members (Kl/r 121 to 200)			
$\dfrac{Kl}{r}$	f_a (ksi)	$\dfrac{Kl}{r}$	f_a (ksi)	$\dfrac{Kl}{r}$	f_a (ksi)	$\dfrac{Kl}{r}$	f_a (ksi)	$\dfrac{Kl}{r}$	f_a (ksi)
1	35.92	41	30.15	81	20.65	121	10.20	161	5.76
2	35.83	42	29.95	82	20.37	122	10.03	162	5.69
3	35.74	43	29.75	83	20.09	123	9.87	163	5.62
4	35.64	44	29.55	84	19.80	124	9.71	164	5.55
5	35.54	45	29.35	85	19.51	125	9.56	165	5.49
6	35.44	46	29.15	86	19.22	126	9.41	166	5.42
7	35.34	47	28.94	87	18.93	127	9.26	167	5.35
8	35.23	48	28.73	88	18.63	128	9.11	168	5.29
9	35.12	49	28.52	89	18.34	129	8.97	169	5.23
10	35.01	50	28.31	90	18.04	130	8.84	170	5.17
11	34.89	51	28.09	91	17.73	131	8.70	171	5.11
12	34.77	52	27.87	92	17.43	132	8.57	172	5.05
13	34.65	53	27.66	93	17.12	133	8.44	173	4.99
14	34.52	54	27.43	94	16.81	134	8.32	174	4.93
15	34.40	55	27.21	95	16.50	135	8.19	175	4.88
16	34.27	56	26.98	96	16.19	136	8.07	176	4.82
17	34.13	57	26.76	97	15.87	137	7.96	177	4.77
18	34.00	58	26.53	98	15.55	138	7.84	178	4.71
19	33.86	59	26.29	99	15.24	139	7.73	179	4.66
20	33.71	60	26.06	100	14.93	140	7.62	180	4.61
21	33.57	61	25.82	101	14.64	141	7.51	181	4.56
22	33.42	62	25.58	102	14.35	142	7.41	182	4.51
23	33.27	63	25.34	103	14.08	143	7.30	183	4.46
24	33.12	64	25.10	104	13.81	144	7.20	184	4.41
25	32.96	65	24.86	105	13.54	145	7.10	185	4.36
26	32.81	66	24.61	106	13.29	146	7.01	186	4.32
27	32.65	67	24.36	107	13.04	147	6.91	187	4.27
28	32.48	68	24.11	108	12.80	148	6.82	188	4.23
29	32.32	69	23.86	109	12.57	149	6.73	189	4.18
30	32.15	70	23.60	110	12.34	150	6.64	190	4.14
31	31.98	71	23.34	111	12.12	151	6.55	191	4.09
32	31.81	72	23.08	112	11.90	152	6.46	192	4.05
33	31.63	73	22.82	113	11.69	153	6.38	193	4.01
34	31.45	74	22.56	114	11.49	154	6.30	194	3.97
35	31.28	75	22.29	115	11.29	155	6.22	195	3.93
36	31.09	76	22.02	116	11.10	156	6.14	196	3.89
37	30.91	77	21.75	117	10.91	157	6.06	197	3.85
38	30.72	78	21.48	118	10.72	158	5.98	198	3.81
39	30.53	79	21.21	119	10.55	159	5.91	199	3.77
40	30.34	80	20.93	120	10.37	160	5.83	200	3.73

TABLE 5-3. (Continued)

Allowable Stress, f_a (ksi), in Steel for Compression Members of 65 ksi Specified Yield Stress

Main Members (Kl/r not over 120)						Main Members (Kl/r 121 to 200)			
$\frac{Kl}{r}$	f_a (ksi)	$\frac{Kl}{r}$	f_a (ksi)	$\frac{Kl}{r}$	f_a (ksi)	$\frac{Kl}{r}$	f_a (ksi)	$\frac{Kl}{r}$	f_a (ksi)
1	38.90	41	32.30	81	21.36	121	10.20	161	5.76
2	38.81	42	32.08	82	21.03	122	10.03	162	5.69
3	38.70	43	31.85	83	20.70	123	9.87	163	5.62
4	38.59	44	31.62	84	20.37	124	9.71	164	5.55
5	38.48	45	31.39	85	20.04	125	9.56	165	5.49
6	38.37	46	31.35	86	19.70	126	9.41	166	5.42
7	38.25	47	30.92	87	19.36	127	9.26	167	5.35
8	38.13	48	30.68	88	19.02	128	9.11	168	5.29
9	38.00	49	30.43	89	18.67	129	8.97	169	5.23
10	37.87	50	30.19	90	18.32	130	8.84	170	5.17
11	37.74	51	29.94	91	17.97	131	8.70	171	5.11
12	37.61	52	29.69	92	17.62	132	8.57	172	5.05
13	37.47	53	29.44	93	17.26	133	8.44	173	4.99
14	37.32	54	29.18	94	16.90	134	8.32	174	4.93
15	37.18	55	28.92	95	16.55	135	8.19	175	4.88
16	37.03	56	28.66	96	16.20	136	8.07	176	4.82
17	36.87	57	28.40	97	15.87	137	7.96	177	4.77
18	36.72	58	28.14	98	15.55	138	7.84	178	4.71
19	36.56	59	27.87	99	15.24	139	7.73	179	4.66
20	36.40	60	27.60	100	14.93	140	7.62	180	4.61
21	36.23	61	27.33	101	14.64	141	7.51	181	4.56
22	36.06	62	27.05	102	14.35	142	7.41	182	4.51
23	35.89	63	26.78	103	14.08	143	7.30	183	4.46
24	35.71	64	26.50	104	13.81	144	7.20	184	4.41
25	35.54	65	26.21	105	13.54	145	7.10	185	4.36
26	35.36	66	25.93	106	13.29	146	7.01	186	4.32
27	35.17	67	25.64	107	13.04	147	6.91	187	4.27
28	34.99	68	25.35	108	12.80	148	6.82	188	4.23
29	34.80	69	25.06	109	12.57	149	6.73	189	4.18
30	34.60	70	24.76	110	12.34	150	6.64	190	4.14
31	34.41	71	24.47	111	12.12	151	6.55	191	4.09
32	34.21	72	24.17	112	11.90	152	6.46	192	4.05
33	34.01	73	23.87	113	11.69	153	6.38	193	4.01
34	33.81	74	23.56	114	11.49	154	6.30	194	3.97
35	33.60	75	23.25	115	11.29	155	6.22	195	3.93
36	33.39	76	22.94	116	11.10	156	6.14	196	3.89
37	33.18	77	22.63	117	10.91	157	6.06	197	3.85
38	32.96	78	22.32	118	10.72	158	5.98	198	3.81
39	32.75	79	22.00	119	10.55	159	5.91	199	3.77
40	32.53	80	21.68	120	10.37	160	5.83	200	3.73

TABLE 5-3. (Continued)

Allowable Stress, f_a (ksi), in Steel for Compression Members of 90 ksi Specified Yield Stress

| \multicolumn{6}{c|}{Main Members (Kl/r not over 120)} | \multicolumn{4}{c}{Main Members (Kl/r 121 to 200)} |

$\frac{Kl}{r}$	f_a (ksi)	$\frac{Kl}{r}$	f_a (ksi)	$\frac{Kl}{r}$	f_a (ksi)	$\frac{Kl}{r}$	f_a (ksi)	$\frac{Kl}{r}$	f_a (ksi)
1	53.84	41	42.39	81	22.76	121	10.20	161	5.76
2	53.68	42	42.00	82	22.21	122	10.03	162	5.69
3	53.51	43	41.59	83	21.68	123	9.87	163	5.62
4	53.33	44	41.19	84	21.16	124	9.71	164	5.55
5	53.15	45	40.78	85	20.67	125	9.56	165	5.49
6	52.95	46	40.36	86	20.19	126	9.41	166	5.42
7	52.75	47	39.94	87	19.73	127	9.26	167	5.35
8	52.55	48	39.51	88	19.28	128	9.11	168	5.29
9	52.33	49	39.08	89	18.85	129	8.97	169	5.23
10	52.11	50	38.65	90	18.44	130	8.84	170	5.17
11	51.89	51	38.21	91	18.03	131	8.70	171	5.11
12	51.65	52	37.77	92	17.64	132	8.57	172	5.05
13	51.41	53	37.32	93	17.27	133	8.44	173	4.99
14	51.17	54	36.86	94	16.90	134	8.32	174	4.93
15	50.92	55	36.41	95	16.55	135	8.19	175	4.88
16	50.66	56	35.94	96	16.20	136	8.07	176	4.82
17	50.39	57	35.47	97	15.87	137	7.96	177	4.77
18	50.12	58	35.00	98	15.55	138	7.84	178	4.71
19	49.85	59	34.52	99	15.24	139	7.73	179	4.66
20	49.56	60	34.04	100	14.93	140	7.62	180	4.61
21	49.28	61	33.56	101	14.64	141	7.51	181	4.56
22	48.98	62	33.06	102	14.35	142	7.41	182	4.51
23	48.68	63	32.57	103	14.08	143	7.30	183	4.46
24	48.38	64	32.07	104	13.81	144	7.20	184	4.41
25	48.07	65	31.56	105	13.54	145	7.10	185	4.36
26	47.75	66	31.05	106	13.29	146	7.01	186	4.32
27	47.43	67	30.53	107	13.04	147	6.91	187	4.27
28	47.10	68	30.01	108	12.80	148	6.82	188	4.23
29	46.77	69	29.48	109	12.57	149	6.73	189	4.18
30	46.43	70	28.95	110	12.34	150	6.64	190	4.14
31	46.09	71	28.41	111	12.12	151	6.55	191	4.09
32	45.74	72	27.87	112	11.90	152	6.46	192	4.05
33	45.39	73	27.32	113	11.69	153	6.38	193	4.01
34	45.03	74	26.77	114	11.49	154	6.30	194	3.97
35	44.67	75	26.21	115	11.29	155	6.22	195	3.93
36	44.30	76	25.65	116	11.10	156	6.14	196	3.89
37	43.93	77	25.08	117	10.91	157	6.06	197	3.85
38	43.55	78	24.50	118	10.72	158	5.98	198	3.81
39	43.17	79	23.92	119	10.55	159	5.91	199	3.77
40	42.78	80	23.33	120	10.37	160	5.83	200	3.73

TABLE 5-3. (Continued)

Allowable Stress, f_a (ksi), in Steel for Compression Members of 100 ksi Specified Yield Stress

Main Members (Kl/r not over 120)						Main Members (Kl/r 121 to 200)			
$\dfrac{Kl}{r}$	f_a (ksi)	$\dfrac{Kl}{r}$	f_a (ksi)	$\dfrac{Kl}{r}$	f_a (ksi)	$\dfrac{Kl}{r}$	f_a (ksi)	$\dfrac{Kl}{r}$	f_a (ksi)
1	59.82	41	46.12	81	22.76	121	10.20	161	5.76
2	59.62	42	45.64	82	22.21	122	10.03	162	5.69
3	59.42	43	45.16	83	21.68	123	9.87	163	5.62
4	59.21	44	44.67	84	21.16	124	9.71	164	5.55
5	58.99	45	44.17	85	20.67	125	9.56	165	5.49
6	58.76	46	43.67	86	20.19	126	9.41	166	5.42
7	58.53	47	43.17	87	19.73	127	9.26	167	5.35
8	58.28	48	42.65	88	19.28	128	9.11	168	5.29
9	58.03	49	42.14	89	18.85	129	8.97	169	5.23
10	57.77	50	41.61	90	18.44	130	8.84	170	5.17
11	57.50	51	41.08	91	18.03	131	8.70	171	5.11
12	57.22	52	40.55	92	17.64	132	8.57	172	5.05
13	56.93	53	40.00	93	17.27	133	8.44	173	4.99
14	56.64	54	39.46	94	16.90	134	8.32	174	4.93
15	56.34	55	38.90	95	16.55	135	8.19	175	4.88
16	56.03	56	38.35	96	16.20	136	8.07	176	4.82
17	55.72	57	37.78	97	15.87	137	7.96	177	4.77
18	55.39	58	37.21	98	15.55	138	7.84	178	4.71
19	55.06	59	36.63	99	15.24	139	7.73	179	4.66
20	54.72	60	36.05	100	14.93	140	7.62	180	4.61
21	54.38	61	35.46	101	14.64	141	7.51	181	4.56
22	54.03	62	34.87	102	14.35	142	7.41	182	4.51
23	53.67	63	34.26	103	14.08	143	7.30	183	4.46
24	53.30	64	33.66	104	13.81	144	7.20	184	4.41
25	52.93	65	33.04	105	13.54	145	7.10	185	4.36
26	52.55	66	32.42	106	13.29	146	7.01	186	4.32
27	52.17	67	31.80	107	13.04	147	6.91	187	4.27
28	51.78	68	31.16	108	12.80	148	6.82	188	4.23
29	51.38	69	30.52	109	12.57	149	6.73	189	4.18
30	50.97	70	29.88	110	12.34	150	6.64	190	4.14
31	50.56	71	29.22	111	12.12	151	6.55	191	4.09
32	50.15	72	28.56	112	11.90	152	6.46	192	4.05
33	49.72	73	27.90	113	11.69	153	6.38	193	4.01
34	49.29	74	27.22	114	11.49	154	6.30	194	3.97
35	48.86	75	26.54	115	11.29	155	6.22	195	3.93
36	48.42	76	25.85	116	11.10	156	6.14	196	3.89
37	47.97	77	25.19	117	10.91	157	6.06	197	3.85
38	47.51	78	24.54	118	10.72	158	5.98	198	3.81
39	47.05	79	23.93	119	10.55	159	5.91	199	3.77
40	46.59	80	23.33	120	10.37	160	5.83	200	3.73

Example 5-2. Given a steel compression member with the characteristics below, find the maximum compression load the member can carry.

Material: A-36 steel (F_y = 36 ksi)
Length: 10 ft
Structural shape: structural tubing TS 6 × 6 × 0.50
Cross-sectional area: A = 10.1 in.²
Minimum radius of gyration: r_y = 2.19 in.

Solution: compute the slenderness ratio:

$$\frac{l}{r_{min}} = \frac{(10)(12)}{2.19} = 54.79.$$

For f_y = 36 ksi and l/r_{min} = 54.79, find by interpolation in Table 5-3: f_a = 17.92 ksi.

Thus, substituting in formula 5.7, we find

$$P = (10.1) \times (17.92) = 180.99 \text{ K}.$$

STEEL TENSION MEMBERS

Tension members in steel are designed according to the formula 5.12.

$$N = A f_t \tag{5.12}$$

where:

N = maximum axial tension load (kips),
A = cross-sectional area of the member (in.²),
f_t = allowable tension stress (ksi).

The allowable tensile stress according to the AISC is given by formula 5.13.

$$f_t = 0.60 f_y \tag{5.13}$$

where:

f_y = yield strength of the particular type of steel being considered (ksi).

The maximum slenderness ratio for tension members in accordance with the AISC specifications should not be larger than 240.

Example 5-3. Given the same data as in the previous example, calculate the maximum tensile load the member can carry.

Solution: From formula 5.13, the allowable stress is

$$f_t = 0.60 \times 36 = 21.6 \text{ ksi.}$$

Therefore, substituting in formula 5.12, we obtain

$$N = (10.1) \times (21.6) = 218.16 \text{ K.}$$

LATTICEWORK OR LACING

When trussing is applied as a subsystem for the construction of so-called "built-up" members, it is usually referred to as "latticework" or "lacing." As in any truss, all members are stressed only in tension or compression, but the design of latticework is different from that of conventional trusses, as illustrated below.

Let us consider built-up laced columns. Their typical construction consists of two, three, four, or more longitudinal members that carry the compression load. It also includes many diagonal members that lace the main members together. The cross-section of the built-up column can be triangular, square, or any other desired configuration. Sometimes, with only two main members, one can obtain a square cross-section, as in the case of lacing together two channels. The material generally used for built-up members is steel. The shape of the members can be any one of a large variety of available configurations, varying from rolled to tubular sections for the main members, while plates or angles are most common for the lacing members. According to the AISC, the longitudinal members are designed as struts, whose unsupported length is the distance between two consecutive points of intersection with the lacing members.

Lacing members must be designed as struts that carry an axial compressive force calculated as follows. Consider 2% of the axial load on the whole built-up column. Divide this force by the number of lacing planes. Take this last force and apply it perpendicularly to the longitudinal member at a nodal point. Compute now the projection of this force in the direction of the diagonal lacing member to be designed. This is the axial compressive force to be used for the design of the member.

Example 5-4. Given: an 18 ft long, single laced column, made of three longitudinal steel pipes arranged at the corners of an equilateral triangle whose sides are 12 in. long; the lacing members form a 60° angle with the longitudinal pipes; and the axial compression load on the laced column is 300 K. Determine the length of the lacing members and the individual longitudinal main members. Compute also the axial compression force N in the lacing members.

Solution: The length of the lacing members is

$$l = \frac{12}{\cos 30°} = 13.86 \text{ in.}$$

The length of the longitudinal members is

$$l' = 2 \times 12 \times \tan 30° = 13.86 \text{ in.}$$

The force F that is perpendicular to the longitudinal member is

$$F = 0.02 \times \frac{300}{3} = 2 \text{ K.}$$

N is the component of F in the direction of the lacing member and its value is

$$N = \frac{2}{\cos 30°} = 2.31 \text{ K.}$$

REINFORCED CONCRETE COMPRESSION MEMBERS

Following the code of the American Concrete Institute (ACI), reinforced concrete members in compression which have a square or rectangular cross-section and are reinforced with longitudinal bars and ties are designed in accordance with the following formula:

$$P = 0.85\,(0.25\,A_g f'_c + f_s A_s)/1000$$

where:

P = axial load capacity (kips),
A_g = gross area of concrete section (in.²),
f'_c = ultimate compressive strength of concrete at 28 days (psi),
f_s = allowable stress in the reinforcement (psi); assume $f_s = 0.4\,f_g \leq$ 30,000 psi,
A_s = cross-sectional area of steel reinforcement (in.²); notice that A_s can vary between the following limits: $0.01\,A_g \leq A_s \leq 0.08\,A_g$.

To facilitate the calculations, the previous formula has been tabulated for several concrete and steel strengths. The axial capacity in kips has been calculated for many cross-sectional sizes with minimum and maximum percentages of reinforcement (see Table 5-5).

Example 5-5. Given a reinforced concrete compression member with characteristics below, compute the maximum compression load the member can carry.

f'_c = 4000 psi
f_s = 20 ksi
A_g = 12 in. × 12 in. = 144 in.²
A_s = 0.04 A_g = 5.76 in.²

Solution: Using Table 5-5, we find that the concrete alone can carry 122 K, and just by simple interpolation (24 + 196)/2, we find that the reinforcement can carry 110 K. Therefore, the maximum total load is 232 K.

352 SIMPLIFIED TRUSS DESIGN

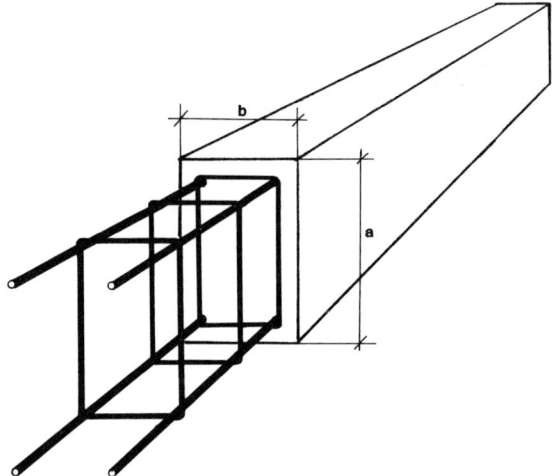

Figure 5-1.

TABLE 5-4. Properties of Steel Bars

Bar Designation	Diameter (in.)	Cross-sectional Area (in.²)	Perimeter (in.)	Weight (lb/ft)
#2	0.250	0.05	0.786	0.167
#3	0.375	0.11	1.178	0.376
#4	0.500	0.20	1.571	0.668
#5	0.625	0.31	1.963	1.043
#6	0.750	0.44	2.356	1.502
#7	0.875	0.60	2.749	2.044
#8	1.000	0.79	3.142	2.670
#9	1.128	1.00	3.544	3.400
#10	1.270	1.27	3.990	4.303
#11	1.410	1.56	4.430	5.313
#14S	1.693	2.25	5.320	7.65
#18S	2.275	4.00	7.090	13.60

TABLE 5-5. Axial Load (kips) on Tied Rectangular Reinforced Concrete Columns

Column Cross Section, $a \times b$ (in.)	Cross-Sectional Area, A_g (in.²)	Load Carried by Concrete Only (kips)			Load Carried by Steel Reinforcement Only (kips)								
					$f_s = 16$ (ksi)		$f_s = 20$ (ksi)		$f_s = 24$ (ksi)		$f_s = 30$ (ksi)		
		$f'_c = 3$ (ksi)	$f'_c = 4$ (ksi)	$f'_c = 5$ (ksi)	Minimum (1%)	Maximum (8%)	Minimum (1%)	Maximum (8%)	Minimum (1%)	Maximum (8%)	Minimum (1%)	Maximum (8%)
---	---	---	---	---	---	---	---	---	---	---	---	---
10 × 10	100	64	85	106	14	109	17	136	20	163	26	204
10 × 12	120	77	102	128	16	131	20	163	24	196	31	245
10 × 14	140	89	119	149	19	152	24	190	29	228	36	286
10 × 16	160	102	136	170	22	174	27	218	33	261	41	326
10 × 18	180	115	153	191	24	196	31	245	37	294	46	367
12 × 12	144	92	122	153	20	157	24	196	29	235	37	294
12 × 14	168	107	143	179	23	183	29	228	34	274	43	343
12 × 16	192	122	163	204	26	209	33	261	39	313	49	392
12 × 18	216	138	184	230	29	235	37	294	44	353	55	441
12 × 20	240	153	204	255	33	261	41	326	49	392	61	490
14 × 14	196	125	167	208	27	213	33	267	40	320	50	400
14 × 16	224	143	190	238	30	244	38	305	46	366	57	457
14 × 18	252	161	214	268	34	274	43	343	51	411	64	514
14 × 20	280	179	238	298	38	305	48	381	57	457	71	571
14 × 22	308	196	262	327	42	335	52	419	63	503	79	628
16 × 16	256	163	218	272	35	279	44	348	52	418	65	522
16 × 18	288	184	245	306	39	313	49	392	59	470	73	588
16 × 20	320	204	272	340	44	348	54	435	65	522	82	653
16 × 22	352	224	299	374	48	383	60	479	72	574	90	718
16 × 24	384	245	326	408	52	418	65	522	78	627	98	783

TABLE 5-5. Axial Load (kips) on Tied Rectangular Reinforced Concrete Columns

Column Cross Section, $a \times b$ (in.)	Cross-Sectional Area, A_g (in.²)	Load Carried by Concrete Only (kips)			Load Carried by Steel Reinforcement Only (kips)							
		$f'_c = 3$ (ksi)	$f'_c = 4$ (ksi)	$f'_c = 5$ (ksi)	$f_s = 16$ (ksi)		$f_s = 20$ (ksi)		$f_s = 24$ (ksi)		$f_s = 30$ (ksi)	
					Minimum (1%)	Maximum (8%)	Minimum (1%)	Maximum (8%)	Minimum (1%)	Maximum (8%)	Minimum (1%)	Maximum (8%)
18 × 18	324	207	275	344	44	353	55	441	66	529	83	661
18 × 20	360	230	306	383	49	392	61	490	73	588	92	734
18 × 22	396	252	337	421	54	431	67	539	81	646	101	808
18 × 24	432	275	367	459	59	470	73	588	88	705	110	881
18 × 26	468	298	398	497	64	509	80	636	95	764	119	955
20 × 20	400	255	340	425	54	435	68	544	82	653	102	816
20 × 22	440	281	374	468	60	479	75	598	90	718	112	898
20 × 24	480	306	408	510	65	522	82	653	98	783	122	979
20 × 26	520	332	442	553	71	566	88	707	106	849	133	1061
20 × 28	560	357	476	595	76	609	95	762	114	914	143	1142
22 × 22	484	309	411	514	66	527	82	658	99	790	123	987
22 × 24	528	337	449	561	72	574	90	718	108	862	135	1077
22 × 26	572	365	486	608	78	622	97	778	117	934	146	1167
22 × 28	616	393	524	655	84	670	105	838	126	1005	157	1257
22 × 30	660	421	561	701	90	718	112	898	135	1077	168	1346
24 × 24	576	367	490	612	78	627	98	783	118	940	147	1175
24 × 26	624	398	530	663	85	679	106	849	127	1018	159	1273
24 × 28	672	428	571	714	91	731	114	914	137	1097	171	1371
24 × 30	720	459	612	765	98	783	122	979	147	1175	184	1469

CONNECTIONS

Without a doubt, connections are the critical part in the construction of trusses. They should be so designed that the axes of all members converging at a joint meet each other at a single point. If this does not occur, the forces are no longer axial and their eccentricity generates unwanted bending stresses at the ends of the members. All this can be avoided only with accurate design of the connection details.

Another critical aspect of connections is that they determine the final size of the members. For instance, when mechanical fasteners are used, their penetration into the members reduce the members' cross-section to a smaller net area that governs their final size.

Regardless of the fastening method employed, connections are a fundamental part in the design of trusses. They are definitely not secondary details to be solved at the end. Consider, for instance, Leonardo's drawings in Plate 5-1. They illustrate the lashing of heavy timbers for a temporary truss bridge. Although experienced carpenters and master builders would have known how to fasten the timbers together, Leonardo's care in showing the exact lashing technique he desired indicates how important he considered these connections to be.

Another conceptual point in the design of connections involves their durability; they are free from inspection, maintenance, and other conditions. This was indeed a major concept in the Renaissance and also in Roman and Greek times, for artwork and building components in general. Bronze hardware for timber connectors was popular in Roman technology. We remember, for instance, that in the bridge over the Rhine, Caesar's legions used bronze hardware produced in Spain for assembling the structure. Our technology, of course, has much to offer in this respect.

356 SIMPLIFIED TRUSS DESIGN

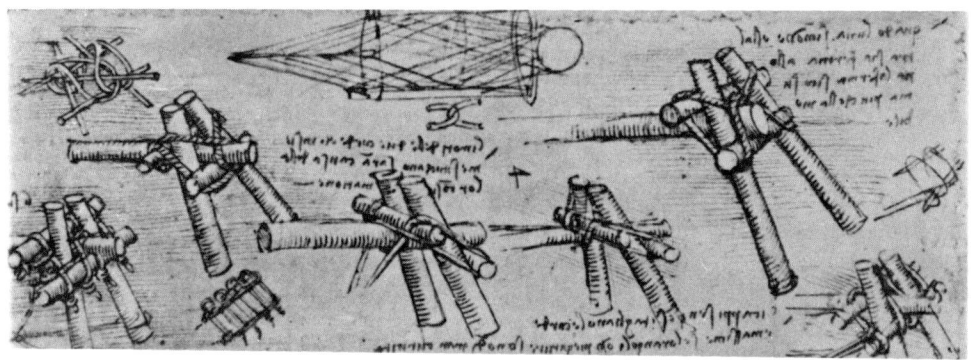

Plate 5-1. Connections are the critical part of the design and construction of trusses. Drawing by Leonardo da Vinci specifies lashing details for timber members of a temporary truss bridge. Notice the minute care with which the great master illustrated how the lashing had to be executed. Although experienced master builders would have known how to fasten the timbers together, Leonardo's intentions were very specific and went beyond conventional techniques.

Wood connections. Heel joints at the supports for heavy timber trusses, reminiscent of the past, can still have a significant influence on the design of timber trusses today. Various examples of these connections are illustrated in Figure 5-2. Additional connection details for timber and steel members are illustrated in Figure 5-3. Present construction for timber trusses employs metal fasteners of different types (see Figure 5-4).

Steel connections. New welded connections have totally eliminated at present all the rivets, bolts, and heavy plates of the old trusses. A new generation of steel trusses has emerged. Lightweight, smooth, and with an intriguing new look, these trusses have conquered a dominant place in architecture.

Direct connections of angles, tees, wide flanges, and round and rectangular tubing produce geometrical composition of linear shapes that flow visually as a continuum. The rational configuration of the geometric forms, which give a sense of order and logic, is carried down to the detail of the connection, where a similar geometric logic is applied to the fitting of the joining members.

Members can be overlapped and welded. This creates a small eccentricity, since the members are not exactly in the same plane.

Members can be cut and fitted so that they can be welded and be in the same plane without the necessity of gusset plates. Gusset plates, however, may be necessary at times. When used, they can be plain and smooth and harmonize with the simplicity of the structure, unlike the old types.

Tubular truss connections. Tubular trussing of the present day was first made possible by the acceptance of welded connections, originally in Europe and later in the United States. The tube, therefore, could finally be used and its structural efficiency, especially for compression members, could then be exploited. Moreover, the absence of gusset plates, usually the most common case, enables the smooth tubular shapes to flow, without discontinuities, throughout the whole structure.

Sometimes, however, when loads are particularly large, gusset plates must be used. This is to allow longer welds for additional strength or to solve erection problems. In fact, using gusset plates permits some welding to be done in the shop rather than in the field. Furthermore, the length of the tubular members in these cases is no longer a critical factor since it can be compensated for by the plate itself. A major inconvenience of gusset plates is that they stiffen the ends of the tubes and produce stress concentration in the pipes. To reduce this phenomenon, the ends of the gusset plates must be tapered.

When gusset plates are not used and several smaller members are welded to a main member, it is advisable at times to reinforce the main member by welding a short sleeve around it at the joint.

358 SIMPLIFIED TRUSS DESIGN

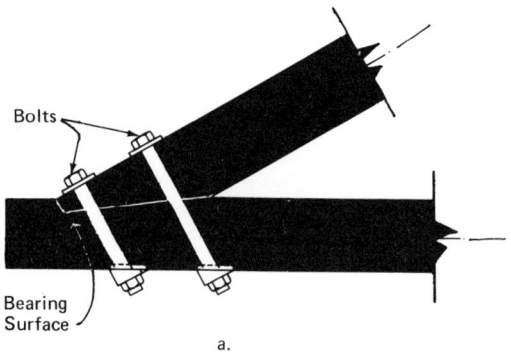

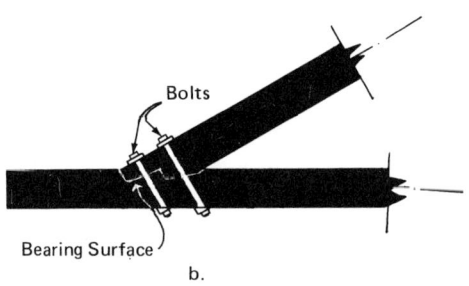

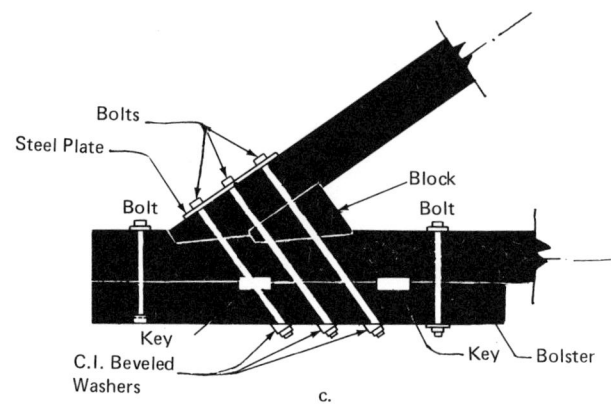

Figure 5-2. Heel joints at supports for timber trusses.
    a. Compression from the inclined strut is transferred to the horizontal tie, in part by direct bearing and in part by friction between the two members clamped together by bolts.
    b. This joint is an improvement over (a) because it increases the bearing area.
    c. If the indentation for the bearing surfaces removes too much material from the horizontal tie, this must be reinforced by an additional member called the *bolster*, which is connected to the tie with timber keys and bolts (note the use of a timber block that allows additional clamping bolts to increase the frictional persistence of the joints).

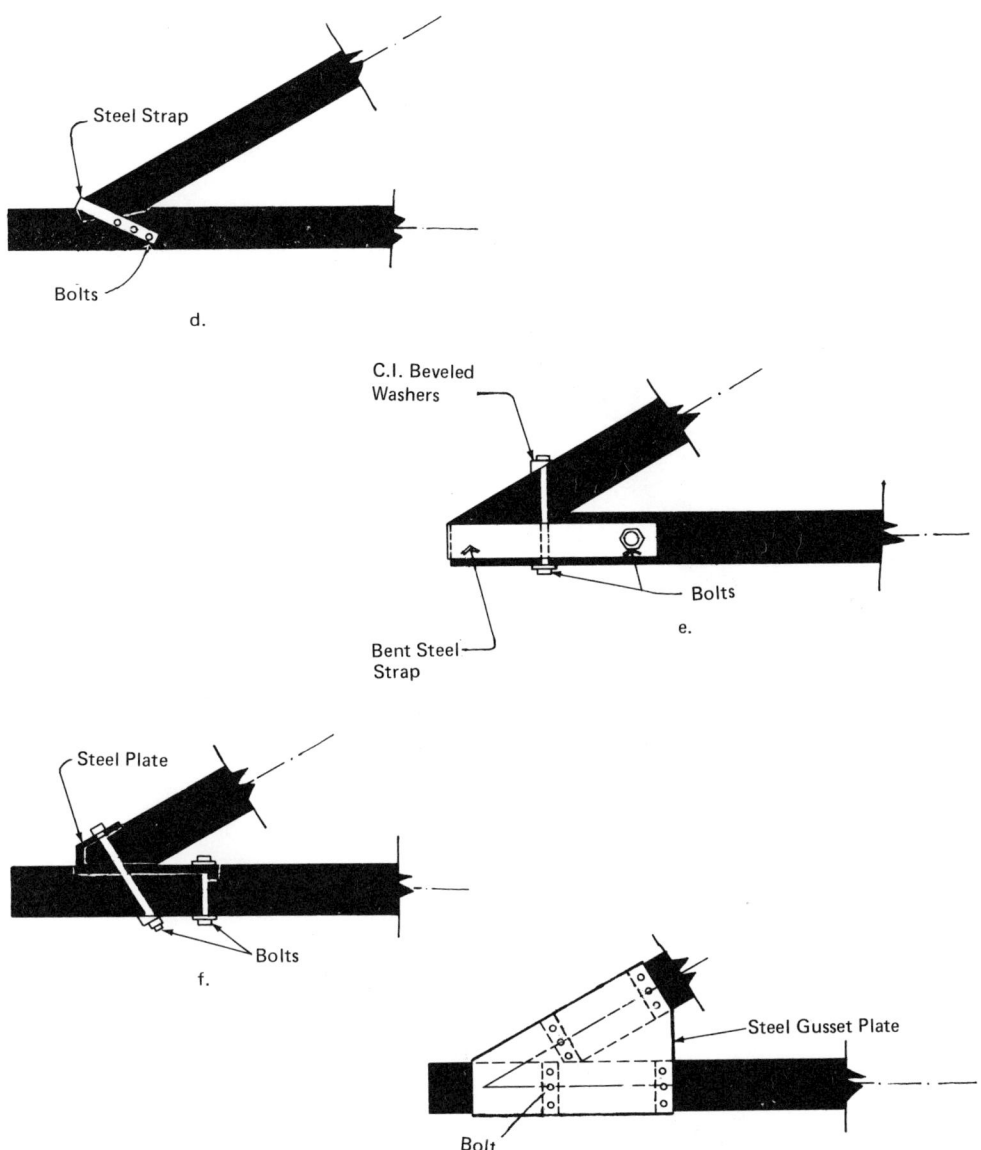

d. The steel strap is used for the overall stability of the joint, but its clamping force is not adequate to provide friction.
e. A bent steel strap receives the compression force from the inclined strut bearing against it, and it receives a tension force from the horizontal tie through the bolt perpendicular to the strut.
f. The forces are actually transferred from one member to the other through the interposition of a steel plate, properly shaped.
g. The forces are transferred from one member to the other by means of a steel gusset plate.

360 SIMPLIFIED TRUSS DESIGN

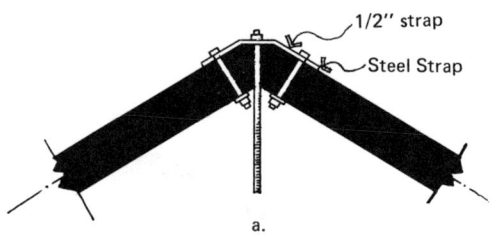

a.

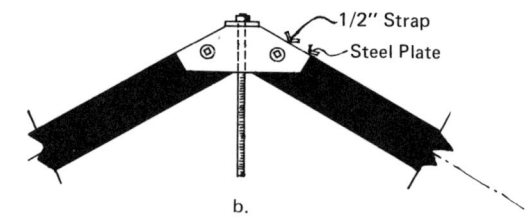

b.

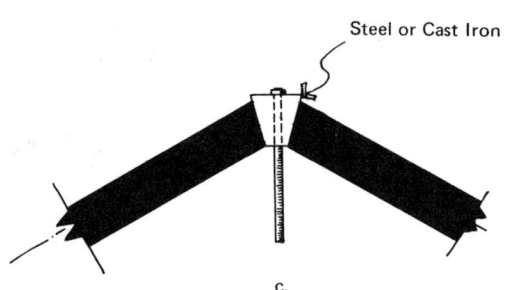

c.

Figure 5-3. Connections for single-piece wood struts and steel ties.
    a. Steel strap distributes the force from the tie to the struts and keeps the struts in place.
    b. Steel strap distributes the force from the tie to the struts. A gusset steel plate keeps the struts in place.
    c. A steel or cast iron stirrup transfers the force from the tie to the struts and keeps all the members in place.

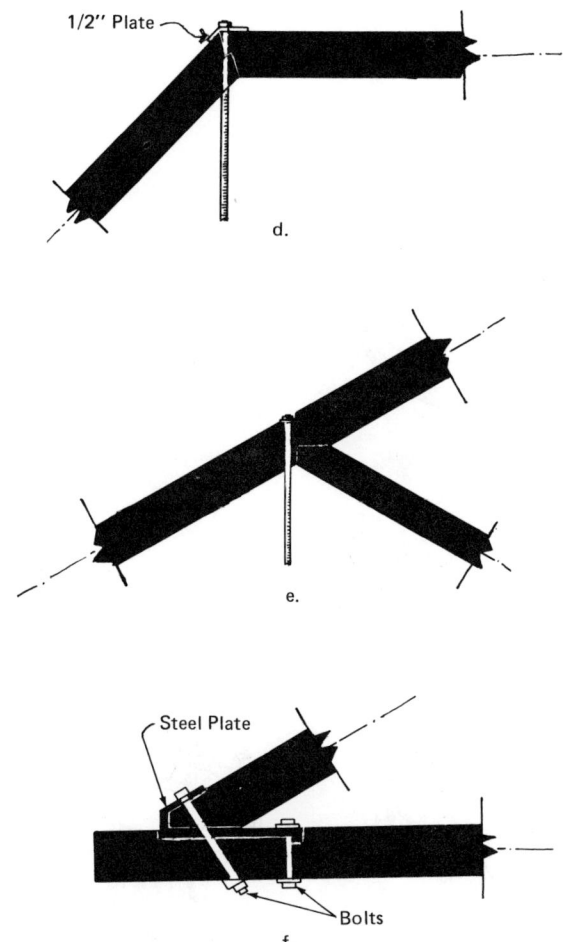

d. Steel plate distributes the force from the tie to the struts.
e. A washer distributes the force from the tie to the struts when the force is small.
f. Anchoring a timber tie to other members can be done by using steel straps.

362 SIMPLIFIED TRUSS DESIGN

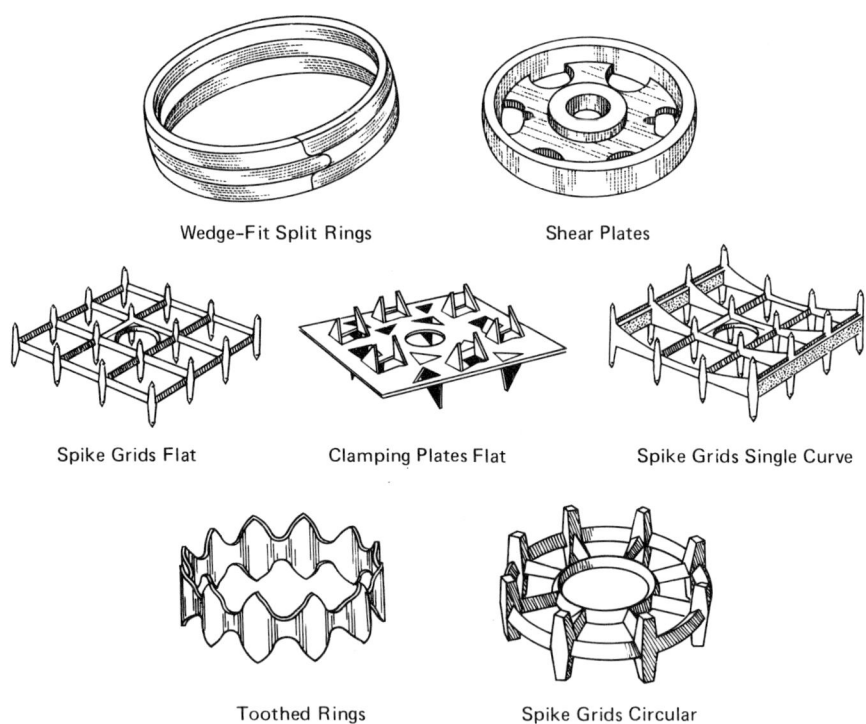

Figure 5-4. Metal connectors currently used for fastening timber members.

MEMBER AND CONNECTION DESIGN 363

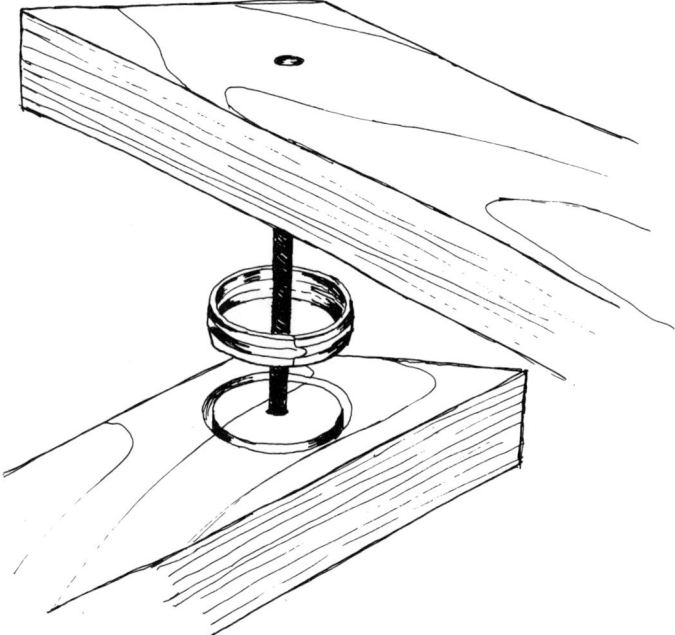

Figure 5-5. Efficient and simple, split ring connectors imbedded between two timbers develop a type of joint of a certain elegance. Such a joint does not violate the integrity of the wood to any great extent and does not torture the material as nails and spikes do. Shown in the figure is the setting of the wedge-shaped split ring in a circular groove cut in the face of the two timbers which have to be connected.

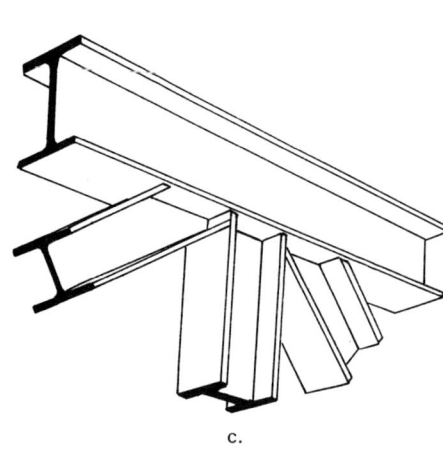
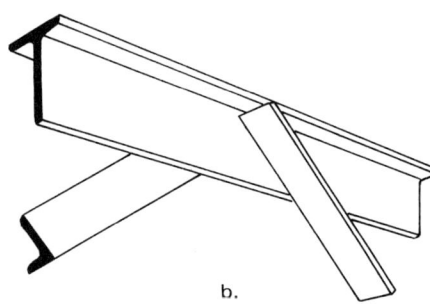
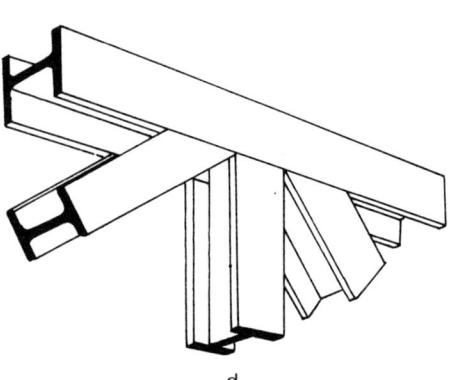

Figure 5-6a. Typical use of gusset plates that, unlike the old generation of connections, do not clutter the joint but leave it crisp and uncomplicated.
    b. Simple and clean connection of diagonal angles, welded to the sides of the T's stem.
    c. Connection of wide-flange, t- and double-angles shapes showing a rational fitting of the members, including the slotting of the wide-flange intersected by the tee.
    d. Connection of heavy wide-flange shapes that, because of the clean fitting of the members, retain a sense of lightness in spite of the size of the members and the magnitude of the axial forces.
    e. Wide-flange members connected to the flange of a top chord. The sharp fitting of the members keeps the joint simple and uncluttered.

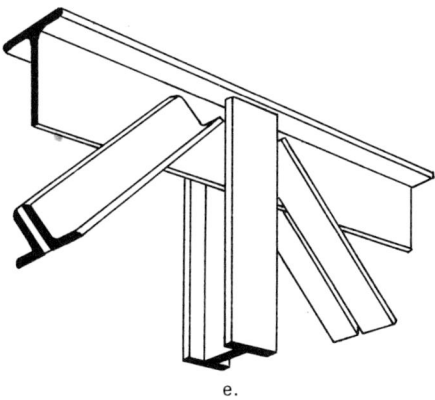

e.

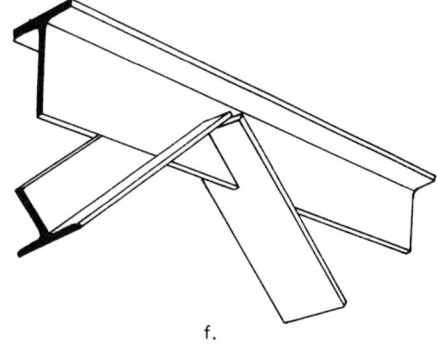

f.

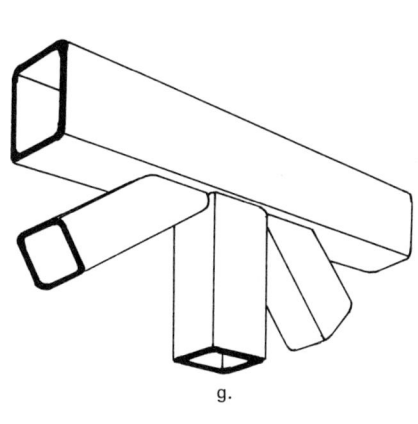

g.

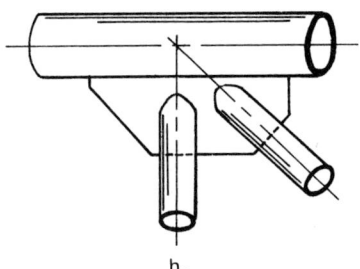

h.

f. Wide-flange members connected to the web and flanges of a chord member. Notice the sense of lightness the connection retains, in spite of heavy members, due to the clean fitting of the members.
g. Clear and smooth connection of rectangular tubular members that create a spatial linear continuum.
h. When gusset plates are necessary in tubular truss construction, it is advisable to taper the edges to reduce the stiffness of the pipes welded to them. This will reduce the concentration of stresses that occur in the pipes at the point where they meet the edges of the plate.

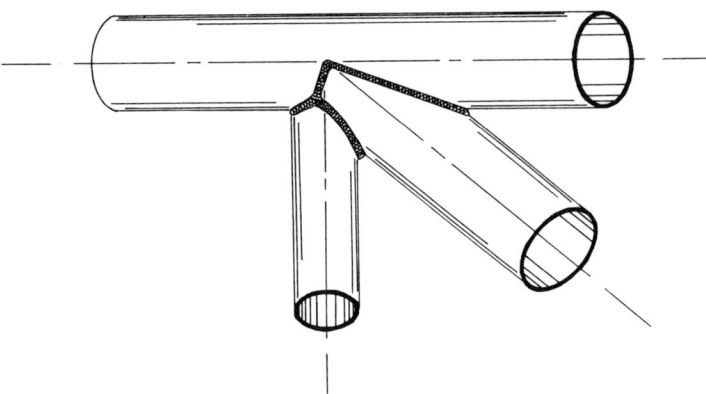

Figure 5-7. Fitting and welding detail of pipes of different sizes at a typical welded joint without gusset plates or other reinforcements.

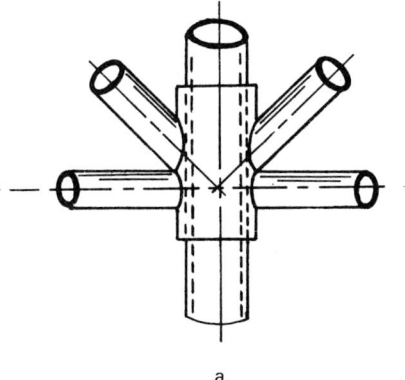

a.

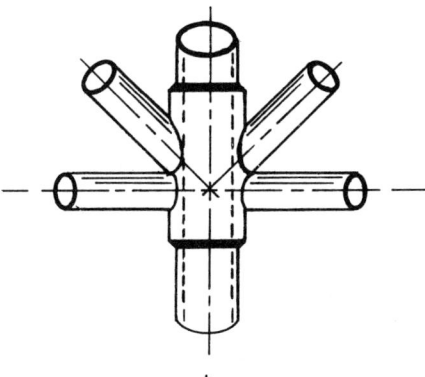

b.

Figure 5-8. When the wall thickness of the major member (the most stressed one) is not adequate, the pipe can be reinforced by either of the two methods illustrated in the figure.
a. A short pipe is placed around the pipe to be reinforced, acting as a sleeve that increases the pipe's thickness.
b. The pipe is cut and an intermediate piece of pipe of adequate thickness is butt-welded between the two ends of the pipe.

Notice that the connection of tensile members is critical and should be treated with much more care than that required for compressive members. When smaller members join a major one, it is suggested that the small tensile member be welded to the major one first, then the compression member with a tapered end should be added.

The eccentricity of members at a connection is a major point. In particular, notice that the eccentricity in the connection of a tensile member has the following effects on the capacity of the joint:

(a) Positive eccentricity reduces the capacity up to 65%.

(b) Negative eccentricity increases the capacity up to 132% of the capacity without eccentricity. (See Figure 5-9a, b, c.)

Notice also that all tubes must be carefully sealed by welding their ends so that air does not enter. This guarantees rust protection inside the tubes, a point of major importance for structural integrity.

Energy absorbing connections. First adopted for the truss connections of the World Trade Center in New York, neoprene pad connectors absorb most of the energy that wind transfers to the building. The vibrations produced in the building by buffeting wind are dampened by the visco-elastic deformations of the 10,000 pads installed at the connections between floor trusses and columns. Interposed between steel plates, the neoprene undergoes deformations under strain and, due to its elasticity, returns to its original shape when the force is removed. This concept, having been proven effective, will probably become a trend.

Reinforced concrete connections. All joints in reinforced concrete trusses must act as hinges. Typical reinforced concrete hinges are considered flexible enough to be acceptable, although not completely frictionless. Note that the criterion of acceptability for hinges depends not on the flexibility of the

MEMBER AND CONNECTION DESIGN 369

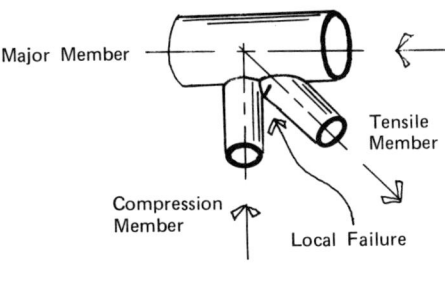

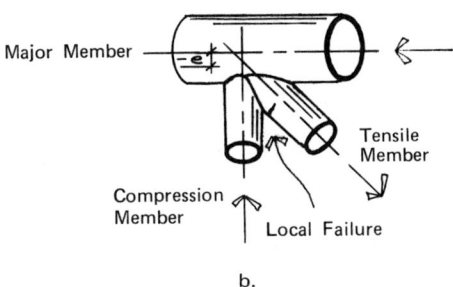

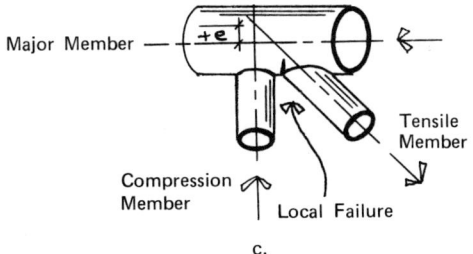

Figure 5-9. The effect of eccentricity on the capacity of tubular tensile member is illustrated as follows:
a. Tensile member with zero eccentricity and tensile capacity 100%.
b. Tensile member with negative eccentricity having an increased capacity of 132%.
c. Tensile member with positive eccentricity having a reduced capacity of 65%.

Notice location of local failure when stressed to rupture.

370 SIMPLIFIED TRUSS DESIGN

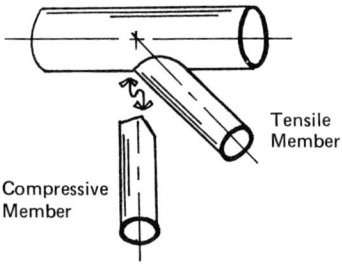

Figure 5-10. Illustration of recommended sequence of welding. Tensile members, whose connection is more critical than that of compression members, should be welded first. Compression members with tapered ends are welded second.

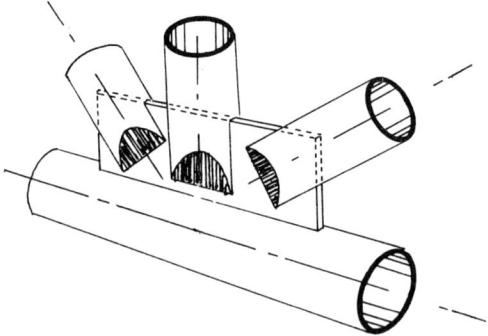

Figure 5-11. Typical connection of tubular members through a gusset plate, showing the plate fitting through the slots at the end of each pipe and the welds that seal the pipe ends. Notice that sealing the pipes is essential to prevent interior oxidation and corrosion.

MEMBER AND CONNECTION DESIGN 371

Figure 5-12. Mastodontic tubular truss of exceptional size, used at Eagle Mountain mine in California for an ore reclaimer. The tridimensional truss with a triangular cross-section, consisting of welded tubular members, shows the wide range of size applications of this type of structure.

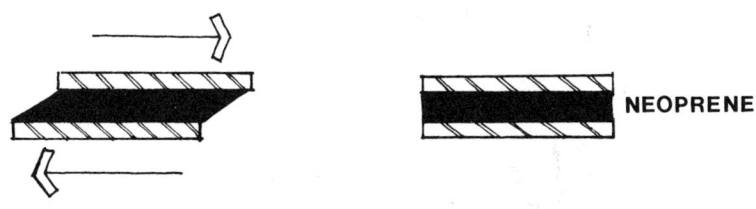

Figure 5-13. Neoprene pads interposed between steel plates are the basic components of special connections designed to absorb and dissipate energy. Used in the World Trade Center to join the floor trusses to the supporting columns, these connectors depend on the visco-elastic deformation of the neoprene to dampen the vibrations induced by the dynamic wind forces.

372 SIMPLIFIED TRUSS DESIGN

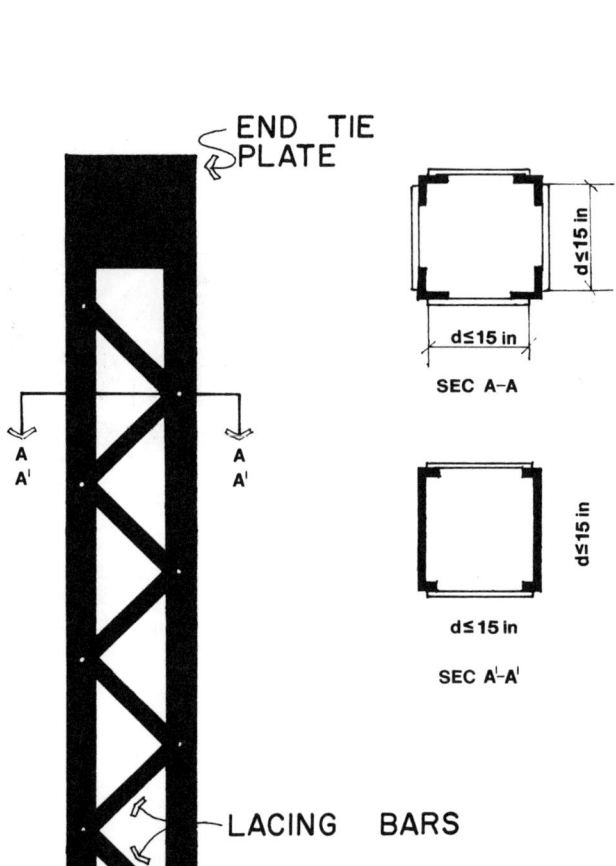

Figure 5-14. Typical built-up steel column with single lacing. Sections A–A and A'–A' show, respectively, the longitudinal members to be four angles and two channels, as two of the many alternatives possible.

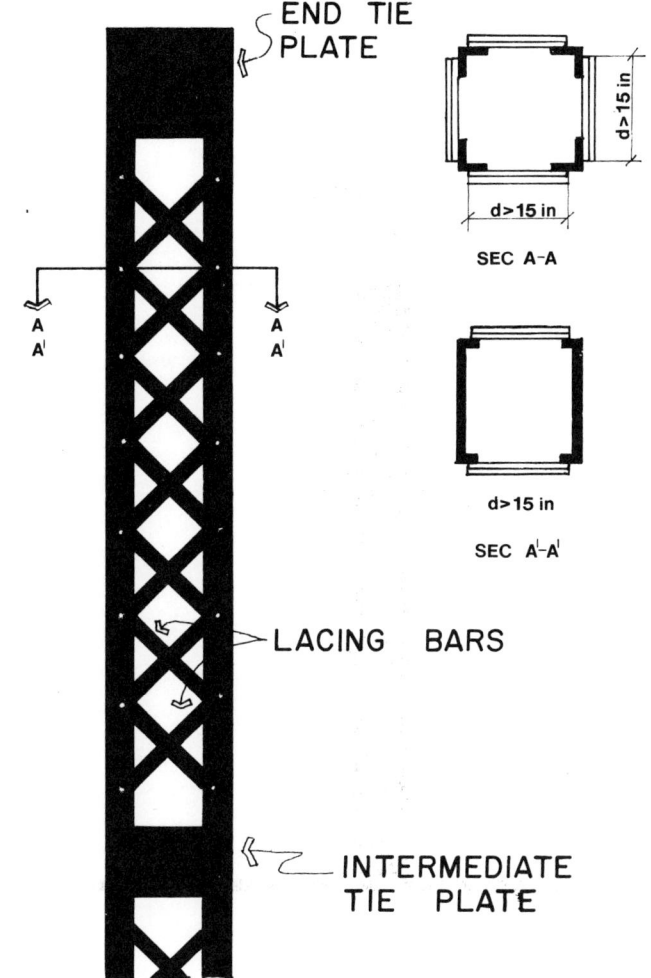

Figure 5-15. Typical double-laced column. Double-lacing is prescribed by AISC specifications when $d > 15$ in. Cross-sections A-A and A'-A' indicate the two of the many alternative choices for the shape of longitudinal members.

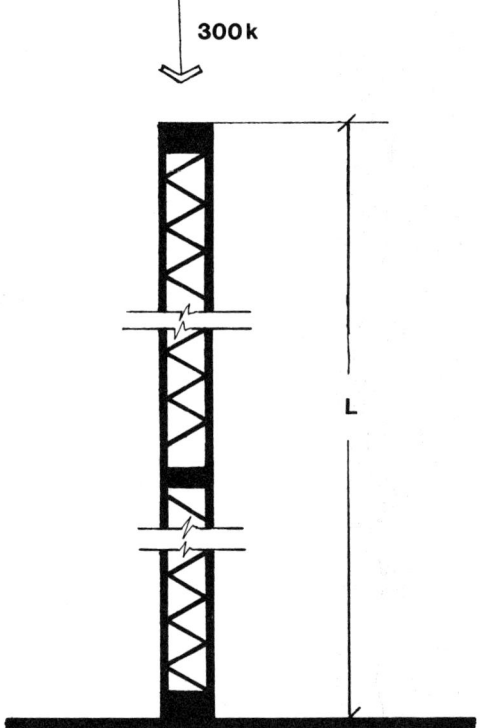

Figure 5-16.

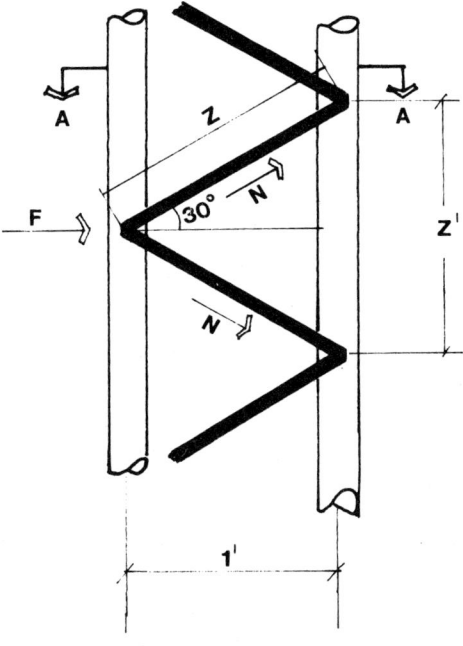

a.

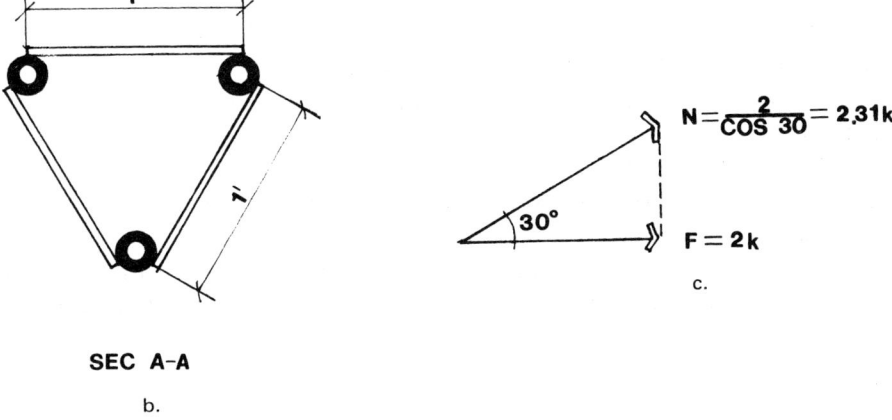

SEC A-A

b.

c.

Figure 5-17.

hinge *per se*, but on the fact that the hinge is a point where the flexibility is much higher than the flexibility at any other point in the structure.

Basic construction elements of concrete hinges include:

(a) The continuity of steel reinforcement from one member to the other through the hinge axis, and the convergence of the steel rods at the hinge.

(b) The indentation in the concrete in order to allow rotation.

While (a) is essential for any moment-free connection, (b) is essential only when large rotations are expected, such as in the case of hinged bridge supports, when the hinge moves as the live load is applied or removed. For roof structure, indentations are usually omitted.

A fundamental design criterion for the concrete hinge is that the axial force and the eventual shear force are carried entirely by the steel reinforcement without any contribution from the concrete.

Typical concrete hinges are of two types, both developed in France and carrying the names of the designing engineers: Mesnager and Considere. As indicated by Figure 5-18 the Mesnager hinge is more flexible than the Considere, because of the convergence of the reinforcement, and it is also more popular.

For the design of the Mesnager hinge, it is important to notice that the gap between the two concrete faces in the indentation is suggested to be

$$g = s \quad \text{or} \quad g \leqslant 1.3s$$

where:

g = gap between concrete faces at the indentation and
s = width of concrete at the indentation.

MEMBER AND CONNECTION DESIGN 377

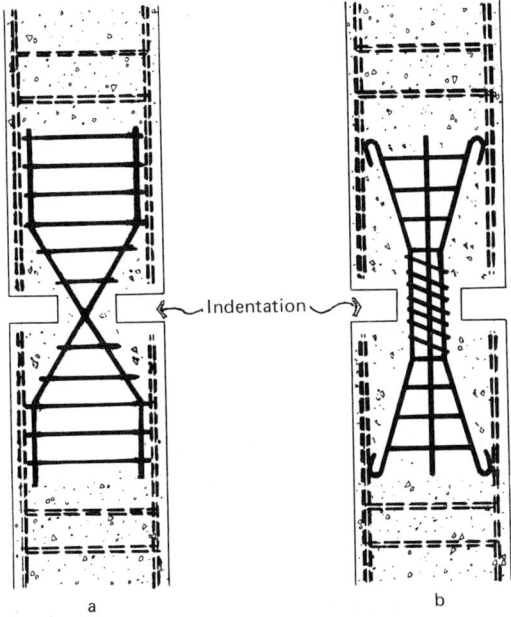

Figure 5-18. Typical reinforced concrete hinges: (a) Mesnager hinge; (b) Considere hinge.

The slenderness ratio l/r of the bars going through the hinge should be contained within limits to avoid their buckling.

$$20 \leqslant l/r \leqslant 40$$

Knowing that:

$$l = g/\cos \varphi \quad \text{and} \quad r = D/4$$

where:

l = length of the individual crossing steel bars,
r = radius of gyration of the cross-section of the individual crossing steel bars,
φ = angle between the axis of the member and the crossing steel bars, and
D = diameter of the individual crossing steel bars.

The total cross-sectional area of the crossing reinforcement bars required at the hinge is determined by

$$A_s = (N/f_s \cos \varphi) + (V/f_s \sin \varphi)$$

where:

A_s = total cross-sectional area of the crossing reinforcement,
N = axial force,
V = shear force,
f_s = allowable stress in the reinforcement, and
f_y = yield stress for the reinforcing steel being used.

assuming

$$f_s = 0.30 \, f_y.$$

The length of the steel bars past the hinge is determined by checking that they develop enough bond to carry the axial force. Ties or stirrups around the crossing steel rods must be installed within a distance m from the concrete faces. The total cross-sectional area of ties or stirrups within the distance m can be determined from

$$A'_s = N \tan \varphi / 2 f'_s + Vm/jDf'_s - 0.005mt.$$

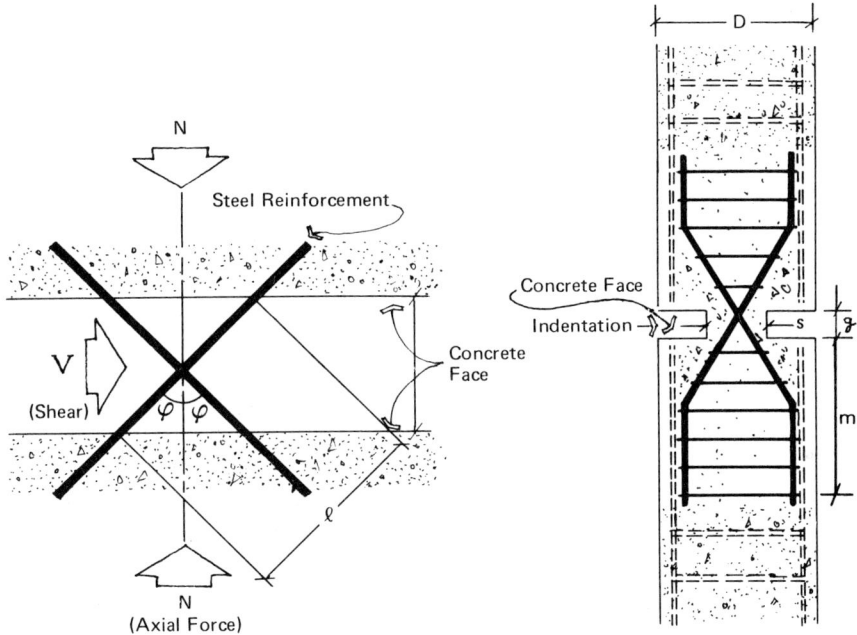

Figure 5-19.

Assume the following:
$$\text{and } \begin{array}{l} m = 8d \\ j = 0.9D \end{array}$$

where:

A'_s = total cross-sectional area of ties or stirrups within a distance m from the concrete face,

m = distance from the concrete face to where ties or stirrups are no longer needed,

D = distance indicated in Figure 5-19,

j = coefficient, assumed to be $0.90D$,

f'_s = allowable stress in ties or stirrups, and

t = thickness of the concrete at the hinge measured perpendicular to the plane of the drawing in Figure 5-19.

d = diameter of the bars of the crossing reinforcement.

380 SIMPLIFIED TRUSS DESIGN

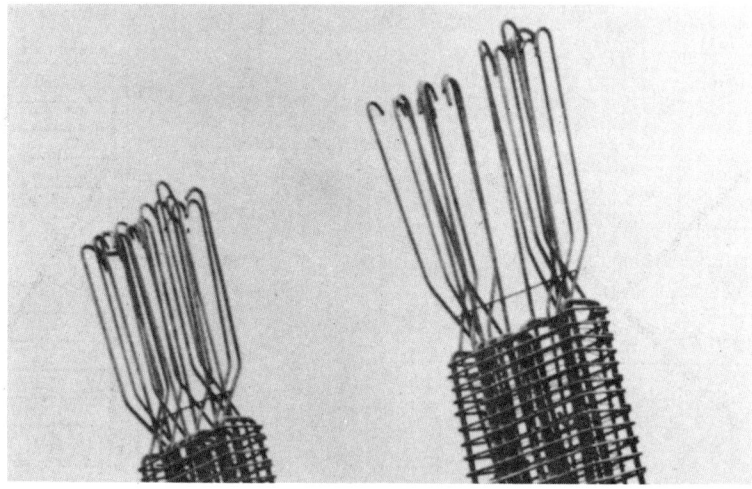

Plate 5-1a. Typical reinforcement for reinforced concrete hinges. Detail from the Vella Bridge at the Via Olimpica in Rome. Engineer: Riccardo Morandi.

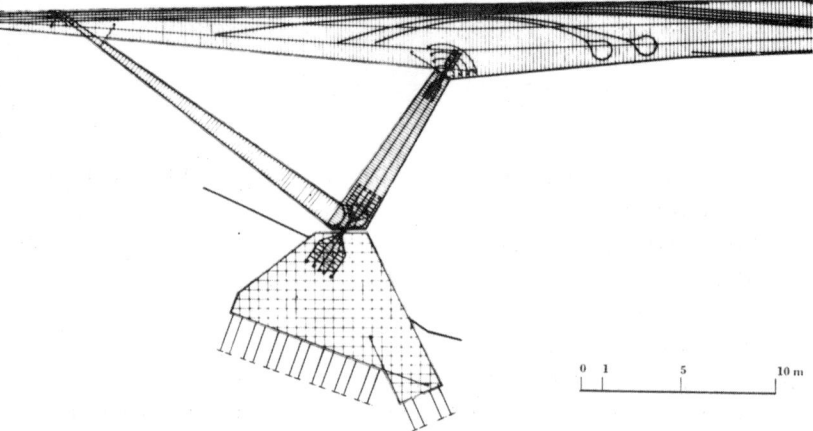

Plate 5-1b. Construction drawing.

MEMBER AND CONNECTION DESIGN 381

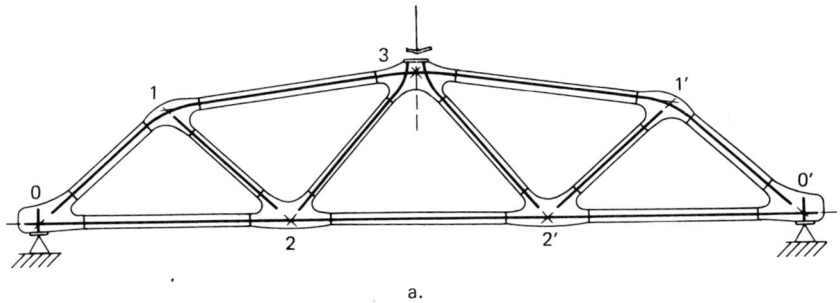

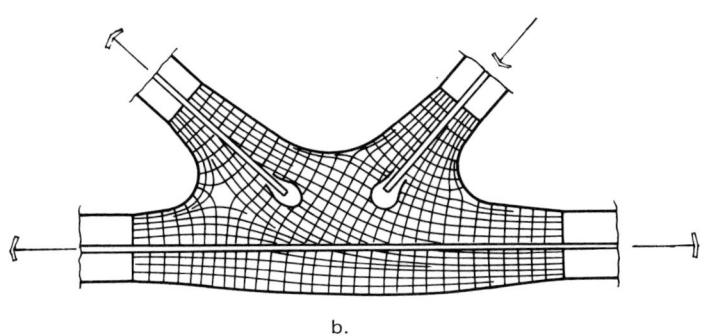

Figure 5-20. Actual stresses present in a model truss with rigid connections, as determined experimentally at Zurich Polytechnic Institute.

Appendix A

The following illustrations are part of a poster which presented a compendium of truss types for roofs and bridges. Credit for the poster goes to the Historic American Engineering Record, National Park Service, U.S. Department of the Interior.

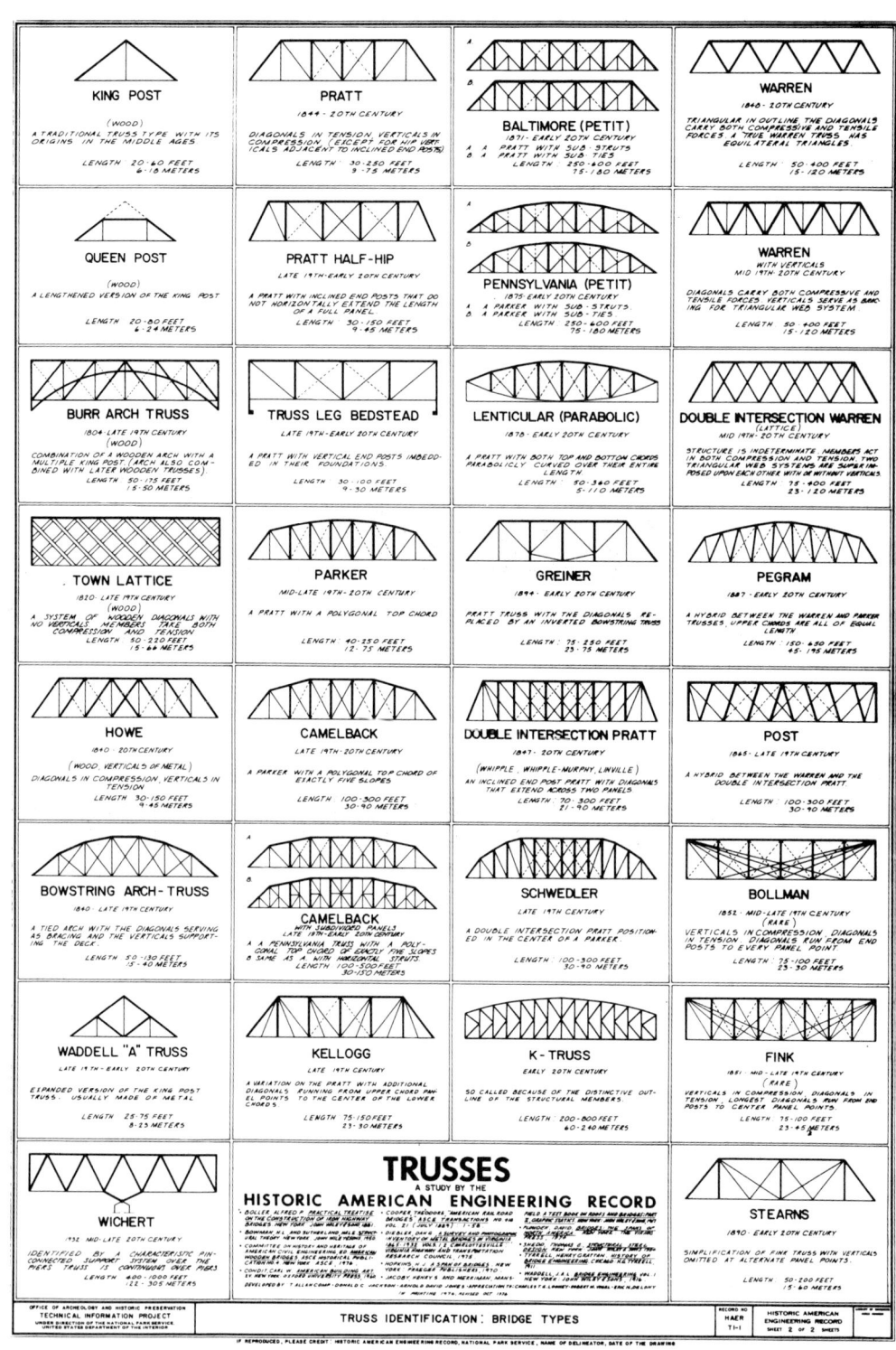

Figure A-1. Cremona's diagram.

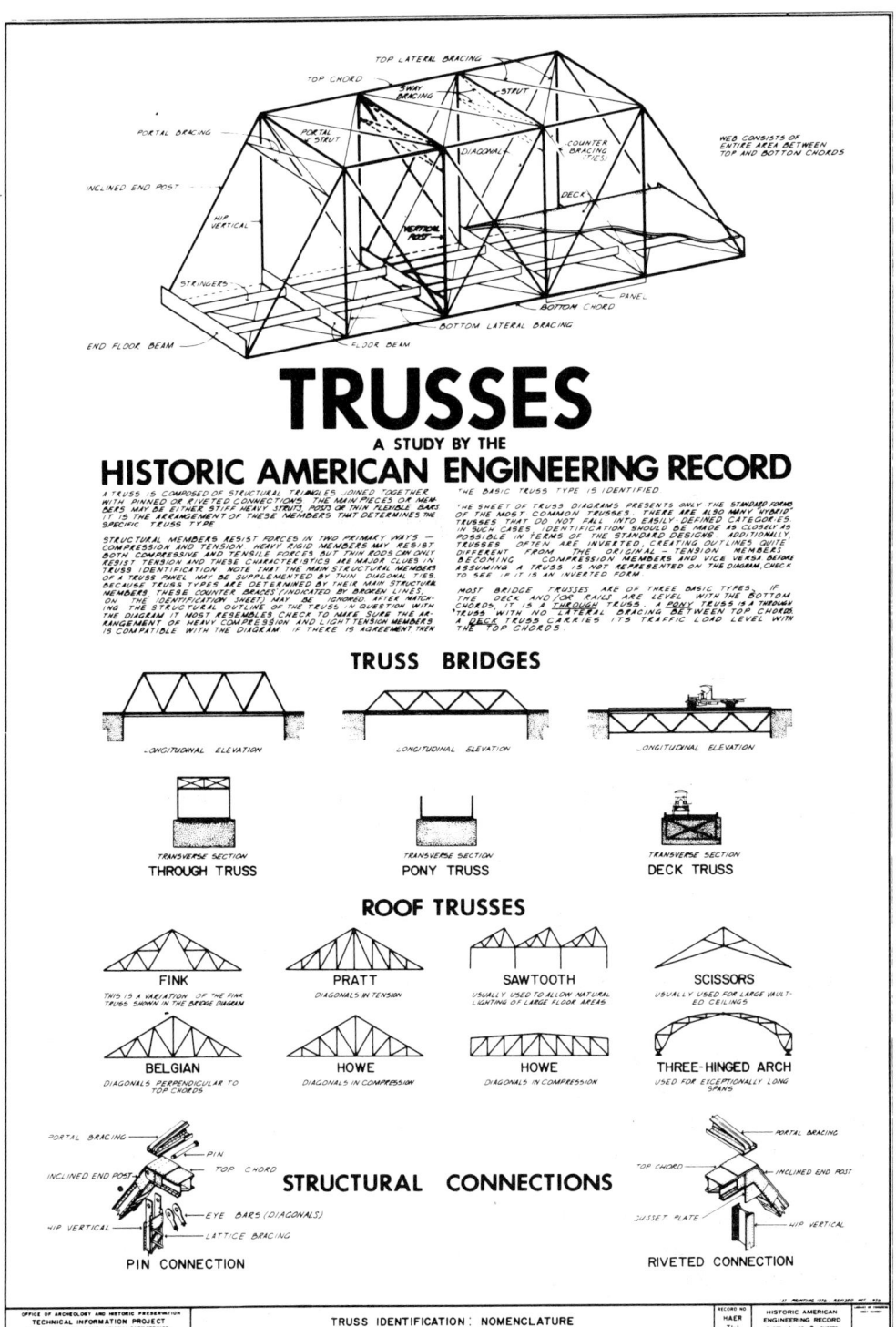

Appendix B

As previously mentioned (in Chapter 3), if the span is kept constant and the height, H, of a truss is increased (i.e., decreasing n), the axial forces in most of the members decrease to various extents, or, in a few instances, remain constant. However, they never increase.

The degree of change in the axial forces can be observed in the tables given in Chapter 3; however, the following examples can help to visually clarify the effects of geometry on force magnitude.

Example B-1. Consider a six panel Flat Pratt with $n = L/H = 7$ (see Figure B-1a) and another similar truss that has an equal span but is shallower; i.e., $n = L/H = 13$ (see Figure B-1b).

If the trusses are loaded only on the top chord with loads P, as indicated in the figure at the top of Table 3-32, the axial force coefficients K_1 for the thirteen members are shown in said table. Observing the formulas for the K_1 coefficients, we see that the forces in members 1, 2, 3, 7, 8, 9, 10, and 11 increase as n decreases, while in members 4, 5, 6, and 13 (the vertical struts), they remain constant for any value of n.

If we plot the values of the axial loads in each member for values of n varying between 7 and 13, inclusive (see Figure B-2), the diagrams show clearly the linearity of the variation and the rate of change. Member 3, with the steepest slope, is affected more than any other member by the change of n.

$n = \frac{L}{H} = 7$

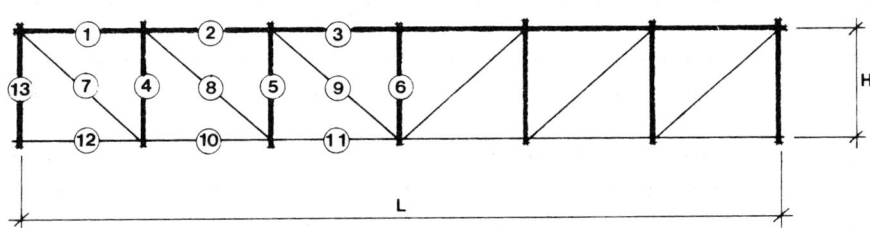

$n = \frac{L}{H} = 13$

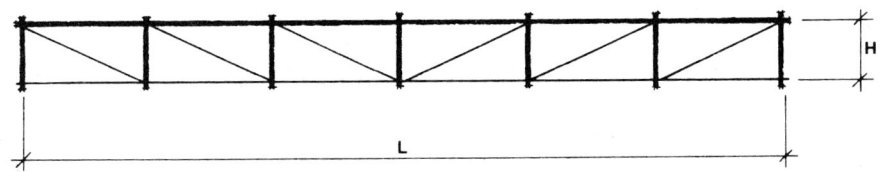

388 SIMPLIFIED TRUSS DESIGN

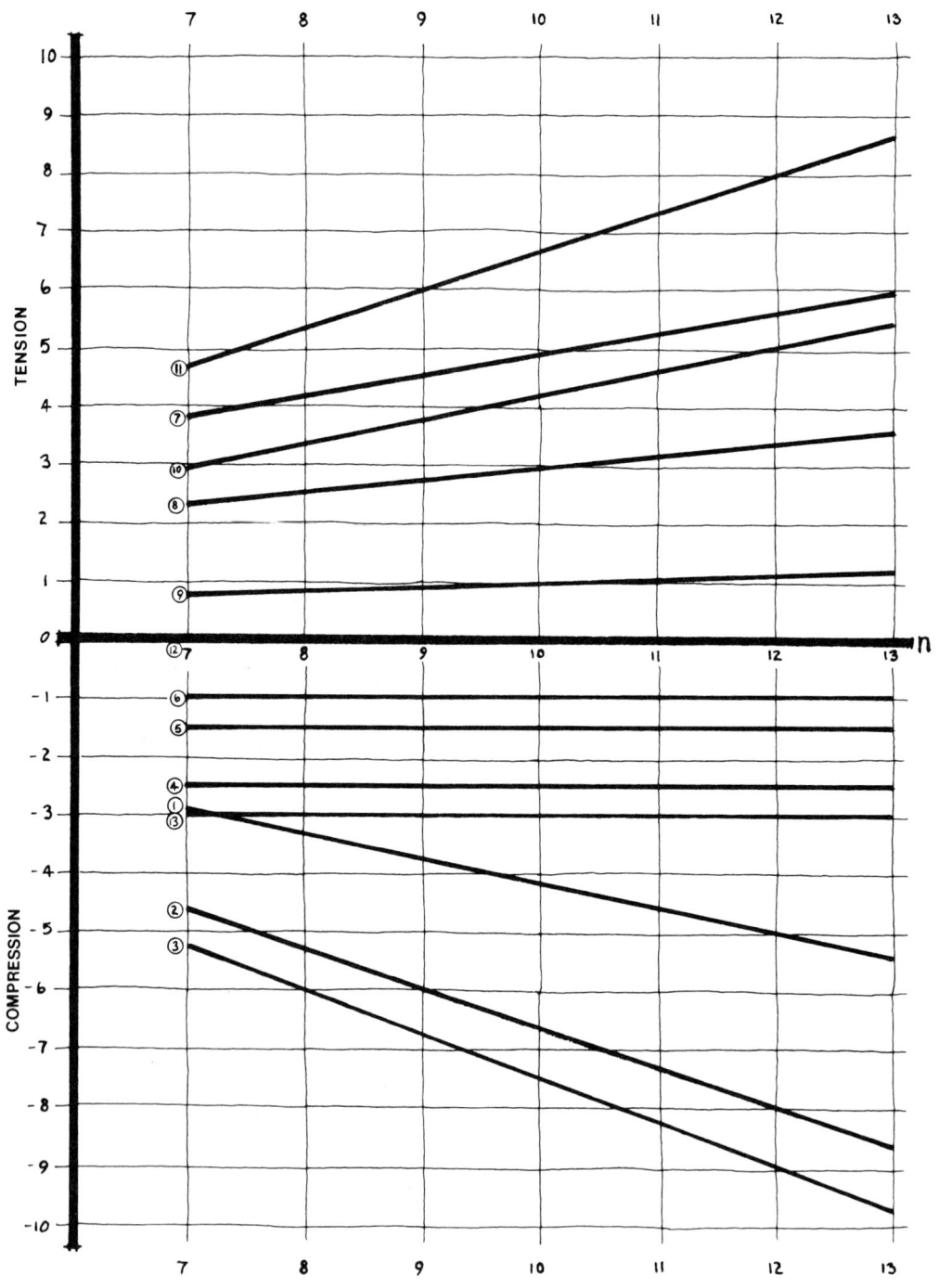

Example B-2. The procedure in this example is similar to that in the previous example; however, the truss is pitched instead of flat.

Consider a Polonceau (or Fink) with $n = L/H = 7$ (see Figure B-3a) and another similar one with $n = L/H = 2$ (see Figure B-3b). Notice that as the depth, H, increases (i.e., the slope increases and n decreases), the configuration varies so much that for $n = 2$ the truss practically becomes a Belgian. In fact, member 6 becomes zero and member 4 coincides with 7.

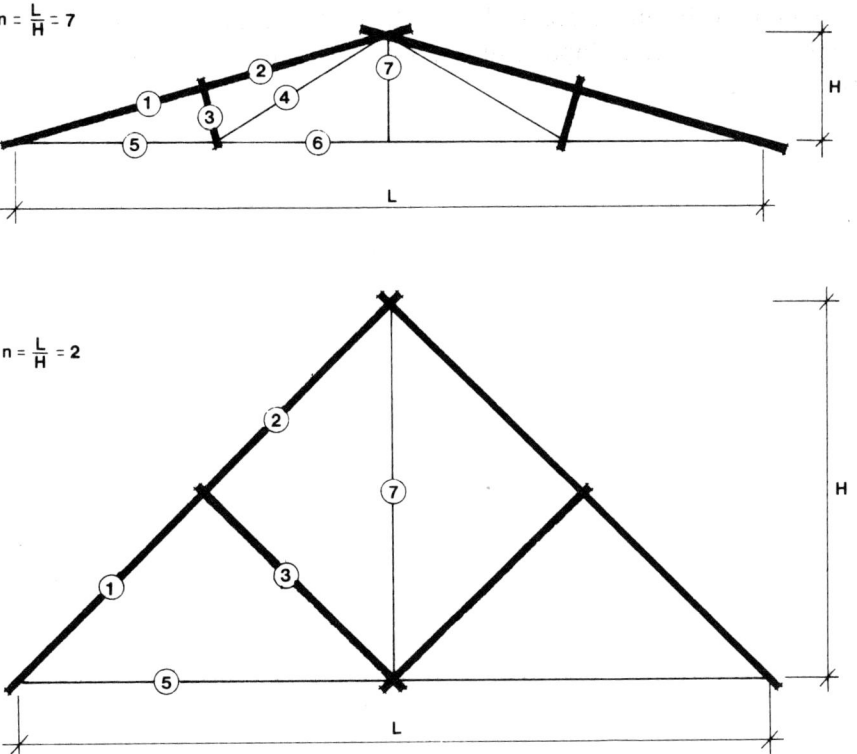

If the trusses are loaded only on the top chord with loads P, in the manner indicated in the figure at the top of Table 3-11, the axial force coefficients K_1 are indicated in the same table. Observing the formulas for K_1, we can see that, in this case, the forces increase in all members as n decreases. (Notice that in the previous example the forces in some members did not change as n varied.)

The rate of change of the axial forces in the members, as n varies from 2 to 7, is graphically indicated in Figure B-4. Notice, for instance, that member 3 has the highest slope; thus, it is the member most affected by the change of n.

APPENDIX B 391

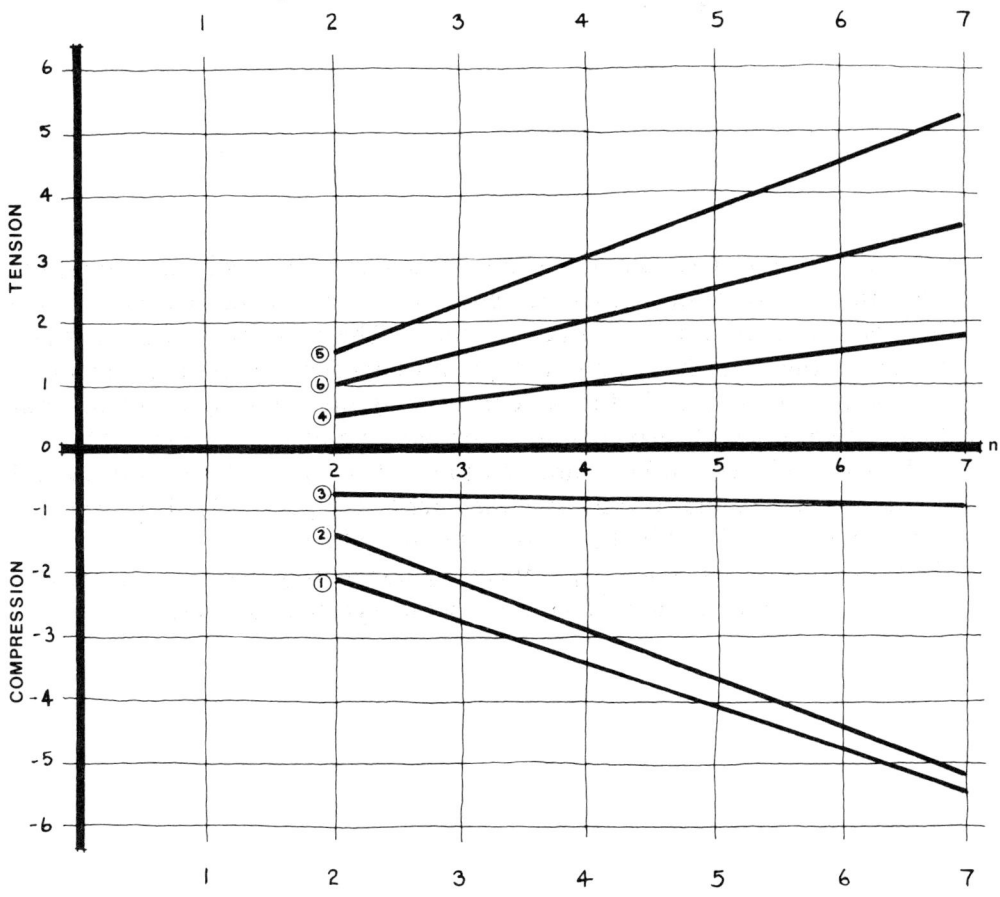

Appendix C

To visualize the generation of a statically determinate (isostatic) truss, the reader can start from one of its members and to its two ends he can attach a third point by means of two other members. The result is an isostatic triangle. He can then attach to it a fourth point by means of two other members. The process can continue in the same manner by the addition of one point at a time, using a set of two members for each point (see Figure C-1). The members in each set can be connected to any two joints of the truss, even if they are not at the ends of the same member (see Figure C-2). However, the two members should not be lined up on the same line.

Isostatic trusses can also be produced by connecting two isostatic trusses, linking them with three non-concurrent members (see Figure C-3).

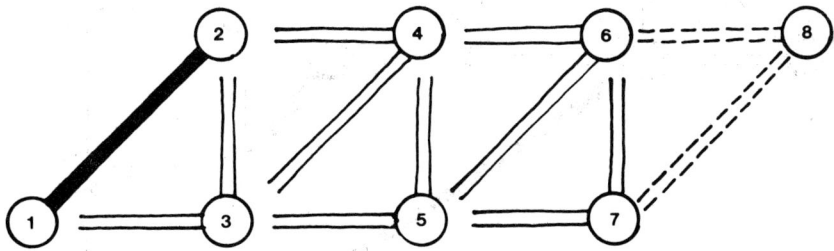

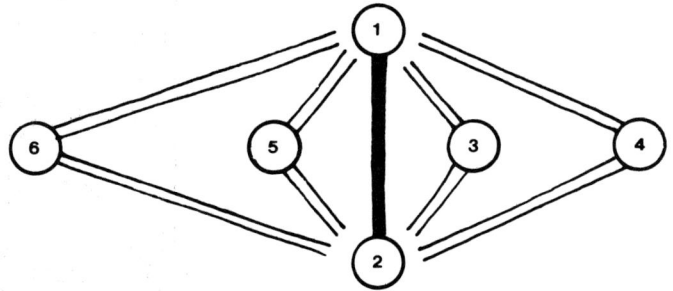

394 SIMPLIFIED TRUSS DESIGN

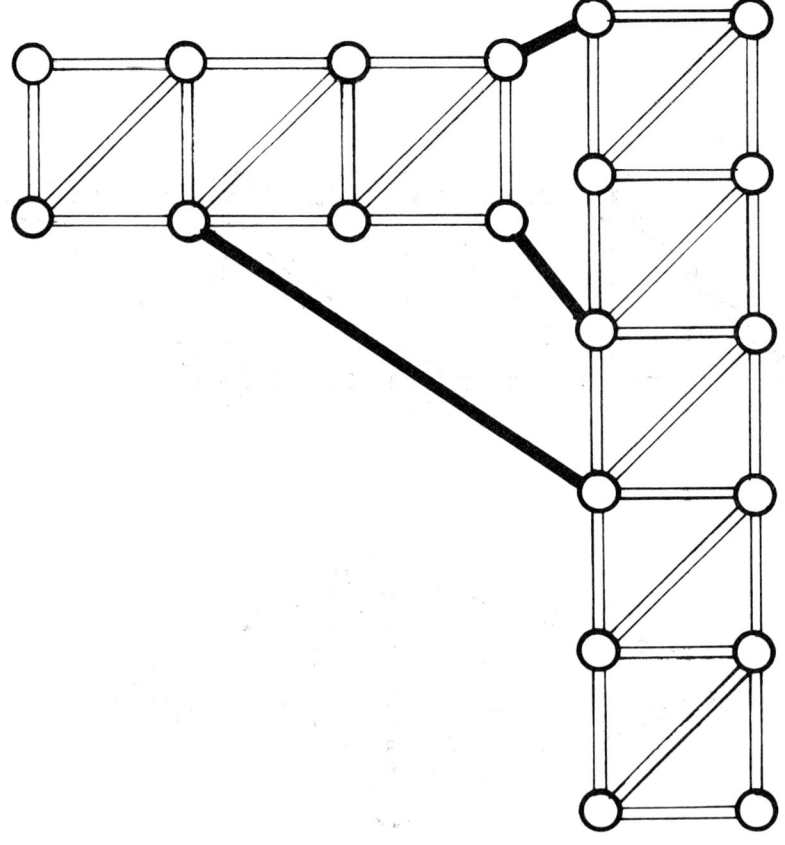

Appendix D

CONVERSION TABLES

To convert from	To	Multiply by
Centimeters	Inches	0.39370079
Centimeters^4	Inches^4	0.610237
Cubic Centimeters	Cubic Inches	0.061023744
Cubic Inches	Cubic Centimeters	16.387064
Feet	Meters	0.3048
Inches	Centimeters	2.54001
Inches	Feet	0.083333
Inches	Meters	0.0254001
Inches^4	Centimeters^4	41.623798
Kilograms	Pounds	2.2046226
Kilograms/square centimeters	Pounds/square inches	14.223343
Kilograms/square meters	Pounds/square feet	0.20481614
Kilometers/hour	Miles/hour	0.62137119
Meters	Feet	3.2808399
Meters	Inches	39.370079
Square Centimeters	Square Inches	0.15500031
Pounds	Kilograms	0.45459237
Pounds	Tons (Metric)	0.00045359237
Pounds/square feet	Kilograms/square meters	4.8824276
Pounds/square inches	Kilograms/square centimeters	0.070306958
Tons (Metric)	Pounds	2204.6226

Bibliography

Acland, James H., *Medieval Structure: The Gothic Vault.* Toronto and Buffalo: University of Toronto Press, 1967.
Allen, Richard S., *Covered Bridges of the Northeast.* Brattleboro, Vermont, 1957.
Bigelow, Jacob, *Elements of Technology*, 2nd ed. Boston, 1831.
Boubée, P., *Le Costruzioni in Legno.* Napoli, 1892.
Breymann, G. A., *Costruzioni in Legno, Costruzioni in Ferro.* Milano, 1927.
Brock, F. B., "Truss Bridges," *Engineering News* **9**: 371 *et seq.*, (1882) and **10**: 5 *et seq.* (1883).
Cable Construction in Contemporary Architecture. Bethlehem, Bethlehem Steel Corporation.
Calhoun, Daniel H., *The American Civil Engineer, 1792-1843.* Unpublished doctoral dissertation, Johns Hopkins University, Baltimore, Maryland, 1956.
Casati, E., *Applicazioni Pratiche della Scienza della Costruzioni.* Torino, 1924.
Clarke, Thomas C., "American Iron Bridges," *Scientific American*, Supplement **32**, August 5, 1876.
Condit, Carl W., *American Building Art: The Nineteenth Century.* New York: Oxford University Press, 1960.
Cooper, Theodore, American Railroad Bridges. *Transactions of the American Society of Civil Engineers* **21**: 1-60.
Corkill, P. A., Puderbaugh, H., and Sawyers, H. K., *Structure and Architectural Design*, 2nd ed., Iowa City: Sernoll Division, Effective Communications, Inc., 1964.
Cremona, Luigi, "Corso di Statica Grafica," Milano, 1868-1869; "Element di Calcolo Grafico," Torino, 1874; "Le Figure Reciproche nella Statica Grafica."
Culmann, Karl, "Der Bau der eisernen Brücken in England und Amerika," *Allgemeine Bauzeitung* **16**: 69-129 (1851) and **17**: 163-222 (1852).

De Camp, Sprague L., *The Ancient Engineers.* Garden City, New York: Doubleday and Company, Inc., 1963.

Dinsmoor, William Bell, *The Architecture of Ancient Greece.* New York: Biblo and Tannen, 1973.

Donghi, D., *Manuale dell'architetto.* Torino, 1906.

Edwards, Llewellyn N., "The Evolution of Early American Bridges," *Transactions of the Newcomen Society of England* **13**: 1-60.

Enciclopedia Italiana di Scienza, Lettere et Arti. Italy: Istituto della Enciclopedia Italiana, founded by Giovanni Treccani, 1949.

Engle, Heinrich, *The Japanese House: A Tradition for Contemporary Architecture.* Rutland, Vermont: Charles E. Tuttle Co., 1964.

Finch, James Kin, *The Story of Engineering.* Garden City, New York: Doubleday and Company, Inc., 1960.

Fleming, Robins, "Weight of Roof Trusses by Empiric Formulas," *Engineering News Record*, 1919.

Fletcher, Robert, and Snow, J. P., "A History of the Development of Wooden Bridges." *Trans. A.S.C.E.* **99**: 314-408.

Föppl, A., *Vorles. über technische Mechanick.* Lipsia, 1912.

Förster, M., *Die Eisenkonstruktionen der Ingenieur-Hochbauten.* Leipzig, 1910.

Gies, Joseph, *Bridges and Men.* Garden City, New York: Doubleday and Company, Inc. 1963.

Grages, F., *Zahlenbeispiele zur statischen Berchnung von Brücken und Dächern, rivisto da Barkhausen.* Wiesbaden, 1900.

Grinter, Linton E., *Design of Modern Steel Structures*, 2nd ed. New York: The MacMillan Co., 1964.

Grinter, Linton E., *Theory of Modern Steel Structures*, 3rd ed. New York: The MacMillan Co., 1962.

Guidi, C., *Lezioni sulla scienza delle costruzioni, parte 3^a.* Torino, 1928.

Haupt, Herman, *The General Theory of Bridge Construction.* New York, 1851.

High-Rise Housing in Steel: The Staggered Truss System. A research project sponsored by USSC, Departments of Architecture and Civil Engineering, and M.I.T., Cambridge, Massuchusetts, January, 1967.

Kirby, Richard S. and Laurson, Philip G., *The Early Years of Modern Civil Engineering.* New Haven, 1932.

Kirby, Richard S., Withington, Sidney, Darling, Arthur B., and Kilgour, Frederick, *Engineering in History.* New York: 1956.
Landsberg, T., *Die Statik der Hochbau-Konstructionen.* Stoccarda, 1909.
Masi, F., *La pratica della construzioni metalliche.* Milano, 1931.
Maxwell, James Clerk, "On Reciprocal Figures and Diagrams of Forces," *Philosophical Magazine*, London, 1864.
Mazzocchi, L. *Trattato delle costruzioni in legno.* Milano, 1893.
McCormac, Jack C., *Stuctural Analysis*, 2nd ed. Scranton, Pennsylvania: International Textbook Co., 1967.
McCormac, Jack C., *Structural Steel Design.* Scranton, Pennsylvania: International Textbook Co. 1966.
Morison, George S., *Bridge Construction.* Ithaca, 1893.
Mörsch, E., *Teoria e pratica del cremento armato.* Milano, 1930.
Otzen, R., *Zahlenbeispiele zur statischen Berechnung v. Brücken und Dächern.* Weisbaden, 1908.
Palladio, Andrea, *The Four Books of Architecture.* New York: Dover Publications, Inc., 1965.
Parker, Harry, *Simplified Design of Roof Trusses for Architects and Builders*, 2nd ed. New York: John Wiley and Sons, Inc., 1953.
Parker, Harry, *Simplified Design of Structural Steel*, 3rd ed. New York: John Wiley and Sons, Inc., 1965.
Parsons, William Barclay, *Engineers and Engineering in the Renaissance.* Cambridge, Massachusetts: M.I.T. Press, 1939.
Pizzamiglio, *Costruzioni metalliche.* Milano, 1911.
Pope, Thomas A., *A Treatise on Bridge Architecture.* New York, 1811.
Rempel, John I., *Building with Wood, and Other Aspects of Nineteenth-Century Building in Ontario.* Toronto and Buffalo: University of Toronto Press, 1967.
Resal, J., *Constructions métalliques.* Paris, 1892.
Ritter, A., *Dach-und Brücken-Construktionen.* Leipzig, 1894.
Sandstrom, Gösta, *Man the Builder.* New York: McGraw-Hill Book Co. Inc., 1970.
Saviotti, C., *Le travature reticolari a membri caricate.* Roma, 1878.
Scharowsky, C., *Musterbuch für Eisenkonstr.* Leipzig, 1908.
Smith C. Shaler, *A Comparitive Analysis of the Fink, Murphy, Bollman, and Triangular Trusses.* New York, 1865.

Straub, Hans, *A History of Civil Engineering.* Cambridge, Massachusetts: M.I.T. Press, 1964.

Torroja, Eduardo, *Philosophy of Structures.* Berkeley and Los Angeles, California: University of California Press, 1967.

Town, Ethiel, *A Description of Ithiel Town's Improvements in the Construction of Wood and Iron Bridges.* New Haven, 1821.

Trusses: A Study by the Historic Engineering Record. National Park Service, Washington, D.C.

Tyrrell, Henry G., *The History of Bridge Engineering.* Chicago, 1911.

Whipple, Squire, *An Elementary and Practical Treatise on Bridge Building.* New York, 1872.

Wood Structural Design Data **1**, 3rd ed. Washington, D.C.: National Lumber Manufacturers Association, 1958.

Zimmermann, H., *Über Raumfachwerke.* Berlin, 1901.

Index

Index

allowable stresses for steel, 338-47
allowable stresses for wood, 328-32
Apollodorus, 9
arsenal at Piraeus, 5
Avery, G.S., 40
axial forces, 114-40

Belgian truss, 127, 141-57
Bollman, W., 38
bottom chord, 114
bronze truss, 3
Burr, T., 33

Canfield, A., 36
chords, 114
colossus, 34
compression members in
    reinforced concrete, 352-4
    steel, 337-57
    wood, 327-37
Concord temple, 1
    crystal palace, 47
    Culmann, K., 45
connections, 355
    reinforced concrete, 368
    steel, 356-7, 368
    wood, 356
Considere, 376
conversion table, 403
Cremona, L., 81
Cremona's diagram, 102-3

Eccentricity in steel connections, 368
economy, concepts of, 86
Eikenberry, L., 40
energy-absorbing connections, 368
expediency, structures of, 136

fan truss, 124, 166-8
Fink, A., 38
Fink truss, 123, 157-65
flat Howe, 194-203
flat Pratt, 204-23
flat Warren, 224-41
Fuller, R.B., 134

generation of trussing systems, 400-402
geometrical instability, 97
German truss, 120
Goubermann, H., 32
Goubermann, J., 32
Greek truss, 1
Gridley, J.B., 40

hammer beam, 122
heel joints, 356
hinges in reinforced concrete, 368, 376-9
historical classification of trusses, 391-3
Howe truss, 125, 169-81, 194-203
Howe, W., 34

instability, 97
interpretation of trussing systems, 86
Italian truss, 118

king post, 116
Kpellbrücke, 32

lacing, 349-50
lattice work, 349-50
length of members, 140-1
Leonardo da Vinci, 23
lions' door, 1
Long, S.L., 34

medieval trusses, 9
Mesnager, 376
Mobius, Augustus, 90
modulus of elasticity of wood, 328-32
Mogine, 47
Mohr, O.C., 45
Monzani, W., 46
Murphy, J., 38

natural forms of trusses, 81-2
Nervi, P.L., 86
Neville, 46

Palladio, 7-8, 20-3
palladrian truss, 118-20
Palmer, T., 33
Pauli, F.A., 46
Pegram, G.H., 40
Pisca Tagua bridge, 33
Polonceau, J.B., 46
Polonceau truss, 123, 157-65
Pratt, C., 36
Pratt, T., 36
Pratt truss, 126, 182-193, 204-23

queen post, 118

Radius of gyration, 334-5
Renaissance, trusses in the, 20
reinforced concrete
    compression members, 351
    hinges, 376-9
    trusses, 66, 109
reinforced triangle, 118
Ritter, W., 45
Roman trusses, 7

Schwedler dome, 133
scissors truss, 120
secondary stresses, 114
    struts, 114
simple triangle, 115

slenderness ratios for steel members, 339-47
Spreuerbrücke, 32
staggered truss system, 75
Stanley, E., 40
statical determinacy, 8
statically indeterminate trusses, 100
steel compression members, 337-48
Strudl, 243
Swartz, A.S., 40

teaching aids, 100
tensegrity structures, 134
tension members in steel, 348-9
terminology, 114
Thayer, G.W., 40
Theseion Temple, 1
Thomas, A.S., 40
ties, 114
top chord, 114
Torraja, E., 128
Trajan column, 9
triangular Belgian, 141-57
triangular fan, 166-8
triangular Howe, 169-81
triangular Pratt, 182-193
triangulation, 1
tridimensional trussing, 128
Trombull, E., 36
trussed structures, 84
trussing systems, 84
tubular truss connections, 357
types of trusses, 115

Vitruvius, 6-7

Warren, J., 46
Warren truss, 82, 128, 224-41
weight of trusses, 104-5-6
Wernwag, L., 34
Whipple, S., 38
wood compression members, 327-37
wood species, 328-32